Growth Cultures

Growth Cultures investigates the rapidly growing phenomena of biotechnology and sets this study within a knowledge economy context. Philip Cooke proposes a new knowledge-focused theoretical framework, 'the New Global Bioeconomy', against which to test empirical characteristics of biotechnology.

In this timely volume, Cooke unifies concepts from the sociology of science, economic sociology and evolutionary economic geography to focus on the problems and prospects for policy agencies worldwide trying to build 'biotechnology clusters'. He develops a superior policy approach of thinking in terms of *platforms* that integrate *proximities* and *pipelines*, which will be of significant interest for the scientific and technological communities as well as economic development policy communities.

Growth Cultures will make fascinating reading for students, policy makers and researchers across management and business studies, innovation and knowledge studies, sociology, science and technology policy, applied economics, development studies and regional science.

Philip Cooke is University Research Professor in Regional Development and founding Director of the Centre for Advanced Studies at the University of Wales, Cardiff. He is Adjunct Professor in Spatial Planning at the University of Aalborg, Denmark. His research interests lie in studies of economics of biotechnology, regional innovation systems, knowledge economies, and policy actions for business clusters and networks. He is author of *Knowledge Economies: Clusters, learning and cooperative advantage* (Routledge, 2002).

Genetics and Society

Series editors:
Paul Atkinson, *Associate Director of CESAGen, Cardiff University*;
Ruth Chadwick, *Director of CESAGen, Lancaster University*;
Peter Glasner, *Professorial Research Fellow for CESAGen at Cardiff University*;
Brian Wynne, *member of the management team at CESAGen, Lancaster University.*

The books in this series, all based on original research, explore the social, economic and ethical consequences of the new genetic sciences. The series is based in the ESRC's Centre for Economic and Social Aspects of Genomics, the largest UK investment in social-science research on the implications of these innovations. With a mix of research monographs, edited collections, textbooks and a major new handbook, the series will be a major contribution to the social analysis of new agricultural and biomedical technologies.

Series titles include:

Governing the Transatlantic Conflict over Agricultural Biotechnology (2006)
Contending coalitions, trade liberalisation and standard setting
Joseph Murphy and Les Levidow
978-0-415-37328-9

New Genetics, New Social Formations (2007)
Peter Glasner, Paul Atkinson and Helen Greenslade
978-0-415-39323-2

New Genetics, New Identities (2007)
Paul Atkinson, Peter Glasner and Helen Greenslade
978-0-415-39407-9

The GM Debate (2007)
Risk, politics and public engagement
Tom Horlick-Jones, John Walls, Gene Rowe, Nick Pidgeon, Wouter Poortinga, Graham Murdock, Tim O'Riordan
978-0-415-39322-5

Growth Cultures (2007)
The global bioeconomy and its bioregions
Philip Cooke
978-0-415-39223-5

Local Cells, Global Science (2008)
Embryonic stem cell research in India
Aditya Bharadwaj and Peter Glasner
978-0-415-39609-7

Growth Cultures

The global bioeconomy and its bioregions

Philip Cooke

Routledge
Taylor & Francis Group
LONDON AND NEW YORK

First published 2007 by Routledge
2 Park Square, Milton Park, Abingdon, Oxon, OX14 4RN

Simultaneously published in the USA and Canada
by Routledge
270 Madison Avenue, New York, NY 10016

Routledge is an imprint of the Taylor & Francis Group, an informa business

Typeset in 10/12pt Sabon by Graphicraft Limited, Hong Kong
Printed and bound in Great Britain by MPG Books Ltd, Bodmin

British Library Cataloguing in Publication Data
A catalogue record for this book is available from the British Library

Library of Congress Cataloging in Publication Data
Cooke, Philip (Philip N.)
Growth cultures : the global bioeconomy and its bioregions / Philip Cooke.
p. cm.
ISBN 978-0-415-39223-5 (hard cover)
1. Biotechnology—Economic aspects. 2. Pharmaceutical biotechnology—Economic aspects. I. Title.
TP248.23.C68 2007
338.4'76606—dc22

2007001186

ISBN10: 0-415-39223-3 (hbk)
ISBN10: 0-203-08730-5 (ebk)

ISBN13: 978-0-415-39223-5 (hbk)
ISBN13: 978-0-203-08730-5 (ebk)

Contents

List of tables

List of figures

Preface

I first became seriously interested in biotechnology in about 1985, nearly ten years after it started in its modern, commercial guise with the formation of the first firm, Genentech, in California. There was a good deal of academic and social interchange among a community of radical *regional* political economists associated with the Union of Radical Political Economists (URPE) in the US and their counterparts, the Conference of Socialist Economists (CSE), especially its Regionalism group in the UK. The peculiarity of this community was, first, a shared interest in the theory of small-scale, collaborative, flexibly specialised and embedded industrial organisation. Authors such as Becattini (2001) had earlier pioneered research in Italy as he analysed the evolution of industrial districts. The second shared interest was in the prospect that new forms of high technology industry emerging first in California and, in the UK, the so-called M4 Corridor shared this Marshallian form of industrial organisation embeddedness, as it seemed to do rather than that of the characteristically large, corporate, multinational. The latter was clearly becoming almost hegemonic in those early years of modern globalisation. The conjoining interest was in the potential for liberated forms of work organisation that this more embedded model seemed to offer, compared to the rigid hierarchies and regimented forms of occupational position imposed by the corporate model. Avoiding the clutches of such behemoths as RCA seemed also to be a key motivation of the networked entrepreneurs of interest in these new agglomerations, a matter of no little extra interest to we observers, soon, we hoped, to be participant ones (Suchman, 2000).

In 1985, at Cardiff University, I managed to get a research council studentship from ESRC to work on a study of the emergent Californian 'retro-model' of high technology capitalism. The researcher was one Graham Sewell, student deejay and serious Beastie Boys and Cramps fan, for whom months and years in UC Santa Cruz proved a fine platform for future academic success. His host was my late friend Rick Gordon, the classic Mustang-driving Australian and founder of the Silicon Valley Research Group. It was Rick who told us about biotechnology and its propensity to be as protean in economic impact as the semiconductor industry on which

Silicon Valley had been built and which had so attracted the brightest young talents to study under the tutelage of other friends and acquaintances of the time. Notable among these were, of course, Ed Soja, Rebecca Morales, Allen Scott and Michael Dear in southern California, and Dick Walker, Ann Markusen, Peter Hall, Manuel Castells and Michael Burawoy in Berkeley. The rising stars included, mostly, economic geographers like AnnaLee Saxenian, Meric Gertler, Michael Storper, Erica Schoenberger and Amy Glasmeier among others. I was hosted by many of these kind people and was delighted to reciprocate. It was through Rick, the political scientist, though, that in 1987 Graham and I were invited to present our project's first biotechnology paper at Philippe Schmitter's then new Centre for European Studies at Stanford University. There I first heard the likes of Mark Granovetter and David Teece talking a more economic sociologically nuanced version of what we regional scientists and economic geographers, or in Manuel Castells' case, urban sociologist had also been thinking and exploring. There I also met Nicola Bellini, shortly to become political adviser to the president of Emilia-Romagna in Italy, subsequently a long-term Pisa, Tuscany-based academic colleague, and expert on localised enterprise and innovation services, policies and institutions. In one way or another, all these thinkers and writers were influential upon me and a host of others, but to my surprise they were not, except for a few and later, interested in biotechnology. Rather it was ICT innovation from the silicon chip and its computing and software applications that, in technological terms, seemed then to hold for the majority the greatest interest.

Discussing this much later, the consensus was that the intellectual investment of 'tooling-up' to research a wholly new, technologically challenging field was something of a disincentive compared to realising the academic returns from having got to grips with nascent ICT. This was a much larger and, in terms at least of the main focus of theoretical economic geography, which is *agglomeration*, a more appealing intellectual focus. By March 1987, as well as Graham's work in California, I had started on a study of the dynamics of biotechnology in Minnesota, whence I had been invited to be a visiting professor at the Humphrey Institute of Public Affairs at the University of Minnesota, Minneapolis. The only fragment of that research into some thirty Minnesota biotechnology firms to be published was in Cooke (1988) in a special issue of *Environment and Planning D. Society & Space* co-edited with Allen Scott. The research and the excellent resources of the Wilson Library, where I read one of the most famous US Office of Technology Assessment briefings on the nature of biotechnology, gave me a reasonably clear outline of its three branches at the time. These were *fermentation*, some of the products of which I had become extremely familiar with over preceding decades as well as in the delightful Minneapolis microbreweries; *monoclonal antibodies* (Mabs) by which genetic diseases could be inhibited through being 'switched-off'; and DNA, the code of life, within which genetic mutations that were the cause of some diseases could be identified and manipulated. Most firms I interviewed were involved in Mabs and/or DNA studies, hunting for molecules that had the chance to become a life-saving medicine or therapeutic treatment. But the strongest lesson I learned was about American technological entrepreneurship: for whereas naïve old me was fascinated by the intellectual and technical skills involved, I lost count of the number of times my question as to the ambition of the numerous biotechnologists then completing their PhDs received the answer – to find a successful molecule, sell it to Eli Lilly or some other pharma for a few million, retire and go fishing in Florida!

All this stood me in good stead when, in 1999, ten years after my second stint in Minnesota, I got an out-of-the-blue invitation to serve on the UK Minister of Science and Technology, Lord Sainsbury's Biotechnology Clusters Task Force (DTI, 1999). This was a swiftly executed mission, beginning in April, finished by September 1999 and reporting within the calendar year. But the ground covered, the intensity of the investigation undertaken and need to absorb new knowledge instantaneously, while still remembering to ask the right questions, were exhilarating at best, exhausting at worst. The Task Force researched many firms, official bodies, research institute heads, experts and observers in the UK and US, the two leading biotechnology commercialisation economies. It was – and here I must be careful – my introduction to the *modus operandi* of New Labour's technology policy formation process. In sum this involved privileged access at the highest institutional levels to a non-random sample of entrepreneurs and industry experts, some subsequently well known to be enthusiastic supporters and, in various ways, beneficiaries of the New Labour project. Nevertheless, and despite the idiosyncrasies of what was clearly an exercise not in scientific but *policy* research – a crucial distinction in these days of pervasive image marketing and the application of spin – the knowledge gained by this researcher, trying his best to adhere to aspects of a Mertonian interrogative agenda explored more fully in Chapter 1, was illuminating. Aspects of it inform parts of the early historical accounts in Chapter 7 of the localised biotechnology agglomerations in Cambridge, Massachusetts, and Cambridge, Oxford and Scotland (UK).

Stimulated somewhat by this new edge on the evolution of biotechnology at a buoyant time, before the stock-market decline of 2000–2001, I embarked on a knowledge 're-tooling' process of my own, involving intensive secondary knowledge-mining, aided considerably by the vast on-line information resources for medical healthcare generally and biotechnology specifically. I cajoled old friends Gerd Fuchs, Gerd Krauss and Thomas Stählecker, of the now sadly defunct Centre of Technology Assessment for Baden-Württemberg at Stuttgart, Germany that biotechnology clustering was a suitable theme for one of their excellent symposia, on at least two of which my research centre and theirs had successfully collaborated before (regional innovation systems in 1995, and multimedia in 1997). In 1999, when we discussed this, Germany had not long embarked on its ambitious

BioRegio cluster-building initiative for which seventeen sub-regional contestant groups had been whittled down to three main winners, Munich, Heidelberg and Rhineland, with Jena in former East Germany receiving a special award. Thus policy interest in biotechnology was high at the time; accordingly a symposium was held at CTA in 2000 and resulted in two stimulating international collections in *Small Business Economics* (2001) and Fuchs (2003). Thereafter commissions to undertake comparative European, followed by more global analyses of key biotechnology agglomerations arrived in approximately equal measure with requests to report such results at academic conferences. Thus a period of high demand-led productivity ensued – including partnering winning policy commissions (mainly EU, foreign and UK DTI or Regional Development Agency), and ESRC funding bids, notably our joint Cardiff–Lancaster Universities Centre for Economic and Social Analysis of Genomics (CESAGen) research centre, whose book series with Routledge is the proximate cause of this volume, and a separate ESRC project on Collective Learning in the Knowledge Economy comparing business performance of biotechnology and ICT firms within and outside cluster settings in the UK and, in a separately funded project, Austria (Cooke *et al.*, 2007). I wish to thank and acknowledge the multitude of named and un-named partners and colleagues involved in this more recent enterprise.

Perhaps not surprisingly, some of the published and more unpublished academic output of this intensive period of research into the economic geography and sociology of biotechnology finds its way into this book. However, this proved to be far less than might have been the case. To explain the etymology of the chapters briefly and in sequence, Chapter 1 is wholly new, the kick-off for which, it becomes clear from the first sentence of Chapter 1, was my stint as adjunct professor at the University of Aalborg, Denmark in September–October 2006. I thank Anne Lorentzen and Søren Kerndrup for the invitation and their hospitality, Bent Flyvbjerg for interest and stimulus, and Bent and Beata Dalum for accommodation and hospitality beyond the call of duty! The content of Chapter 2 was largely written in 2006 as a work package report for EURODITE, an EU Framework 6 research project on local–global knowledge flows in key industries, ably co-ordinated by Stewart MacNeill of Birmingham University, who is thanked for his geniality and perseverance. For the global co-publication and co-patenting research results, I was superbly assisted by Ann Yaolu, later of the University of Iowa, and funded by our EU DIME 'Network of Excellence' role of disseminating new kinds of economic development indicator knowledge from academe to the world of policy. Chapter 3 has a lengthy and complicated etymology: it was first presented at the Regional Science Association 42nd European Congress at Dortmund, Germany, 27–31 August 2002 but remained unpublished. It was nevertheless improved by comments from the many participants showing interest, particularly Martina Fromhold-Eisebith, Allen Scott, Robert Hassink, Javier Revilla

Diez, Michael Kiehl, Markku Sotarauta and Gert-Jan Hospers. Thereafter it was re-presented and developed at the International Workshop on 'Regional Learning' at the University of Nijmegen, 26 September 2002 and The Competitiveness Institute conference in Cairns, Australia on 11–12 October 2002. Comments from Eike Schamp, Ben Dankbaar, Arnoud Lagendijk, Jan Lambooy, Don Di Salle, Ifor Ffowcs-Williams and Erik Von Bahr were attended to. Roel Rutten and Frans Boekema edited an early version of Chapter 3 for a Special Issue of *European Planning Studies*, Vol. 12, No. 5, July 2004 for which the publishers are acknowledged and thanked for allowing this updated version to be published.

A version of Chapter 4 was presented at the Fourth Triple Helix Conference, 'Breaking Boundaries – Building Bridges', Copenhagen, Denmark – Lund, Sweden, 6–9 November 2002 but was not published. Elements were inserted from what also began life as a conference paper, the origins and substance of which are delineated in reference to Chapter 6. It was also then further refined for a presentation to the Netherlands Ministry of Economics, organised by Bart Nooteboom, then of Erasmus University, Rotterdam, followed by a joint commission with PriceWaterhouseCoopers from the Merseyside Biotechnology Council and North West England Regional Development Agency. It was published in *Environment and Planning C: Government and Policy* in its 'biotechnology clusters' theme issue. I thank the editors and, among others, Mats Benner and Luis Suarez-Villa for comments on its original formulation, Robert Kaiser and Tiago Santos Pereira for responding to its second incarnation, Poul op ter Boom, Bart Nooteboom and Paiivi Oinas, its third and Rob Arnold, Ken Poole and other reviewers, its fourth. Finally, Mika Kautonen of Tampere University, Finland also helped with information and comments. Chapter 5 began life when Rikard Stankiewicz kindly invited me in 2005 to present at the European Forum Conference on 'The Role of Universities in Innovation Systems', 17–18 June, at the Robert Schuman Centre for Advanced Studies, European University Institute, Fiesole, Florence, Italy. Comments from David Mowery, Paul David, Stuart Blume, Aldo Geuna and Henry Etzkowitz added value. Thanks are due once more to Ann Yaolu for further assistance with the scientometrics. Thanks also to Lennart Stenberg, VINNOVA, for discussing various concepts and hypotheses on globalisation processes in biotechnology. Finally, thanks to colleagues in Sweden in the CIND (Uppsala) and CIRCLE (Lund) innovation research centres and in Canada, the ISRN network who commented on these observations in preliminary form and encouraged their development.

Chapter 6 was first commissioned and presented at the conference on 'Rethinking Science Policy: Analytical Frameworks for Evidence-Based Policy', held at the Science Policy Research Unit, University of Sussex, 21–23 March 2002. I am grateful to the late Keith Pavitt, Ammon Salter, Edgar Grande, Svend Remøe, Fumi Kitagawa and the above-mentioned Robert Kaiser and Tiago Santos Pereira for comments. Thanks also to

Meric Gertler, Mika Kautonen, Stu Rosenfeld and David Wolfe for assistance, especially on information about biotechnology in their countries at that time. In 2003, it appeared in pristine form in David Wolfe's edited book on *Clusters Old & New*, published in Canada by McGill-Queens University Press. Later on, different versions for distinctive audiences were invited for publication in June 2004 by Bill Page, editor of *Science & Public Policy*, and as a broader Life Sciences version by the 'Clusters' Special Issue editors of *Urban Studies*. All editors are acknowledged for their assistance in enabling this updated variant to be included in this volume. Chapter 7, the longest in the book, contains much hitherto unpublished material. Most of it arose from private commissions on global bioregional comparisons by DTI in 1999, VINNOVA, the Swedish Innovation Systems Development Agency in 2004 and the German Ministry for Science and Technology (BMBF) in 2006. All are thanked for enabling extracts to be utilised in this chapter. My colleague on the EURODITE biotechnology brief, Robert Kaiser, both noted above, researched the tabulated data on up-to-date German and French biotechnology activity and I here acknowledge his contribution and that of Lennart Stenberg for his assiduous reading, and correction, of the relevant text.

Chapter 8 is a new chapter that, nevertheless, draws extensively on the pioneering *Nature Biotechnology* Supplement of late 2004, to which my attention was drawn by its main editor, Halla Thorsteinsdóttir. Chapter 9 is also largely new although some tabulated data fragments appeared in my article for *Research Policy* in 2005, whose editors are acknowledged in this regard. Once again VINNOVA was important as a stimulus to various stages of writing this chapter. Parts of it were written at the behest of the unpublished report I wrote for Lennart Stenberg and, in 2006, I was invited to be on the evaluation panel for the *VinnVäxt* programme and its first three project initiatives. One of these involved two day-long meetings with the main actors in the *Skåne* 'Food Innovation at Interfaces' agro-food project. Here I learnt a great deal, again very swiftly, about the challenges and perceived opportunities of agro-food biotechnology. I am grateful to VINNOVA and the project principals for this opportunity to reflect upon and share *my* perception of the status and key processes involved, but this can be checked against the highly informative etymology of the evolution to date of Scanian Agro-Food and Functional Food contained in Lagnevik *et al.* (2003). Chapter 10 on the complex financial dimensions and instrumentation of biotechnology is also new and unpublished except as an internal work package document to the EU Framework Programme 6 project on access to finance for biotechnology (AFIBIO). This is most ably co-ordinated by Olivier Kitten of Nantes Atlanpole, near Brittany, France with his director Jean-Francois Balducchi and partners from San Raffaele Biopark, near Milan, Italy, Heidelberg's Technology Park in Germany, Cardiff's Gene Park in Wales (UK) and Tartu's Biopark in Estonia, accompanied by numerous partners from, *inter alia*, the European

Investment Fund of the European Investment Bank, the European Business Angels Network, the French Business Angels organisation, Deloitte's PLC and others. I was extremely well supported in the Centre's role of academic observers, documenters and 'critical friends' of the AFIBIO project by research associate Fangzu Zhang with whom the original report was written. It was a real joint effort and I wish to acknowledge her role in this shortened and updated version of the research we conducted.

Chapter 10 also contains a summary of findings from an earlier EU biotechnology project, BIOLINK on bioincubation and co-incubation where, accompanied by colleagues Dan Kaufmann and Chen Levin of the Jerusalem Institute for Israel Studies, and Avri Harmon of the Ben Gurion University of the Negev, Be'er Sheva, we played a comparable critical documentation role of the incubation services at Oxford BioTechNet (UK), Consorzio 21, Sardinia, Italy, BioM, Munich, Germany, Genopole, Evry, south Paris, France, and Hadasit, Jerusalem, Israel. For those interested in these highly 'close to market' aspects of biotechnology, some of the BioLink findings not utilised here were published in Cooke *et al.* (2006). I wish deeply to thank all the mentioned institutes and participants in these two projects for the enlightenment and consternation, in equal measure, that the world of biotechnology intermediary activity and entrepreneurship bestowed upon all of us involved. Finally, Chapter 11 summarises, unifies and reflects upon the content and results of the foregoing ten chapters. Rather like Chapter 1, it snakes around the ethical, the sceptical and the empirical, trying to find a fulcrum for this complex science-based, sometimes breathtakingly roguish, but always remarkably innovative 'growth culture'. It also contains numerous *innovation biographies* as we refer to the target methodology of the EURODITE project. Here they range from organic restaurants and ecological supermarkets to innovations in endoscopy and the treatment of breast cancer. These are carefully selected to stress the underlying economic sociology and evolutionary economic geography findings that embeddedness, socio-spatial interaction and collaborative knowledge interchange remain fundamental to socio-economic endeavour. This is no matter how much emphasis society's masters spin it as unavoidably competition, rivalry and individual struggle. And this reminds us of the importance, at the very least for social scientists, of continuing attachment to the possibly crumbling Mertonian norms, particularly the one that still prevails in science and ironically, among venture capitalists, as the meta-norm of *scrutiny*. The usual disclaimer applies.

Phil Cooke, Cardiff
29 December 2006

1 Growth cultures

Meaning and interpretation in the knowledge age

Introduction

We live in unstable times. I look out of my window as an 'adjunct academic' in building 13 at Aalborg University, at grass being cut by a dark green, unmanned, electronic Husqvarna robot mower. I recall first seeing something like it at EPCOT, Walt Disney's futurama in Florida, around 1995. Now it bumps gently against the wall under my window, reverses slightly and redirects itself to cut the difficult grass closest to the wall. No gardener can be seen, only the small trolley on which it arrived and to which it will return, parked on the pathway. For the university, with an attractive 1970s campus swathed with grass that flows into all the awkward corners produced by its design, modelled on North African village architecture, the robot mower makes economic sense. For the academic labour in the buildings, the Husqvarna is immeasurably quieter and cleaner than its petrol-driven predecessor that in any case does not function well in tricky corners. But for labour it signifies the challenge of growth without the necessity of direct labour application.

On the other hand, as the advertising material says:

> 'the Auto-Mower will constantly and quietly keep the lawn trimmed – even recharging itself – while you find something a little more interesting to do with your weekend leisure time.
>
> This green-thumbed robot is capable of maintaining a lawn area of between 1,200 and 1,500 square metres completely autonomously once the low-current perimeter wire is installed. The boundary wire is stapled or buried in the ground to stop the Auto-Mower from leaving the lawn and wandering into the neighbour's veggie patch.
>
> Available in either a battery or solar powered version, the Auto-Mower parks itself on the charging unit when its power is low by following a 'search wire', or in the case of the solar model, stops moving to allow time for the sun to recharge its batteries.
>
> Virtually silent and able to work in any weather conditions, the Auto-Mower is designed to continuously cut in a random pattern.

> This keeps the grass at a consistent height and clippings are fine enough to be left on the lawn as mulch, avoiding the need to use a catcher. Available from Husqvarna dealers, this escape route from what must be the most hated of Sunday afternoon chores costs €2,034 including the charging station but not installation.
>
> (www.automower.com)

Clearly a small object of desire, allowing the owner to utilise the time saved by this release from physical exertion normally associated with lawn mowing, even of the electrically or petrol supported kind, to go to the gym, go for a swim or hike up a mountain. Yet more opportunity for consumption choices – keeping or getting healthy on our own terms. In spending the €2,000 on the mower, we release ourselves to spend a further few hundred or thousand on sufficient hiking gear and health club plus appropriate gymwear and swimwear to be inconspicuous yet noticeable and possibly ourselves further small objects of desire.

Growth cultures

In such ways is the modern economy a *growth culture* in which time expended performing selfless, unrewarded acts of community or household service may appear to GoreTex or Husqvarna to be consumption 'down-time'. Similarly, other Sunday 'chores' like talking to partners, children or dogs while out on a walk in the park, wearing old jeans and jumpers from the 1960s is, from this perspective, unproductive. Such talk might be about the growing, picking, purchasing and preparing of a Sunday dinner with vegetables from the garden and meat delivered or picked up from the local butcher. If the former, it might come weekly from an organic farm, possibly accompanied by organic vegetables. Or their organic or non-organic equivalents might arrive, similarly delivered in a small white van, from the supermarket responding to our Internet order. Alternatively, the plan might not involve home-cooking but a meal out. This is something that some 24 million Britons had done at least once per year by 1996, and many much more according to Warde and Martens (1998) who noted a doubling since 1993. During approximately the same years the number of Britons actively dieting had risen to 12.8 million. The USA Department of Agriculture (1999) report based on food consumption surveys noted that food obtained by 'eating out' – restaurants, coffee houses, vending machines, cafeterias and fast food combined – had less fibre (25 per cent less), iron levels 29 per cent below recommended daily intakes, and calcium 20 per cent below recommended minimum levels compared with food obtained by 'eating in'. The frequency of eating out has almost doubled in the last 20 years. About 30 per cent of meals were eaten out, a trend expected to continue, primarily due to low prices for such food, aided by dual-income families with little time and energy left to prepare food at home. Thus one rather pessimistic scenario that could be constructed from such data, is that far from utilising time saved from such 'chores' as mowing lawns by exercising, substantial numbers of the population would use it for eating relatively unhealthily. This is the essence of 'growth culture'.

Healthcare

This brings us, conveniently, to the subject of healthcare, since a major Western concern directly connected to the foregoing is the dramatic rise in the numbers of people who are negatively affected by eating disorders, notably those that result in obesity. To quote British statistics once more, during the last 25 years, the rate of obesity has quadrupled in the UK. In England alone, 22 per cent of men and 23 per cent of women were classified as clinically obese in 2002, while 43 per cent of men and 34 per cent of women were overweight. This meant that over half of all adults weigh more than their recommended weight. In the USA obesity rates increased from 15 per cent of the population in 1980 to 31 per cent in 2000. In Canada, almost 50 per cent of the population is either overweight or obese. Fifty-six per cent of adults and 27 per cent of children in Australia are either overweight or obese. Obesity can lead to a variety of health problems, including high blood pressure, coronary heart disease, osteoarthritis and diabetes. A UK National Audit Office study from 2002 estimated that obesity cost the National Health Service €750 million a year, and the cost to the country overall has been put at up to €10 billion per annum.

On a far larger scale than that implied in the Robot Mower illustration, these statistics bear out the defining aspects of *growth culture* whereby scale economies applied relentlessly over decades in the world food industry have resulted in systems of production, distribution and consumption that may be literally lethal to consumers. Just a few examples of the perversity of scale production in agro-food can illustrate the key issues here. One concerns rising perinatal mortality among Holstein cattle breeds, the world's greatest milk producers, inbred for decades to achieve massive milk yields per cow. Adamec *et al.* (2006) recorded rising stillbirth scores from producers on 120,434 US Holsteins 1985–1996. Stillbirth scores 3 to 5 were coded as difficult births leading to deaths within the first 48 hours after birth. The observed effects of inbreeding were consistently unfavourable. In other words, Holstein inbreeding to increase milk yields has given rise to barriers of a genetic and physical kind that substantially constrain natural birth processes. The limits have clearly been reached of scale-methodologies typically associated with generic, massified and cheap food production based on successive rounds of inbreeding. A different illustration concerns the long-observed problem of Danish pigmeat, produced under some of the most intensive, indoor production programmes in the world. This is the economy that pioneered mass food-production following the agricultural crisis of the 1880s and the opening of large-scale dairy farms in Jutland. Such is the crowding of animals in production that pigs are routinely injected with penicillin to counter disease. As immunity builds up, more and more massive doses have to be given. Consumers of Danish pigmeat absorb the injected penicillin in above-normal quantities and if unlucky to catch a disease, especially a foreign strain such as *staphylococcus* from China, they prove remarkably resistant to treatment due to the penicillin absorbed in eating pork products. Hence, massification in the second case directly affects the health balance not only of the animals but of consumers of intensively farmed food products.

However, in the face of such difficulties for animals and humans occasioned by mass food production, upon which the profits of agrobusiness and supermarkets largely depend, come a range of responses. In the food sector, one approach that is taking on the character of an embryonic new social movement is that which rejects mass food production and favours its small-scale, localised, seasonal, often organic opposite. Growth in the sale of organic produce has run at 10–15 per cent in Europe and 20 per cent in the US in the first decade of the twenty-first century. In Europe this has been stimulated by the lethal nature of some food production causing deaths by 'mad cow' disease, *E-coli* bacteria, especially among the aged and young children, and more specific 'food scares' such as the US study showing that consumption of processed meats increased the risk of pancreatic cancer (Nothlings, 2005). According to the research findings, people who consumed the most processed meats (hot dogs and sausage) showed a 67 per cent increased risk of pancreatic cancer over those who consumed little or no such meat products.

A different approach concerns 'functional foods' which are biotechnologically derived and promise, as *nutraceuticals*, to bring health gain to customers through their health-enhancing contents. Perhaps the best-known are the *lactobacter* and related 'healthy bacteria', dairy products like Yakult, Benencol, Activia and the lesser-known though pioneering ProViva. Activia by Danone, for example, is a low fat yogurt that contains *Bifidus regularis*, a natural probiotic culture that helps regulate the digestive system by helping reduce long intestinal transit time. *Bifidus regularis* is a live, 'friendly' bacteria claimed to be beneficial when eaten daily. ProViva, invented by Swedish company Probi, a progenitor of this kind of digestive support ingredient, failed to market its juice product as well as its competitors, possibly through entering the market too early and with insufficient preparation of consumer product consciousness. The firm was founded in 1991 and the health drink product was launched in partnership with Skåne Dairy in 1994, *Lactobacillus plantarum 299v* being the active ingredient in this case. Yakult is the oldest dairy product and its Japanese founder Minoru Shirota's goal was to contribute to human health by acting upon Nobel laureate Elie Metchnikoff's idea of 1900 that a person's health could depend on the condition of the intestines. Metchnikoff, a researcher at the Pasteur Institute in Paris, developed a fermented milk drink as a carrier for the *Lactobacillus casei*, influenced by the remarkable life expectancy of Bulgarians and their common practice of drinking 'fermented' milk. Through Yakult, Shirota introduced one of the world's first probiotics. Benecol is a well-marketed Finnish variant on this general theme. Research published in 2004 showed that most of the claims of beneficiary health effects from these products were unfounded and that, at best, they do no known harm and at worst can induce negative reactions (Adams, 2004).

Hence, we return once more to the *growth culture* tendency to operate along lines reminiscent of Bertolt Brecht's *The Resistible Rise of Arturo Ui* in which, it will be recalled, a gangster mob threatens the peace of the city of Chicago and then offers to police the resulting social disorder, an offer readily accepted by Mayor Dogsborough, thus ushering in the rise of the gangster to power in the city. Modern market systems induce perverse over-consumption effects on the way up, as it were, earning profits in so doing. Among such negative effects can be important incidents in which consumer health is threatened in the ways outlined. Then they make possibly even greater profits by concocting apparent remedies to the pathologies they have induced in the identical consumers. This may be even better expressed by reference to the contribution of the Nobel laureate in economics, George Akerlof, who introduced the important concept and subsequent theory of *asymmetric information*. Akerlof's metaphor, embodied in the title of his prize-winning contribution, was that of the used-car dealer, knowledgeable, as few customers could ever be, of the likelihood that the car in question was a dud. 'The market for "lemons"' was the memorable title of his first scholarly article, which was turned down for publication by the first four leading journals he tried, on grounds of 'triviality'. But it gave rise to numerous fields of economics, including studies of market failure, adverse selection, the nature of contracts, and principal-agent relations, some of which have also exercised the minds of economic sociologists (Edling and Stern, 2003; Shapiro, 2005).

In the 'knowledge economy' the opportunities for such chicanery have expanded exponentially given that there is vastly more knowledge and its codified form – information – available in society. However, it is by no means ubiquitously available and, in the variety of forms knowledge may take, may appear as an array of impenetrable knowledge monopolies. One important field, which is defined by such concerns, forms the focus of this book, which is the modern pharmaceuticals industry. For in a massified world of production of generic drug solutions to healthcare problems associated with the health deficits noted above, pharmaceuticals companies have become modern behemoths. But in the face of new knowledge deriving from the research and innovation associated with biotechnology, they can also look dinosaur-like in their inability to adapt to the biotechnology paradigm. Hence, pharmaceutical biotechnology is instructive from both the perspective of the sociology of consumer knowledge and, less commonly, the knowledge revolution associated with the rise of microbiology and more recently genomics within business. In this respect, biotechnology has produced remarkable and even pioneering shifts in established power structures. These have limited potential to moderate some of the knowledge asymmetries noted above. This is mainly because the force propelling new biotechnology knowledge is coming from the *public* financing of *public* research. The market sits atop this knowledge driver facilitating but not originating commercialisation of new knowledge in the form of innovative therapeutic treatments. And of course, as Owen-Smith and Powell (2004) underline, for all the caveats about confidentiality regarding pipeline agreements expressed in tightly worded contracts between public researchers and private innovators, public science is *open science* or what, in sociological terms, is normally referred to as following *Mertonian* scientific norms. This makes it likely that there will be leakage of such supposedly privately appropriated knowledge in normal discourse between public scientists.

Biotechnology is in the front line of the debate about private appropriability of publicly funded knowledge. Academic entrepreneurship has risen in profile and even when exaggerated claims are made regarding, for example, the number of new firms arising annually from it, there is nevertheless a seemingly desperate urge by university administrations and governments to expand it. Statistics show that in most advanced economies over half of academic entrepreneurship is accounted for by biotechnology (Shane, 2004). In the US, scientists are routinely required to make disclosures to their university knowledge transfer office if they make a discovery that might have the chance to become a commercial innovation capable of earning returns for both university and academic. Accordingly, the decision as to whether or not and how much patent variety should be contemplated may be taken increasingly away from the scientist. The implications for patenting, especially broad patenting, which can cover a sizeable knowledge sub-field in an innovative area, may also be subject less to scientific and more to economic calculation, as we shall see later in this introduction. Licensing of patented knowledge to pharmaceuticals firms who innovate new treatments as expensive as the affluent US healthcare market will bear then creates difficulties for poorer countries who may need but cannot afford such treatments. This has led to disaffection, critique and deligitimisation of the patenting concept accordingly.

Merton and biotechnology

Before reviewing and introducing the intellectual property miasma that biotechnology has given rise to, particularly since genome decryption and encryption became its leading edge scientific impulses, a reminder of the norms of Mertonian science is in order. They are the five that follow, mnemonised from the capital letter of each heading as CUDOS:

Communalism – The findings of academic science are public knowledge rather than private knowledge.

- secrecy is prohibited in the sense that it cannot carry weight or be given credit in scientific discourse;
- dishonesty is not tolerated and mutual, personal trust is the norm;
- findings are accepted as primary literature in the public record only as a result of peer review;
- the ethic of communalism helps to explain the importance of empiricism and intersubjectivity in scientific epistemology; epistemic strategies include quantification, measurement, standardisation (and calibration), instrumentation, experimentation, and verification

Universalism in science has two faces:

- contributions to science cannot be excluded because of race, nationality, religion, social status, gender, sexual preference, or other irrelevant criteria (i.e. science is multicultural); science strives to be a meritocracy; the scientific community attempts to be democratic and fair to its citizens;
- empirical 'facts' have to be consensible (clear and unambiguous) and the scientific 'theories' that interpret them have to be consensual; the categories of 'fact' (representation of experience), 'taxonomy' (classification of facts), and 'theory' (explanation of the classification) have to be shared among the scientists.

Disinterestedness – Objectivity is an attempt to separate the political, religious, economic, personal and vested interests of scientists from their findings. All reference to economic, political, religious or other social interests is routinely and systematically excluded.

- Scientists are motivated by the same things that motivate non-scientists, but their work must reflect their disinterestedness in any particular outcome. They must be open to conclusions that are most consistent with their results.
- Academic science tries to attain consensual objectivity or *intersubjectivity* by merging individual interests into a collective enterprise; the purpose of the knowledge is often said to be 'for knowledge's sake', as if there were no interest in the uses to which the knowledge could be put.
- When scientists present themselves or their work to the scientific community, they are 'deferential' to their own observations and results and to previously reported observations and results.
    - Scientific papers are written in a neutral, impersonal, passive voice to minimise the personal and social context of the research.

- An impression of humility is conveyed by systematically citing formal scientific sources for everything that is not entirely their own work.
- Although impersonally written, research papers are never anonymous; credibility is the prime personal asset and reliability the prime social asset.

- Thus, one role science has to offer a democratic society is as independent arbiter of social questions (of a scientific nature) that otherwise could only be resolved by reference to political, religious, or economic institutions.

Originality – Scientists are self-reliant, independent thinkers who are trained to be original – whether choosing a research question, deciding on an approach to address it, or finding a way to convince others of their findings. Scientists must display novelty in order to gain maximum credit; when they publish, something about the research has to be new; thus, freedom or independence is a necessity in science, generally referred to as *academic freedom*. Moreover, the right to dissent is absolutely critical; this need for originality explains the emphasis on research rather than, say, scholarship and the drive towards specialisation. Research areas stimulating projects and proposals have to be formulated and in some sense discovered.

Scepticism – This is an attitude of questioning and doubt that defers judgement until supporting evidence is available. Scrutiny of research claims is a hallmark of good science. The scientific culture is an institutionalised context for argumentation.

- Peer review allows the most important moment for scepticism to be exercised and is the key institution in scientific culture. It allows scientists to be held accountable to a community, rather than a superior authority.
- Peer review occurs at both ends of the process: in funding research proposals and in accepting research reports for publication. Credibility and relevance are two important criteria when reviewing research; verification (via reproducibility) is a 'triple point', where psychological, social and epistemic considerations come together in a critical arena. A further degree of confidence is achieved by 'validation by triangulation' when different approaches give the same answer.

Consider, in light of these norms, the following four recent experiences arising in biotechnology, and more broadly in the interface between biotechnology research and entrepreneurship, particularly that involving contemporary commercial pharmaceuticals business practices.

Communalism – 'The findings of academic science are public rather than private knowledge'

The first genetically engineered organisms were created in the United States in 1973 by academic scientists Herbert Boyer and Stanley Cohen. With venture capitalist Robert Swanson they formed Genentech, the world's first modern biotechnology company, to exploit for profit the results of publicly funded research. The genie was out of the bottle. By this time, mainstream sociologists of science had largely abandoned models of scientific research based on Merton's theory of scientific norms. The reason relates to the historical development of the discipline: early philosophers and sociologists of science regarded science as unique among intellectual activities, and in particular, they thought that science as a discipline was defined by a high level of agreement among scientists about assertions of fact; for this reason early sociologists of science, including Merton, were chiefly concerned with constructing models to explain the phenomenon of scientific consensus. However, during the 1960s and 1970s – influenced by developments in the history and philosophy of science – sociologists of science began to take a more cynical view of competition and collaboration within scientific communities. Questioning the existence of any distinctive scientific ethos, they turned away from an idealised picture of consensus among scientists and instead became preoccupied with studying scientific debate and disagreement. As a result of this shift in focus, many of what had been central issues in the classical sociology of science came to be generally neglected, among them issues relating to intellectual property and the openness of scientific communication. Moreover, because the new 'sociology of scientific knowledge' incorporates a diverse range of theoretical approaches, its insights have not been as easily accessible to scholars outside sociology as those of Merton and his colleagues.

Today, most sociologists of science doubt the reality of normatively controlled behaviour, preferring instead to treat references to norms in the course of scientific debate as either mere rhetorical tools or rationalisations for interest-driven behaviour. In light of actual scientific practice, this idea may be unbalanced, as the following analysis indicates. Conflict surely exists, as does fraud, but in general terms certain Mertonian norms – notably that concerning scrutiny through replicability of experiments – must remain an abiding norm for deviance to be exposed as frequently as it is. In other words, despite infractions the system still works. The idea of scientific norms remained influential both within and outside the discipline of sociology, and when questions about the impact of commercialisation on scientific research attracted the attention of intellectual property lawyers in the late 1980s, possible effects on the normative structure of science provided a natural starting point for discussion (see, for example, Kellogg, 2006; Shorett *et al.*, 2003). This period, during which the economic returns to patented biotechnological knowledge began to be seen as remarkably abundant for some of the first-mover 'academic entrepreneurs', heralded a change in institutional mentality towards privatisation of the profits from exploiting publicly funded science. Universities, with few if any regulations until the 1980 Bayh-Dole Act in the US, began to see an attractive new source of income, while some 'star' scientists in leading bioscientific research centres added the aspiration of wealth to the nobler one of a Nobel laureateship in their 'ambition portfolio', as we might term it. Governments, often looking for less 'feather-bedded' university research, notably in the UK, began to assess performance and cut and reward accordingly in ways that, for the first time, prioritised 'enterprise' and bemoaned the failure to patent breakthrough discoveries from public science by the likes of Milstein and Köhler at the Medical Research Council laboratory in Cambridge, home to some thirteen Nobel laureates. Industry, in the guise of 'big pharma' came to see that its interests were also served by outsourcing R&D to such knowledgeable 'academic entrepreneurs', their laboratories *and* their 'dedicated biotechnology firms' (DBFs).

What some perceived as a benign 'triple helix' among the three key interests; academe, government and industry (Etzkowitz and Leydesdorff, 1997) that sprang first from biotechnological commercialisation in the 1970s and Genentech, could appear more like a 'witches brew' when such firms began to file for 'broad patents' covering, for example, good prospect gene sequences in the early 2000s. In hindsight these transgressed the patenting norms of precision and originality (novelty) and should never have been allowed, as they threatened to paralyse academic research in those patent fields. Thankfully, following Cohen (2005) we find that such broad patents are either known about and ignored by a minority of academics or not known about and thus ineffective for the majority of academic researchers in the patent field. However, it is a test of the 'triple helix' that has yet to materialise in practice, though it clearly exists in principle and a single challenge in the courts by a patent-holder against the university in question would produce a radical re-appraisal of such insouciance.

Universalism – 'Science strives to be a meritocracy; the scientific community attempts to be democratic and fair to its citizens' and 'Categories of "fact", "taxonomy" and "theory" have to be shared among the scientists'

Regarding both of these, probably the most egregious case of science *not* being driven by meritocracy concerned the *Heliobacter pylori* – the cause of peptic ulcer. It had been known since 1874 when Arthur Boettcher reported on the small, curved bacteria found repeatedly in stomach ulcers. Subsequent research by many scientists confirmed this. Law (2006) refers not to 'evidence-based' practice but to 'eminence-based' practice as one explanation for the way outmoded standards caused non-epidemic infections to be overlooked. More suspiciously, antibiotics by the 1940s had begun to work on stomach ulcers. Pharmaceuticals companies were complicit with medical professionals for the disappearance of references to the Bottcher work on the infectious origins of stomach ulcers from the medical textbooks. For Glaxo, for example, Zantac was the first blockbuster drug, a stomach ulcer treatment that was the world's highest earning drug at $2.4 billion in 1990. The same year Zantac was launched, Dr Barry Marshall, a scientist at Royal Perth Hospital in Australia, swallowed a cocktail of *H. pylori* from a dyspeptic patient, developed stomach ulcers and then treated them with antibiotics, recovering in twenty-four hours. However, it was not until 1995, after thousands of patients had suffered and died, that Glaxo produced the first *H. pylori* eradicator drug. Little-known researchers in far away places with inconvenient facts were disenfranchised from scientific *universalism* with positively lethal effect. Subsequent studies to Marshall's, even on patients with 'intractable' ulcers, showed them cured by antibiotics. As Le Fanu (1999) noted, 'There could be no more powerful illustration of the importance of knowing the cause of a disease' (p. 184). For his trouble, Marshall was awarded the Nobel Prize for Medicine in 2005.

Disinterestedness – 'To separate the political, religious, economic, personal and vested interests of scientists from their findings' and Scepticism – The attitude of questioning and doubt that defers judgement until supporting evidence is available

A curious case of most of these infelicities, by both scientist and the peer review system, concerns the case of Dr Jan Hendrik Schön, of Bell Laboratories, New Jersey, who briefly rose to prominence after a series of apparent breakthroughs in research on biomicroprocessors. This use of organic molecules in microelectronics first came under suspicion in 2002 when graphs in two different papers were revealed to be impossibly identical (Brumfiel, 2002). He had been a recipient of the Otto-Klung-Weberbank Prize for Physics in 2001, the Braunschweig Prize in 2001 and Outstanding Young Investigator Award of the Materials Research Society in 2002. These were later discovered to have been based on fraudulent research. The Schön scandal provoked discussion in the scientific community about the degree of responsibility of co-authors and reviewers of scientific journals where peer review is designed to find errors, as well as determine relevance and originality of papers, rather than detect fraud. It was considered seriously disturbing to the scientific community that none of Schön's malfeasances were picked up in the peer review process.

In May 2002, Bell Labs appointed a committee to investigate the possibility of this being a case of scientific fraud. The committee sent questionnaires to all of Schön's co-authors, and interviewed his three principal co-authors (Zhenan Bao, Bertram Batlogg and Christian Kloc). They examined electronic drafts of the disputed papers, which included processed numeric data. They requested copies of raw data but found that Schön had kept no laboratory notebooks. His raw data files had been erased from his computer. According to Schön, the files were erased because his computer had limited hard drive space. In addition, all of his experimental samples had been discarded or damaged beyond repair. In late 2002, the committee publicly released its report. The report contained details of 24 allegations of misconduct. They found evidence of Schön's scientific misconduct in at least 16 of them. It was shown that complete data sets were re-utilised in a number of different experiments. Graphs, purportedly plotted from experimental data, had been produced using mathematical functions. It was reported that all of the misdeeds had been performed by Schön alone. Bell Labs fired Schön on the day they received the report. It was the first known case of fraud in the history of Bell Labs. In 2002 *Science* withdrew eight papers written by Schön and in 2003 *Nature* withdrew seven papers written by Schön. Why did he do it? Career pressure, conviction of the rectitude of his beliefs, and operating in a difficult-to-replicate experimental field are three reasons advanced by close observers. But Schön himself has yet to admit he did anything wrong.

Originality – 'Scientists are self-reliant, independent thinkers who are trained to be original'

The almost identical case of Hwang Woo-Suk illuminates the quest for the kind of originality that will surely result in a Nobel Prize and possibly great riches as well as international celebrity status. Hwang had established a celebrity status in South Korea through his transgenics research. In 2004, he announced the creation of an embryonic stem cell with the somatic cell nuclear transfer method, following up with a paper in *Science*. This was surprising since it was thought that creating a human stem cell was inordinately complex. He explained that his team used 242 eggs to create a single cell line. He was fêted with commemorative Korea Post stamps displaying Lazarus style imagery. In 2005 he claimed to have created 11 stem cells using 185 eggs, publishing the findings in *Science*. This was perceived as a biotechnological breakthrough since the donors were of different ages and gender. If true, it would have meant that any patient could receive designer stem cell treatment. Later that year he claimed to have cloned an Afghan hound. Accordingly he was made head of the World Stem Cell Hub, a leading stem cell research centre, whereupon University of Pittsburgh co-researcher Gerald Schatten announced he was ceasing collaboration with Hwang on ethical grounds relating to the sourcing of eggs for experimentation. It transpired that these had been purchased from women by the head of MizMedi Women's Hospital at $1,400 each, a breach of ethical conduct. Supporters of Hwang's work, including the Korean media, called criticism of his work 'unpatriotic'. Hwang claimed in interview, 'I was blinded by work and my drive for achievement . . .'. At the end of 2005 his university determined that all 11 of Hwang's stem cell lines were fabricated and in 2006 *Science* withdrew his papers based on the fabricated results.

Ghostwriters, Pascal's Quadrant and the Oncomouse

In the 'falsificationist' sense, these cases and a number of others mostly from medical bioscience, including the Abderhalden, Alsabti, Baltimore, Benveniste, Breuning, Czaja, Darsee, McBride, Van Parijs, Poehlman, Ricaurte, Sudbø and Summerlin scandals show the Mertonian norms to have been well and truly destroyed – if indeed they ever existed in the minds of scientists. But from a 'distributionist' perspective, these scandals are statistical outliers in relation to a huge bell-curve inside which ethical science approaching those norms continues to be practised. Moreover, although it took time and raised serious questions about the peer review and 'community of practice' presumptions of collaborators, the system of determining what constitutes scientific truth and fakery worked, usually when someone else tried to replicate the fraudulent results and failed. Of course, there may well be more such fraudulence awaiting discovery, a factor which contributes to a general weakening of trust in the scientific ethics system. However, it is when such *examination* knowledge from the testing and patient trialling process reveals problems, such as dangerous side-effects of drugs upon patients, and is deliberately not disclosed, as has happened with Merck's Vioxx and other cox-2 inhibitors (painkillers) or Glaxo's Seroxat (Paxil) antidepressant, that a crucial ethical line has been crossed. From a UK sample of 2,232 patients, the latter drug was linked to 16 suicides, 47 attempted suicides, 92 patients who thought of self-harm or harm to others, and 19 children who suffered strong side effects. Responding to media pressure, Glaxo in 2003 corrected its initially published estimate of the likelihood of withdrawal symptoms from one in five hundred prescribed the drug, to one in four (Law, 2006).

Two other aspects of the 'culture' of bioscience, enveloped as it seems by the power and influence of multinational pharmaceuticals firms, give cause for anxiety when viewed from a social scientific perspective in which such mega-sponsors are largely unheard of. The first of these is what we shall term the ATBA or 'author to be announced' syndrome. That is the common practice whereby hundreds of articles in medical journals carrying academic authorship have been penned by ghostwriters hired by drug companies. The practice is sufficiently blatant for titles of articles to appear on company websites, carrying in place of the real author, an ATBA notice, inviting academics or doctors interested in selling their name and improving their published output to sign in. In this way an article 'hiding' inconvenient trialling results can clearly be slipped into often highly influential sources based on which doctors will prescribe drugs and hospital treatments. Barnett (2003) found evidence that many articles apparently written by independent academics were authored by writers working for agencies in return for large payments from pharmaceuticals companies to persuade readers of the superiority of their product. Reminiscent of the worst instances of 'tobacco science' whereby tobacco companies gave substantial research grants to life scientists to disprove the link between smoking and lung cancer, these practices reveal the dangers of what Murray and Stern (2005) refer to as 'dual knowledge', namely knowledge that has both pure and applied purposes, thus enabling scientific frontiers to be pushed and commercial gain to be made.

The *New England Journal of Medicine* was forced to retract an article published in 2002 by doctors from Imperial College, London and the National Heart Institute relating to cardiology when it was reported by one of the eight co-authors that they were not responsible for the research and that he had never even seen drafts of the paper before it was published. In the same year, the *Journal of Alimentary Pharmacology* published an article that involved a medical writer hired by AstraZeneca – a fact that was not revealed by the author, who merely acknowledged her contribution. The article plugged AstraZeneca's Omeprazole gastric ulcer treatment, even though it had worse side effects than comparable treatments. A letter posted on the *British Medical Journal* website reported the lengths to which medical writing agencies go to disguise academic papers ghostwritten on behalf of pharmaceuticals companies:

> One standard procedure I have used states that before a paper is submitted to a journal electronically or on disc, the editorial assistant must open the file properties of the Word document manuscript and remove the names of the medical writing agency or agency ghostwriter or pharmaceutical company and replace these with the name and institution of the person who has been invited by the pharmaceutical drug company (or the agency acting on its behalf) to be named as lead author, but who may have had no actual input into the paper.

The European Medical Writers Association (EMWA) now has largely unenforceable 'guidelines' for academic authors that say a 'substantial portion' of the research and written article should have been conducted by them if they are to be credited as authors (Jacobs and Wager, 2005).

The second element of the 'culture' of bioscience's imbrication in commercialisation and pursuit of economic growth – hence its rather unusual 'growth culture' compared to most sciences – concerns, once again, the implications of patenting for research. In this case, the 'predator' was not initially a large pharmaceuticals company but an academic discovery patented and heavily promoted 'pharma-style' by an academic institution, in this case Harvard University, one of the world's leading academies. In Murray and Stern's (2006) summary of the case of the Oncomouse, they illustrate the dilemmas of 'dual knowledge' by reference to Stokes's (1997) analysis of what he termed *Pasteur's Quadrant*. Stokes had noticed the rise of a new kind of knowledge different from the classic 'pure' or 'disinterested' knowledge reified by Merton (1973) as well as 'applied' knowledge capable of commercial exploitation. Nils Bohr is the eponymic occupant of the former quadrant in the two-by-two diagram, while Thomas Edison eponymises the latter. Louis Pasteur's fermentation discoveries were of immediate influence on the development of the germ theory of infection and the French brewing industry. The axes of the diagram are two questions, one about disinterestedness and the other about the utility motivating research, as shown in Figure 1.1.

Operating in Pasteur's Quadrant – in which bioscience lies – poses scientists with three choices or dilemmas:

- Should they operate in the institutional régime of Mertonian 'open science' or should they participate in a 'private property' régime? This is at the heart of the current dilemma that is central to this book, namely the commercialisation of the university and privatisation of knowledge that has been won from public funding sources, including not only the research grants but also the scientific infrastructure. This is where the 'growth culture' has increasingly moved to as university science outperforms corporate research capabilities.
- The 'publish or patent' dilemma arises when a piece of academic research that should ascribe to the Mertonian publication norms may not be published but rather held back from public view before patenting. Patenting does not amount to secrecy as Murray and Stern (2006) imply. Indeed patents are precisely licenses whereby interested observers must pay to utilise the knowledge contained. But not publishing and not patenting in order to gain some future commercial benefit seems well beyond the border marked 'ethical' unless the science is

Quest for Fundamental Understanding?	Considerations of Use? No	Considerations of Use? Yes
Yes	Pure Basic Research (Bohr's Quadrant)	Use-inspired Basic Research (Pascal's Quadrant)
No		Pure Applied Research (Edison's Quadrant)

Figure 1.1 Pasteur's Quadrant

privately contracted. Of course publishing and patenting together are not mutually exclusive.

- Operating in two or more régimes causes potential dilemmas and conflicts regarding disclosure and ownership of research findings. It is not clear whether and what transgressions occur through the individual academic shifting régimes, for example contracting development to a firm of research funded publicly, and what proof of evidence would be legally implicating. Although tested to some extent in law courts, public policy is by no means pellucid in guidance.

Now the context is set, back to the Oncomouse. It was developed by a team led by Philip Leder, head of Genetics at Harvard Medical School and the research published in *Cell* in 1984. It was a mouse that had been genetically engineered by inserting an oncogene into a mouse embryo, thus making the mouse particularly susceptible to cancer. Its use for identifying genetic sources of cancer was varied and extensive. Scientists shared mice and developed their own. In 1988 the Oncomouse discovery was granted a broad and controversial patent, controversial because the licensee Dupont was aggressive in enforcing its property rights. This was the first genetically engineered animal, and as discussed earlier in this chapter, the prospects of researchers being unable to proceed untrammelled by the conditions of a broad patent brought forth anxiety and anger. Dupont also wanted rights to innovations made using the mouse, demanded early sight of publications, prohibited mouse-sharing among scientists, and charged commercial rates for the mice. Scientists resisted Dupont but did not seek the courts to narrow the patent. Rather they lobbied the National Institutes of Health, whose head Harold Varmus negotiated improved but not entirely unrestrictive terms with Dupont. Scientists also learned the power of patenting and filed many more patent applications than hitherto. This was seen as strengthening their hand in commercial negotiations with industry, improving their academic profile towards employers, other academics and industry, and in some cases enabling them to raise their income – although patents can be costly both to acquire and, more importantly, to defend in court.

Thus we have come to the nub of this book. How, it asks, did this heightening in the complexity of research life occur with knowledge capabilities seesawing away from large corporations where hitherto Chandler's (1988) insights into the power and control of peak assets like R&D by firms with globe-spanning scale had reigned supreme? How did university scientists and small 'dedicated biotechnology firms' (DBFs) come to dominance at the innovation end of this particular value chain? And why is the economic sociology of the university-DBF alliance against, but also, paradoxically, in a relation of dependence upon 'big pharma'? Where do the lines and spaces of key knowledge domains lie and are these shibboleth-shaking characteristics spreading from biotechnology to other Chandlerian industries dominated by a few large, global corporations?

The book thus proceeds as follows. In Chapter 2, a theoretical framework is developed that is germane to the idea of 'growth cultures' both in the narrow sense in biotechnology, and in the broader sense of economic growth more generally in the burgeoning 'knowledge economy'. In this, attention will be drawn to aspects of economic sociology, knowledge science and evolutionary economic geography that combine fruitfully to offer a general explanation of the key processes under investigation. This is followed by a chapter (3) in which the implications are traced of the switch from small molecule chemical and biotechnological methodologies to the large molecule interest consequent on the rise of genomics. This is again referred back to theory in seeking to explain the emergence of 'knowledge domains' in which specific institutions like Harvard Medical School so dominate the publication fields associated with biotechnology. Chapter 4 is about the organisation of the biopharmaceuticals industry and its validity as a wider industrial organisational model, rooted in the idea of 'knowledge entrepreneurship'. Then in Chapter 5 the spotlight falls upon academe, posing the question: How did research institutes displace corporate R&D in bioscientific knowledge leadership? Chapter 6 is about shifts in 'big science' and science policies associated with the rise of bioscience and threats from newer social concerns, notably climate change, that may trim its growth. Chapter 7 focuses upon the market sector of DBFs and their economic sociology model of clustering, along with their complex relationships with 'big pharma' on whom they are dependent for so much despite their superior knowledge. Chapter 8 poses questions as to whether bioscience's globalising tendencies have implications that include or exclude developing countries. Chapter 9 begins a re-focus on bioscience in non-pharmaceuticals fields, notably agro-food, followed by Chapter 10 on bioenvironmental and bioenergy and biofuels technologies and growth cultures. The concluding chapter (11) reviews what has been learned from the process termed by that point the 'globalisation of bioregions'.

2 The knowledge economy and growth cultures

A theoretical framework

Introduction

The complexities of where and why economic growth occurs nowadays are difficult to pin down, especially if using growth models that are not accomplished at dealing with the exigencies of the knowledge economy. Hence a first task is to seek to show how 'knowledge economy' conventions create different and distinctive demands of people and places from the prominent, century-long and more, effects of industrialisation and what came to be called the 'Industrial Age'. This is performed in the first main section of Chapter 2, drawing on features introduced in Cooke (2002) and others, like Törnquist (2002) and Burton-Jones (1999) who pay attention to issues concerning human intellectual and scientific performance in identifying growth locations in the contemporary era. These are often distinctively different from those associated either with classic industrial age growth points where heavy industry concentrated, or with those lighter manufacturing belts that succeeded them. We shall see that certain large cities, notably national or regional capitals, are growth engines of the knowledge economy, when they were not necessarily so earlier. Industrial Age belts were highly resource-dependent, so upland areas and valleys with appropriate ores or energy sources prevailed. The ghosts of resource-dependence prevailed to some extent thereafter, although sometimes new 'satellites' suited a 'Fordist' linkage of new production and consumption norms. But by now, these too, while by no means abandoned, are often less than buoyant in their economic life and lose population accordingly. If not capital and otherwise large metropolitan cities, then it is the 'knowledge economies', parts of cities concentrating much knowledge-intensive economic activity, or other knowledge centres like university, science, or creative knowledge locations outside the larger cities, that display growth culture.

After a reprise of the knowledge economy in its most recent forms, socially and economically, the second main section of this chapter turns towards a formulation of the circuitry of growth in the knowledge economy. In doing this, the chapter places a perhaps unexpected emphasis on regions and even smaller locales as fulcrums of growth. It is one of the surprises to many observers of the rise of globalised web-based communication technologies that work and communities have not spread out ubiquitously as a result of the information society and its attendant 'footloose' locational potential for people and jobs. It is not that long ago that imagery of a famous mountaineer being able to e-mail a chapter of her forthcoming autobiography from the summit of Everest awed the literary world. But it is not that comfortable composing text with the jet stream ripping at the portals of the tent, frostbite nipping at the bodily extremities, and altitude sickness rendering memory delusional and subsequently incomprehensible, so most sensible authors do not fit that imagery. We live in a world of floating signifiers whose semantic connections to each other may be as random as those ritually thrown up shotgun-style by Google. Ask Google a question and a shower of responses containing words utilised in composing the question appears instantly. In 2006, research showed that a remarkable 60 per cent of scientific questions posed to Google yielded usably correct answers. But rather than that being considered a remarkable feat, it should rather be seen as defining the limits of its particular search-engine methodology.

The Google-style portal forms no intrinsic meaning because there is no understanding, so the question is 'answered' probabilistically by something approximating luck. Semantic websites are needed for greater accuracy and 'friendliness'. In the same way, floating signifiers tend to have overwhelmed the business of claiming something remotely resembling the truth. While postmodern irony implies that deployment of such a word as 'truth' is accompanied by some kind of health warning, its anti-foundationalism has turned pathological. So, comparable to the 'approved health claim' status of the health drinks discussed in Chapter 1, we might invoke an 'approved truth claim' status for anything said by governments, especially their leaders about, for example, the presence of life-threatening weapons capable of mass destruction abroad, or the healthiness of possibly BSE-infected hamburgers fed publicly to a small family member by a minister under pressure. Or at the mundane level of the 'growth culture', claims by all sorts of places to be 'at the heart of Europe', to be 'nuclear-free', 'sustainable' or to be a 'Science City', 'an innovative cluster' or a 'high-tech region'. Who approves the truth claim? In a media-saturated world, whoever finances the mediated claim. The Internet's pretence to represent reliable truth claims about the world, absent scrutiny, are often as fake as those of the perpetrators of the 'biotechnology scandals'. So why does this happen? Because deep inside such 'approved truth claims' is the aspiration to display an attractive, successful, celebrity status – not at the personal but the governance, associational and institutional levels – apparently enjoyed by a few such glamorous jurisdictions in the world. For these have 'constructed advantage': they do not merely flourish from the accident of resource-endowment and its bestowal of 'comparative advantage', nor necessarily due to their more nuanced possession of 'competitive advantage' where, like Microsoft, they dominate a particular, global market because of entrepreneurial, legal, acquisitional or some other power-asymmetry over others. It is nearly impossible for governances, associations and related institutions to bring off competitive advantage for a firm for long in the knowledge economy because their policy instrumentation is too blunt in the face of liberalising, more open market structures such work increasingly entails. Moreover, unless these are possibly aberrant examples like Enron and WorldCom, themselves fraudulent companies, their transgression of economic let alone legal conventions and rules should, as in science, yield eventually to the application of scrutiny. But regions are different – they are collectivities, not yet registered on stock exchanges, and all sorts of claims may be made without recourse to, or much possibility of, the kind of regulatory scrutiny normal in science and business. This is an attractive option for 'growth culture' contestants. But, except for the mould-breaking originator, their truth claims are difficult because they are frequently hamstrung by a methodology that involves a claim by many or all to be the new 'Silicon Valley' and the attendant problem of selecting from the 'sample of one' (March, Sproull and Tamuz, 1991). Thus places are often highly complex, literally impossible fully to 'clone,' and of course co-evolved with their own intrinsic and the world's extrinsic, space–time characteristics. 'Whatever exists must be necessary' said Hegel (1830), but on the implications of *when* it exists and its *meaning* for practice, he is less clear. Yet in the contemporary era, a certain idealism carries the growth culture through, often in the face of 'objective truth': it is true because we deem it to be so, being at the heart of the 'growth culture's' appetite for 'spin'. Logically, this 'spin'-induced theoretical framework should have the 'Valley' of 'approved truth claims' at its conceptual heart, but it does not. This is because, in its *ex-postmodern* stance, it privileges the absorptive capacity (Cohen and Levinthal, 1989) of the locations of scrutiny – places where congregate communities that in a Rortian manner may determine the validity of truth claims (Rorty, 1980). As Akerlof (1970) showed, such expertises are highly asymmetrically distributed, few in number depending upon the knowledge domain in question, yet extremely powerful in their knowledge attributes. As such they are of vital interest to the project of the book, which is to understand the impulses behind the rise of growth culture in the knowledge economy.

Finally, section three, by way of exemplification of, first, the manner in which biotechnology's evolution and particularly, geographical location, is illuminated by the framework, but second, showing how biotechnology is something of a pioneer, the organisational model for which is increasingly emulated in other industries, not just high-tech, and how the framework takes on something of a generalised model for industrial organisation, economic sociology and geography under knowledge economy conditions. Thereafter the aim is to anatomise and delineate the key institutional actors in this complex evolutionary process by which scientific and financial *scrutiny* of varieties of truth claim drive forward and inform the growth culture of the contemporary era.

Lineaments of proximity in the knowledge economy: aspects of growth cultures

In recent years, with the rise of the knowledge economy, there has been a spotlight on science-based clusters, interactive innovation, and creative, tolerant and talented concentrations of politically desirable economic growth. Surprisingly, despite the exaggerated claims that the Internet brought *ubiquity* and the *death of distance* (Cairncross, 1997) and processes of *ubiquification* (Maskell *et al.*, 1998), much attention has lately been devoted to fascinations with the idea of *proximity*. In well-rooted research on, for example, traditional industrial districts in Italy, one senses a revitalised, albeit economic, recognition and analysis of community (Becattini, 2001). Indeed, such is the dynamism of these agglomerations that many now host in their interstices, communities within communities, as thousands of Chinese entrepreneurs and workers juxtaposed with fewer but still substantial numbers of Pakistanis and North Africans migrated into cities like Prato, near Florence, in the early 2000s (Becattini and Dei Ottati, 2006). Foremost are agglomerative opportunities for tacit, complicit, codified face-to-face and tactile contact and knowledge exchange and of course business. Such socio-economic 'communities of practice' are the *ne plus ultra* of modern economic development (Brown and Duguid, 2001).

It is, of course, astonishing that the role of proximity in human life could have been so occluded by modern economics and other social sciences. It was once said rather frivolously by an anonymous economist that 'the economy is about houses and cars': that is, we have to spend on accommodation (and subsistence) and we will spend what it takes to get close as swiftly as possible to other comparably minded individuals to make life worthwhile. Marx put it more succinctly when he wrote about 'capitalist expansion that annihilates space with time'.[1] Were an historical audit conducted of commercialised new knowledge in the form of patented innovations in recent time, it is probable that a majority would be concerned in some way or other with 'annihilating space with time' to facilitate the achievement of proximity. All the great Kondratieff bursts of innovation, from steam trains to ship technologies, to air transport, the automobile, communications technologies, TV, and the Internet were predicated on enhancing proximity.

Thus the death of distance and the end of geography were rumours much exaggerated upon the implementation of all these innovations, most recently and erroneously with the advent of the Internet (Morgan, 2004). However, what happened as a consequence of these innovations is that *proximity*, the literal meaning of which includes, *inter alia*, nearness, closeness, contiguity and propinquity, all with traditionally geographical connotations, has evolved elaborate and geographically unconfined meanings involving nearness in context, domain and even opinion. Thus chat rooms are quite neighbourly places in virtual space. A multinational company displays characteristics of organisational proximity in all its global operations because of its common rules, conventions and resources, from job titles to the commonalities of its intranet. Zeller (2004) in an interesting article tracing the dependence of Swiss 'big pharma' on innovative biotechnology clusters elsewhere, lists, as well as geographical proximity, the following: institutional (e.g. national laws); cultural (e.g. communities of practice); relational (e.g. social capital); technological (e.g. Linux software users); virtual (e.g. a multinational); internal and external (e.g. firm supply chain management). Actually few feature prominently in such empirical analyses and those that do are usually less important in explaining locational behaviour these days than the core idea of *geographical* proximity. Thus pharmaceuticals firms open R&D 'listening posts' or acquire incumbent firms capable of quarrying American biotechnology clusters. This is a strategy to re-balance the knowledge asymmetries that have arisen as university centres of excellence and DBFs have outperformed 'big pharma', resulting in some re-establishing domestic R&D headquarters abroad in the San Diego and Cambridge, Massachusetts biotechnology 'megacentres'[2] (Cooke, 2005a). Nevertheless, the point has been made that, although for words not to become wholly meaningless concepts should not be over-stretched, the notion of proximity, unlike propinquity, never had any expressly spatial 'nearness in place' meaning. Thus Zeller (2004) performed a useful service in this sense.[3]

However, it is hard to escape an almost 'iron law' that dictates that contemporary knowledge economy development tends to become increasingly urbanised. Indeed, this is the most powerful and determinant fact about the importance of geographical proximity as a means to achieving many, if not all of the other kinds of proximity. While the emphasis in the viewpoint that foresaw the end of geography/death of distance was wrong in respect of the idea of ubiquitous access to the full range of necessities for economic activity over space, it was correct in postulating globally networked information flows. This rests on the observation that globalisation proceeds through networks linking nodes of economic power, mainly cities. The 'spaceless playground' perspective missed the point that such nodes would be the result of increasing returns to urban agglomeration (Sternberg and Litzenberger, 2004). By and large this has meant increasing returns derived from a variety of spillovers, especially knowledge spillovers that tend to concentrate in cities or 'knowledgeable cities' such as university or research towns. This is true for North American and European (and Israeli) cities for which the required analysis has been performed.[4] Clearly, such a wide array of city settings means that the growth process is by no means identical in all cases. Moreover, as Begg (2002) and others show, the competitiveness of cities often accompanies social polarisation. However, this is also a by-product of growth where immigrants are attracted because of perceived economic opportunities absent in their location of origin. Thus in liberal market economies urban sprawl into neighbouring non-metropolitan municipalities is usual, whereas in co-ordinated market economies this is less the case. The chapter proceeds from this point with deeper reflections upon, first, the difference space makes in developmental terms, second, the role of agglomeration and economic diversification for innovation, thought crucial for development nowadays, and third, the way biotechnology embodies this 'knowledge move' in economic sociology and geography.

An emergent pattern in the contemporary variety of proximities, underlined by Begg (2002), is that proximity to knowledge spillovers is crucial to city growth in the knowledge economy era. This harks back to the initial contention of Glaeser *et al.* (1992) that human capital and scarce skills are significant factors in a city's capability to retain and augment its economic growth. This is something of a progenitor of Florida's (2002) talent-led analysis of US city growth in the contemporary era. However, much of the finer detail of variations within growth trajectories is lost in these analyses, not least because of definitional, and even unit of data analysis complexities. One interesting differentiation first hypothesised from a *static* analysis of major concentrations of European knowledge economy sectoral activities, derived from European Union city and region level data on high technology manufacturing and knowledge-intensive business services (KIBS), was that major cities, sometimes also capital cities, accreted much of the KIBS employment. Contrariwise, more specialist satellite cities concentrated high technology manufacturing employment to a greater extent. Live instances of that modern urbanisation process would include, for example, Cambridge and numerous lesser high-tech satellites of Boston such as Waltham, Worcester, Woburn and Andover; San Francisco *vis-à-vis* many such places in Silicon Valley; London in relation to Cambridge, Oxford and the Thames Valley; Stockholm and Uppsala; Helsinki and Espoo; and Copenhagen in cross-border relationship to Lund, the so-called Medicon Valley, traversed by the Øresund bridge.

In a test of this proposition, Cooke and Schwartz (2003) collected appropriate longitudinal statistics for Israel and found support for the proximity to the major city thesis. Statistically significant relationships were found for the growth over time of such knowledge-based urbanisation. But in Israel specialist satellites were growing in employment around the main KIBS city of Tel Aviv rather than the capital Jerusalem. Yet even Jerusalem, for all its political troubles, managed to evolve the country's main biotechnology cluster. Nevertheless in such proximate satellite locations as Rehovot, Herzliah and Ramla higher than average employment growth rates in high technology manufacturing occurred from 1995 to 2002. There was even a biotechnology emphasis to Rehovot's growth and ICT emphases in the other two instances. This suggests that in countries where the main financial centre is not the capital city, the former will exert the stronger proximity effect but that where, as in the UK and, for example, Austria, the capital is also the leading financial services centre, a strong spatial monopoly (or more accurately quasi-monopoly) proximity effect is exerted (Cooke *et al.*, 2007). This is the classic result modelled by Krugman (1995) in applying increasing returns to scale theory, under conditions of imperfect knowledge, to two hypothetically competing candidate cities with the consequence that one always ended up monopolising space. Contemporary city growth theory places knowledge spillovers from (geographical) proximity at the forefront of the explanation for these observed tendencies.

To repeat, this is not to say that geographical proximity determines economic activity to an overwhelming degree. If anything, the implication of what has been concluded here is that the defining feature of knowledge spillovers from geographical proximity is qualitative and quantitative in equal measure. That is, a firm, let us say, located proximately and actively in relation to multiple and varied sources of high-grade intelligence, creativity and connectivity is in principle at an advantage compared to a competitor who is not. However, connectivity to other appropriate knowledge nodes elsewhere in the relevant global knowledge networks is likely to be quantitatively less intensive albeit of qualitative equivalence or even superiority. In their discussion of precisely this geographically proximate as against virtually proximate relationship Owen-Smith and Powell (2004) argued for the superiority of geographical proximity along the following lines. Key processes by which dynamic proximity capabilities are expressed interactively in research or *exploration knowledge* transfer, and commercialisation or *exploitation knowledge* transfer include the following:

- There is a difference between 'channels' (open) and 'pipelines' (closed). The former offer more opportunity for knowledge capability enhancement since they are more 'leaky' and 'irrigate' more geographically proximately. Pipelines offer more confidential, contractual means of proprietary knowledge transfer. This may occur locally or over great geographical distances based on contractual agreements. These are less 'leaky' because they are closed rather than open.
- In high-tech fields, research centres may be a magnet for firms because they operate an 'open science' policy, promising spillover innovation opportunities. These are possible sources of productivity improvement, greater firm competitiveness, accordingly proximate, localised economic growth.
- Such open science conventions influence inter-firm innovation network interactions. Although researchers may not remain the main intermediaries for long as successful firms grow through patenting and commercialisation, they experience greater gains through the combination of proximity and conventions, than through either proximity alone or conventions alone.

These propositions each receive strong support from statistical analyses of research and patenting practices in the Boston regional biotechnology cluster. Thus:

> 'Transparent modes of information transfer will trump more opaque or sealed mechanisms when a significant proportion of participants exhibit limited concern with policing the accessibility of network pipelines . . . closed conduits offer reliable and excludable information transfer at the cost of fixity, and thus are more appropriate to a stable environment. In contrast, permeable channels rich in spillovers are responsive and may be more suitable for variable environments. In a stable world, or one where change is largely incremental, such channels represent excess capacity.'
>
> (Owen-Smith and Powell, 2004)

Finally, though, leaky channels rather than closed pipelines represent also an opportunity for unscrupulous convention-breakers to sow misinformation among competitors. However, the strength of the 'open science' convention means that so long as research institutes remain a presence, as in science-driven contexts they often do, such 'negative social capital' practices are punishable by exclusion from interaction, reputational degrading or even, at the extreme, convention shift, in rare occurrences, towards more confidentiality agreements and spillover-limiting 'pipeline' legal contracts. We note, further, how open science and its scrutiny conventions attract, in further evolutionary rounds, 'open innovation' to such knowledgeable clusters when it might otherwise be assumed that openness should mean knowledge advantage erosion. But likely gains are perceived to outweigh losses by customers taking the plunge. This is a major factor in proximity-based economic growth since knowledge supplier firms garner a substantial share of their income from, especially, R&D outsourcing by larger customer firms (Chesbrough, 2003; Huston and Sakkab, 2006).

Towards a knowledge theoretical framework of economic sociology and geography

These localised knowledge spillovers and proximity questions relate in interesting and complex ways to issues of specialisation versus diversification as wellsprings of innovative capability. At issue in the former debate is whether such localised knowledge spillovers as those discussed, reside in the combination of effects that gives local innovation proactivity to the region or locale itself. Or does the locale derive its extra proximity capabilities from the combined relational effects of the firms that typify it? The latter is argued by Breschi and Lissoni (2001) who see no convincing evidence that non-pecuniary spillovers, the apotheosis of localised knowledge spillovers, have displaced Marshall's classic definition of 'external economies' in terms of pecuniary advantages. In other words firms co-locate because it is profitable to do so.

However, the existence of and contribution to an entrepreneurial social infrastructure of *localised knowledge spillovers* is advocated by most of the academic community interested in the explanation for clustering and, at a higher scale, the formation of regional innovation systems. These are areas of hot debate currently, the outlines of which are explored by Caniëls and Romijn (2005). The proponents of the view that proximity offers innovation advantages in itself begins in relatively recent times with Jaffe *et al.* (1993). The argument here was that R&D in particular constitutes a public good in locations where it concentrates and that this is sufficient to cause firms to concentrate in proximity to such knowledge spillover opportunities to access them as free goods in advance of competitors. This, of course, is entirely consistent with the attractions of 'open science' and subsequently 'open innovation' in proximity to (informal) knowledge spillovers discussed in the previous section of this chapter. Thereafter the likes of Audretsch and Feldman (1996) and Malmberg and Maskell (2002) also argued in favour of the power of localised knowledge spillovers as drivers of innovation, especially in knowledge-based clusters. However, the 'market' dimension of such growth impulses maybe overstated, since if free markets in such knowledge really operated, proximity would not be necessary. In this sense we hypothesise knowledge clusters as more like 'growth culture' clubs than commodity exchanges.

Interestingly, both sides argue their cases in respect of regional analysis. Thus the Caniëls/Romijn critique is of both sides for ascribing too much influence to regional *milieu* and too little to firm *capabilities* or what may also be referred to as *entrepreneurship* and specifically in this vein *knowledge entrepreneurship*. The compromise position suggested as far as the treatment of such externalities (static or dynamic) is concerned is to establish a more penetrative analysis of the firm-level contribution to regional *capabilities*. For now, further work is required on types of agglomeration advantage, ranging from static to dynamic spillovers, pecuniary to non-pecuniary, and pure versus impure knowledge spillovers at the firm level but aggregated up to at least the regional level. This means, of course, stepping further than firms alone in the analysis, since regional capabilities, including knowledge entrepreneurship capabilities, are seldom reducible to those of firms alone. One of the puzzles of knowledge economies is how cognitive boundary-crossing among 'epistemic communities' (Haas, 1992) like entrepreneurs and researchers usually representing the private and public sectors respectively actually occurs. Epistemic communities are *professionally* distinctive associations, gaining identity and economic status precisely from their distinctiveness. Communities of practice, by contrast, derive their specificities from professional, relational combinations defined by *place* rather than professional or even corporate 'culture' (Brown and Duguid, 2001).

This is excellent candidate territory for the exploration of *complicit* knowledge spillovers of the kind conceptualised in Table 2.1. Authors like Galison (1997) refer to the means of explicit–implicit interchange as 'contact languages' and even talk of 'pidgins' as means of communication and scrutiny. Summarising, it is arguable that concentrations of scarce and valuable capabilities among incumbents and intermediaries such as varieties of knowledge to explore an idea, test or examine it and then exploit it as a commercial innovation may be a kind of quasi-monopoly of knowledge it is in the interests of incumbents to protect, as far as they are able, from outsiders. 'Openness' in innovation may clearly have certain 'club' characteristics, as has been suggested. In Table 2.1 the role of *complicit* knowledge is introjected into the processes by which tacit knowledge becomes codified. Thus *places* may 'monopolise' knowledge in certain inter-connected spheres where specific *exploration* knowledge assets exist in geographical proximity. A *geographical knowledge domain* of the kind discussed in relation to Zeller's (2004) study is established. At the cognitive level, such *implicit* knowledge requires extraction if appropriation or ownership is to be established. In this space *complicit* intermediary firms and other actors mediate the 'translation' of such implicit, exploration knowledge by *examination* (scrutiny) of it for validity, reliability, safety and so on, at which point it becomes codified as explicit, appropriable knowledge in the form of say, a copyright, trade mark or patent.

To achieve this, supporting knowledge capabilities are necessary to reproduce and improve the advantageous knowledge asymmetries that proximity has assisted in accumulating in the spatial knowledge domain. Thus talented personnel are trained to conduct research to develop technique that represents the spatial knowledge domain's knowledge 'monopoly' or preferably quasi-monopoly since it is an open though asymmetrically accessible knowledge domain. In institutional terms, these elements are represented in the form of a 'regional innovation system' whose innovation institutions focus network interactions assisting translational research for commercialisation of new knowledge. Once such knowledge exists in appropriable form it may be represented digitally as software or the code that represents, for example, the human genome. In discussing the structure of a classic regional innovation system (Cooke *et al.*, 1998) it could be represented as a binary relation between two crucial sub-systems, one concerned with knowledge exploration, the other with knowledge exploitation. In simple terms, a university

Table 2.1 Knowledge: from implicit domains to digital innovation systems

	Implicit	*Complicit*	*Explicit*
Knowledge domain	Invention	Translator	Appropriation
Knowledge capability	Talent	Research	Technique
Innovation system	Institutions	Networks	Digital

centre of research excellence evolves patentable knowledge, then a firm licenses, exploits and commercialises it. But as Galison (1997) hints, this is too binary and facile. There must be an intermediating, third sub-system centred upon complicit knowledge that 'translates', 'filters', and 'interpolates' implicit knowledge, assisting its appropriation as explicit or codified knowledge in the exploitation phase (Table 2.1).

Thus it is hypothesised that the old binary relations between implicit and explicit 'senses of knowing' (Orlikowski, 2002) are more complex than suggested in the literature begun by Polanyi (1966). As hinted by Zitt *et al.* (2003):

> in many cases both tacit and codified information are linked in a number of ways. For example tacit information may act as a facilitator for the transmission of codified knowledge. If direct information flows matter then the availability of skilled manpower (universities as recruitment base, mobility of researchers) also plays an important role in localised dissemination of knowledge.[5]
>
> (p. 296)

A broad conceptual model of the intermediary cognitive and institutional steps from implicit to explicit knowledge in bioscience, referred to as distinct kinds of *complicit* knowledge, identified at least fifteen of these (Cooke, 2005a). More recently, detailed research of a member firm of only 54 employees revealed the utilisation of at least six different kinds of intellectual property rights (IPR) intermediation assisting the process of translating, assisted by research, implicit *exploration* knowledge through a complicit *examination* knowledge phase to reach the explicit or codified *exploitation* phase in the form of an IPR patent. Such detailed empirical analysis would undoubtedly yield more. Though it is not incorporated in Table 2.1 so as to keep the basic relationships simple, Harmaakorpi (2006) further identified the role of pre-tacit and self-transcending knowledge as a crucial element in the knowledge creation process.[6] To clarify usages in the above analysis further, attention is drawn to Table 2.2, which analyses the exemplifications of the distinctive research knowledge chain in relation to different epistemic categories of knowledge (analytical/scientific, synthetic/engineering, and symbolic/artistic). This is done for purposes of definition rather than any particular relationship such an analytical framework might have for considerations of proximity. The analysis shows that all kinds of knowledge pass through variants of the same three-stage transformation process, as implicit knowledge is complicitly transformed into explicit/codified, marketable new knowledge or innovations.

In connection with this, there is also a further and evolving debate between two perspectives regarding the most propitious setting for the stimulation of innovation. One says that capabilities from sectoral *specialisation* produce the best results, the other says *diversification*. The former position

Table 2.2 Characteristics, forms and stages of knowledge production

	Analytical	*Synthetic*	*Symbolic*
Exploration	Mathematical reasoning	e.g. Genc therapy	Experimental art work
Examination	Theorems to test	Clinical trials	Art exhibition
Exploitation	e.g. Penrose tiles/patterns	Therapeutic treatment	Gallery sale

NB: Penrose tiles are mathematically derived computer graphics

is associated with Glaeser *et al.* (1992) and Griliches (1992) who, as we have seen, envisage specialised knowledge 'spillovers' as key to innovation. The latter view begins with Jane Jacobs (1969) and is supported by, for example, Feldman and Audretsch (1999) who show that sectoral diversity is most strongly associated with regional innovativeness. The first tend to emphasise *markets* while the second give greater weight to the milieu of institutional infrastructure (innovation support system) and microeconomic linkages across agents and firms (networks). So we have a comparable issue as that which cropped up regarding localised knowledge spillovers. However, a slightly different resolution arose when, subsequently, Henderson (2003) showed specialisation effects on knowledge spillovers to have a strong but short-lived impact in high technology industry while *diversification* effects persist far longer. This suggests that as they evolve, such clusters first specialise then later diversify. This makes sense in that science-based clusters are demonstrably formed in historical terms around a single innovation, which swiftly transmutes into families of related products – and processes. One only has to think of the semiconductor that was the trigger for Silicon Valley firms later spawning the integrated circuit, graphic interface, mouse, PC, etc. Similarly, both the San Francisco and Cambridge, Massachusetts biotechnology clusters began with the production by, respectively, Genentech and Biogen, of human insulin.

It is incumbent upon us to try to place these debates in good order insofar as that ambition is feasible in rapidly evolving, empirically informed theoretical fields. But before that attempt, one more, longstanding, binary opposition of relevance to the foregoing discussion requires integration into the debate. This concerns the relationship of Marshall's (1918) contrast between *urbanisation* and *localisation* economies to the foregoing, not least because important elements of the contemporary debate have their origins in Marshall's insights of a century ago. These of course were fully informed by his clinical observation of the industrial districts of Britain, which reached their apogee shortly before the outbreak of the 1914–1918 war. Urbanisation economies are somewhat cognate with Jacobs's notion of the innovative buoyancy of industry diversification, which is naturally far more pronounced in urban than non-urban areas. Localisation economies are more specialised and their externalities equivalently so. Thus Marshall speaks of the manner in which British (Yorkshire) woollen towns specialised in specific kinds of smooth (e.g. worsted) or rough (tweed-type) woollens and (Lancashire) cotton towns Blackburn, Nelson, Oldham and Preston displayed different types of weaving of cotton and 'the constant intercommunication of ideas between machine makers and machine users'[7] that underpinned the evolved specialisations in specific localities. These dynamic, partly non-pecuniary localisation externalities enabled firms to specialise while maximising equipment utilisation. This contrasts with equivalent static, pecuniary externalities derived from access to a highly specialised, localised skills base. It is conceivable, but only just, that such ruminations possibly followed by specifications could be conducted in virtual proximity. But as Sapsed *et al.* (2005) show, tactile meetings are still required for project design even within single distributed corporations engaged in complex projects.

These are crucial concepts regarding varieties of knowledge-based clustering. This is as a prelude to the second, somewhat briefer analysis of contemporary global knowledge flows. Thus we distinguish between firm and milieu characteristics, between diversification/urbanisation and specialisation/localisation innovation contexts and types of externality. We further distinguish static and dynamic externalities associated with urbanisation versus localisation economies. Finally, we categorise the *main* mode by which knowledge spillovers are accessed. Though not alluded to here for the sake of limiting complexity, this can also be done in terms of pure/impure or pecuniary/non-pecuniary spillovers. The account below is intended to be illustrative of theoretical nuances arising from the juxtaposition of conceptual categories discriminating between types of knowledge and types of industry organisation.

For the *firm*, the taker of static externalities in a diversified, urbanised innovation context is not 'networked' but may still be innovative, possibly as described in Cooke (2005b), as a *recombinant* innovator, finding diverse customers for incremental innovations. The milieu most associated with low networking propensity because of its static externalities is the industry agglomeration. The firm exploiting dynamic urbanisation externalities in a diversified innovation context without significant 'distant (knowledge) networking' is a stylised description of the modern, particularly Italian, industrial district firm, capable of incremental innovation, occasionally innovating disruptively, but seldom radically.[8] Hence its typical milieu is a modern industrial district. In contrast, the firm seeking only static knowledge spillover externalities in a localised, specialised innovation context is perhaps 'protectionist', path dependent and capable only of incremental innovation at best – it is reminiscent of firms in mature, classical industrial districts such as the silk or ceramics districts of Great Britain that died from this kind of protectionist disposition (Cooke, 2002).

They equate to the oligopolistic cluster, a category that resonates well with Bottazzi *et al.*'s (2002) typological categorisation. However, we would not necessarily expect only decaying oligopolies to operate in such clusters. In Finland, for example, Nokia operates as a highly innovative company in such contexts, often with its R&D laboratories embedded in university science parks with clusters of spinout suppliers. Finally, we propose the virtually networked firm to be the likeliest candidate for radical innovation, particularly when also engaged in the similarly dynamic knowledge spillover settings of diversified, urbanised economic space. Much research shows such innovators to be highly networked globally and locally (Gans and Stern, 2003). Accordingly, they are likely to be located in innovative business clusters where open science and open innovation opportunities abound. Interestingly, the debate may have been resolved by insights, empirics and scrutiny from evolutionary economic geographers. In particular, research by Boschma (2005) and Cantwell and Iammarino (2003) shows clearly that high innovation and economic growth rates occur in regions displaying 'related variety' of industry mix. By contrast, economies displaying Marshallian 'specialisation' or Jacobsian 'diversification' perform less well. Ketels (2006) has recently argued that this is the case not only in European but also US regions.

The analysis for biotechnology clusters and other sectors

To access the kind of analytical or synthetic knowledge discussed above, which under earlier internationalisation or 'Globalisation 1' (Cooke, 2005b) and 'closed innovation' régimes firms used to generate in-house, larger firms today acquire it from SME firms, or take a majority shareholding if it is a reasonably mature firm, or make 'milestone' payments against IPR as a firm progresses past agreed development and trialling targets with a view ultimately to licensing the technology in question. Penetration of global talent pools, as with Novartis who, as noted, gained proximity to US biotechnology clusters to access leading edge knowledge, marks 'Globalisation 2' where key drivers in the new economic geography and sociology are knowledgeable SMEs and public research institutes or university centres of research excellence. Firms may also offer to buy the technology rather than the licensed service, something Chesbrough (2003) portrays as 'open innovation', an example being the relationships of large corporations like Monsanto and Eli Lilly *vis-à-vis* Millennium Pharmaceuticals' DNA sequencing software. Millennium sold them its proprietary product after much heart-searching because the sale generated sufficient revenue for Millennium to buy three smaller biopharmaceutical companies to become the drug manufacturer its owners had always sought to be. This is today a major means of knowledge transfer in biotechnology and ICT through varieties of often triangular 'partnership' among a large customer, a smart technology SME and sometimes a university or institute researcher or team responsible for the discovery or invention.

As an illustrative indication of this for biotechnology in the single post-boom year of 2003, Table 2.3 is instructive. There we see a major effort by US pharmaceuticals corporations to transfer-in knowledge from European and US dedicated biotechnology firms by forming a variety of partnership agreements with them. Many of these firms might otherwise have been targets for venture capital houses but it is clear that by 2003 pharmaceuticals firms were willing to invest substantial sums in products that in some cases were only at early-stage trialling. It is noteworthy too, that many such deals focused on the end-point of the drug development process, namely commercialisation. For the foreseeable future the large scale of

Table 2.3 Top ten biotechnology partnerships 2003

Biotechnology firm *€ million*	*Partner*	*Transaction*	*Value*
Millennium Pharmaceuticals (US) 425	Johnson and Johnson (US)	Global marketing rights for oncology drug	
Biovitrum (Sd) 414	Amgen (US)	Develop and commercialise Phase 2 diabetes drug	
Regeneron Pharmaceuticals (US) 405	Aventis (F)	Develop and commercialise Phase 1 cancer drug	
Regeneron Pharmaceuticals (US) 278	Novartis (Sw)	Develop Phase 2 rheumatoid arthritis drug	
Actelion (Sw) 216	Merck (US)	Marketing heart, kidney and hyper-tension inhibitors	
Idenix Pharmaceuticals (US) 199	Novartis (Sw)	Licensing hepatitis B drug. Co-develop Phase 1/2 compound	
Corgentech (US) 199	Bristol-Myers Squibb (US)	Co-develop Phase 3 vein graft failure drug	
Medivir (Sd) 158	Boehringer Ingelheim (G)	Out-license of Phase 2 HIV therapy	
Lexicon Genetics (US) 152	Bristol-Myers Squibb (US)	Neurosciene drug R&D, commercialisation	
Flamel Technologies (F) 131	Bristol-Myers Squibb (US)	Licence and commercialise insulin product	

Source: DTI (2005)

retained earnings continuing to be enjoyed by pharmaceuticals companies means that they maintain a commercialisation advantage in respect of distribution and marketing staffs (the latter often in the tens of thousands) that DBFs do not have and possibly would not want. Hence the successful 'pharma' firm of the future is likely to be drawn from those who realise their strengths and weaknesses and act as, fundamentally, bankers and sales engines for the products of DBF and university research power.

Before completing this section on 'pipeline' interactions where the pipeline is the timeline of different stages in the evolution of a potential product as it moves from laboratory bench or *in silico* representation as computer graphics, software or design, to innovation on the market, the other main means of global knowledge transfer involving *research* deserves mention.

In published papers such as Cooke (2004b; see Figure 2.1) evidence is provided of the main global co-publishing patterns among biotechnology 'star' scientists in prestigious research institutes. It is largely pre-private sector entry research and largely funded by research councils, governments and non-profit foundations. It reveals the strength at the centre of the hierarchy of US clusters in California, Massachusetts and New York. In Europe, university towns are often the heart of the excellence network. It is thus no coincidence that the locational preferences of recent pharmaceuticals industry openings have been influenced by the existence of these concentrations of accomplishment.

Finally, closer to the point of exploitation when a large pharmaceuticals firm and other investors, including venture capital, will likely be interested in investing, is the patented outcome of the exploration and examination research stages of the pipeline. Again, we have conducted research into biotechnology co-patenting activity by 'stars' in comparably accomplished contexts and these network hierarchies are comparably structured, though in a more condensed way (Figure 2.2). This reinforces the commercial interest of such clusters for large commercialising companies whose internal knowledge capabilities are attenuated by the problems of dealing with new technologies outside their traditional field of research competence – in the case of pharmaceuticals in synthetic *chemistry*, not bioscience.

The generalised theoretical framework: economic sociology and geography in the knowledge economy

In this section, it is shown how knowledge hegemony in pharmaceuticals has shifted to universities and the cohorts of biotechnology firms that often co-locate with those that are 'ahead of the curve' in biosciences. In the course of this account, a new theoretical framework for explaining spatial industry organisational shifts has to be essayed. It is evolutionary in origin, interested in the economics of *search* and *selection* practices of firms in contexts where *variety* acts as 'evolutionary fuel' in Hodgson's vivid phraseology (Hodgson, 1993). By evolutionary fuel is meant iterative,

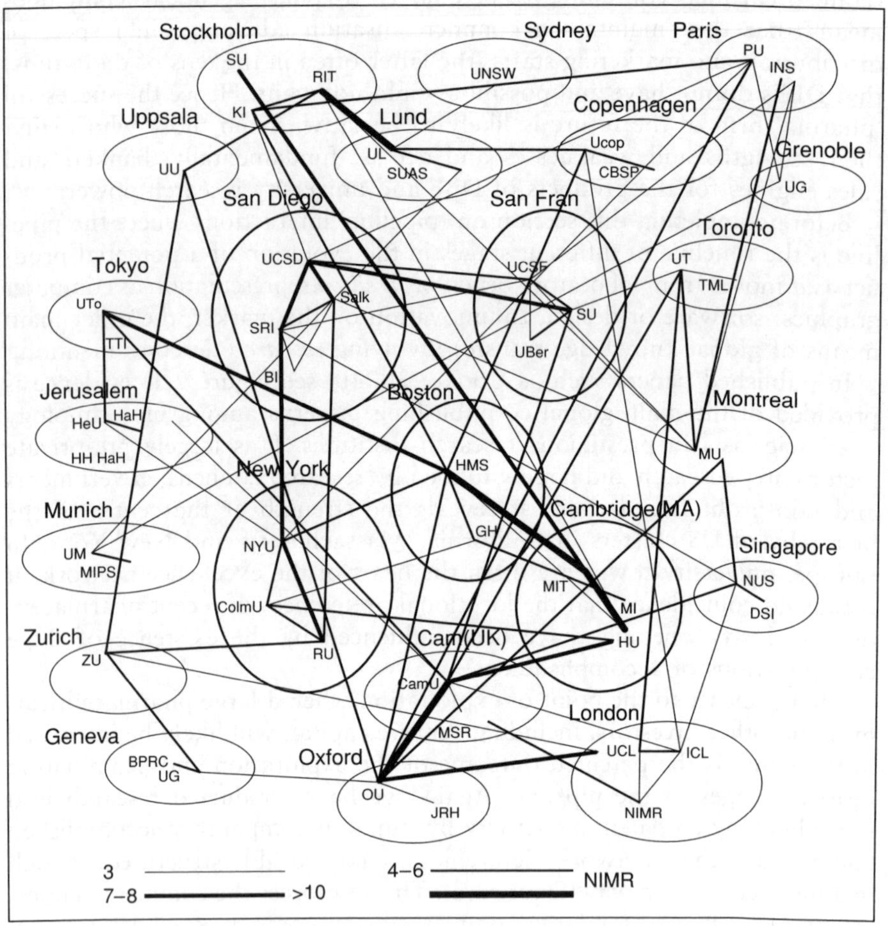

Figure 2.1 Publishing collaborations

NB: Represented are co-publications between 'star' bioscientists in leading research institutes in eight of the top-ten Science Citation Index journals for recent periods: *Cell* (2002–2004); *Science* (1998–2004); *Proceedings of the National Academic of Sciences* (2002–2004); *Genes and Development* (2000–2004); *Nature* (1998–2004); *Nature Biotechnology* (2000–2004); *Nature Genetics* (1998–2004); *EMBOJ (European Molecular Biology Organization Journal)* (2000–2004). In all, 9,336 articles were checked. Abbreviations refer to research institutes.

trial-and-error interactive feedback from experimentation by actors to survive and prosper economically. The greater the variety, the greater the opportunity for innovation arising from interactions with other actors. It has been shown empirically that opportunities for the swiftest innovation occur in conditions of *proximate and related variety* (Boschma, 2005). Cities are one variant of this, but because their variety is often fragmented as well as partly related, they are less fruitful than settings with only related

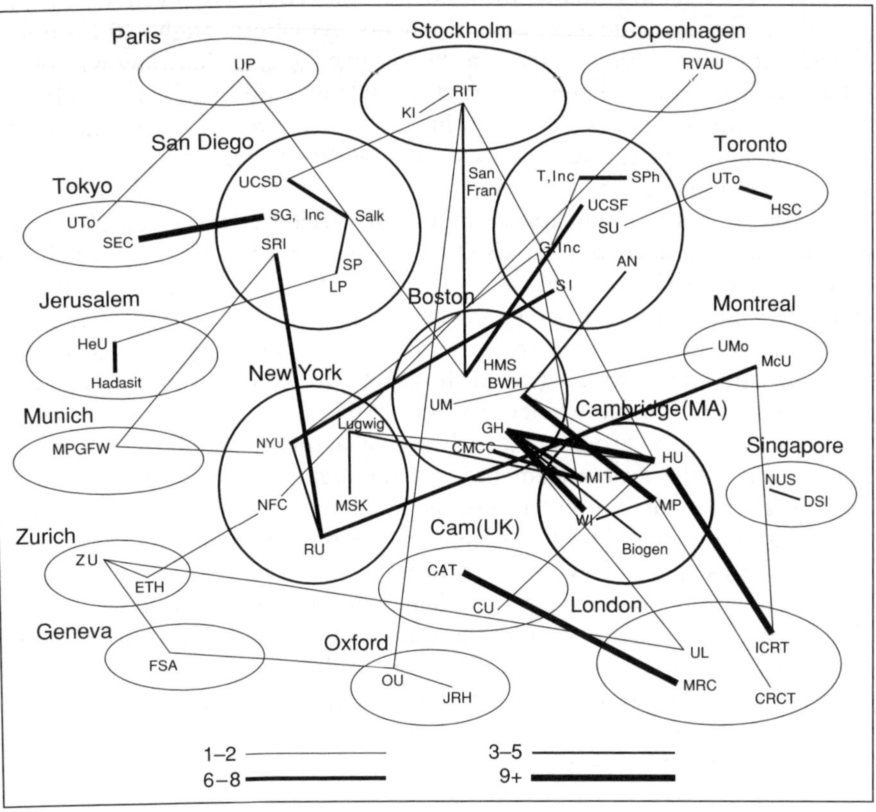

Figure 2.2 Biotechnology co-patenting networks

Source: Derived from US Patent Office database 1998–2004

variety of the kind typically found in regional innovation systems. Hence, as noted, this new perspective settles at the apex of a conceptual triangle between Jacobs (1969) who advocated sectoral *diversification* and Glaeser *et al.* (1992) *specialisation* as key wellsprings of innovative growth. It is post-sectoral, recognising innovative growth to be facilitated through knowledge or technology platforms characterised by openness of knowledge flows and knowledge spillovers. For example, a location specialising in leading edge research in sensors finds numerous applications of such technology in many related yet extensive fields where lateral absorptive capacity is high. *A priori* biotechnology is the exemplar of this mode of industry organisation, but more as pioneer than offshoot now that the model of 'open innovation' building on 'open science' norms is emulated widely, from 'open systems architectures' of various kinds to 'open source' software (Owen-Smith and Powell, 2004). The point here is that 'open science' norms among scientists operate informally through normal 'channels' even if 'formal' confidentiality agreements exist with clients. As Chesbrough (2003) notes, clients know this – complicitly – in the knowledge also that they will themselves receive returns from localised knowledge spillovers in the cluster. Not all of this openness is geographically proximate, as distant networks play a strategic part, and *cognitive* and *relational* proximities come into play as Boschma (2005) stresses.

Nevertheless, the implications of related variety as witnessed in the demise of the *generic* corporate R&D model compared to the *varietal* choice model found in the rich mix of research centres and niche firms in, for example a major biotechnology cluster, is testimony to the attractions to customers of the latter over the former model. These may be measured in terms of capabilities ranging from those relevant to exploration, examination or exploitation knowledge (Cooke, 2005b).

Hence, there are grounds for advancing a theoretical framework that links together these new elements and highlights the role of varieties of knowledge in contributing a testable explanation of regional developmental asymmetries. The key elements are presented in Figure 2.3 and discussed subsequently. We start from the centre of the diagram, denoting a region in which a mix of widely in-demand knowledge capabilities evolves. Connecting to northwest in the diagram, and compared to other regions, this expresses its *asymmetric* knowledge endowment from a variety of knowledge organisations and institutions. *Exploration* knowledge organisations, such as research institutes, knowledge networks among individuals (e.g. 'lunar societies'; Uglow, 2003) and knowledge leadership figures

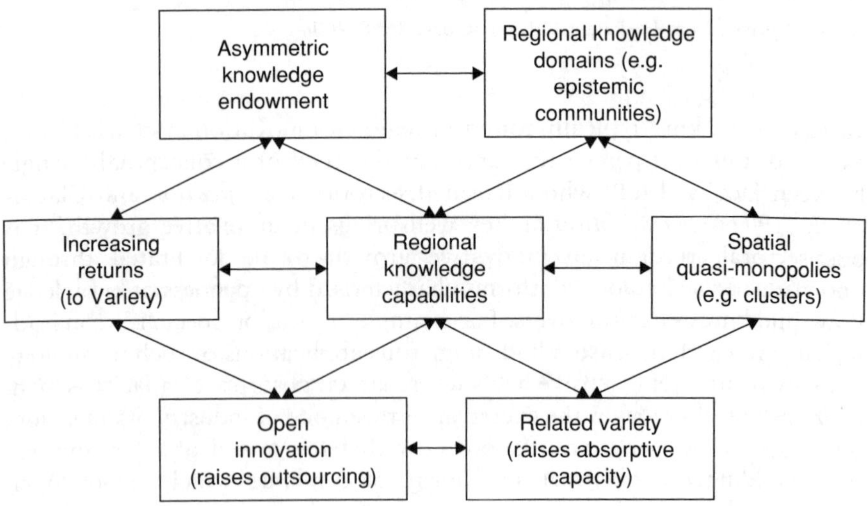

Figure 2.3 Knowledge capabilities and economic geography: a theoretical framework

(e.g. possible future Nobel [or Oscar] laureates) co-exist with *examination* knowledge equivalents for standard setting, trialling, testing and patenting, and *exploitation* knowledge bodies such as entrepreneurs, investors and related professional talent. The evolutionary fuel is supplied (linking westward in the diagram) by the attraction of a variety of imitative and innovative talent to the region, a Schumpeterian 'swarming' realising increasing returns to related variety (south-eastward diagrammatic connection) where innovation may move swiftly through various parts of the innovation 'platform'. Related variety nourishes absorptive capacity because cognitive distance between platform sub-fields is low (think of 'general purpose innovations', after Helpman, 1998).

Moving north-east in Figure 2.3, these processes result in the presence of regional 'knowledge domains'. The dictionary definition of 'knowledge domain' is a region or realm with a distinctive knowledge base, common principles, rules and procedures, and a specific semantic discourse. This naturally fits well with the concept of the epistemic community with its own professional discourse and interests. Such monopolistic features are frequently characteristic of, for example, *clusters* that in regional terms may display related variety (e.g. varieties of engineering expertise in the industrial districts of Emilia-Romagna in Italy in a spectrum from Ferrari cars and Ducati motor cycles (both Modena) to Sasib in packaging machinery (Bologna) and drgSystems machine tools (in Piacenza) (Harrison, 1994). These and other clusters have spatial quasi-monopolistic or 'club' characteristics, exerting exclusion and inclusion mechanisms to aspirant 'members' consequent upon their knowledge value to the club. If such industries operated as markets rather than knowledge quasi-monopolies, it is difficult to see why spatial 'swarming' would occur. But high technology firms at least are willing to pay super-rents of 100 per cent to locate in clusters – even when they are professed non-collaborators – to access anticipated localised knowledge spillovers (Cooke, 2006). Finally, to the south-west of Figure 2.3, it is precisely such localised knowledge spillovers that induce what Chesbrough (2003) calls 'open innovation' whereby large firms outsource their R&D to purchase 'pipeline' knowledge, and access via 'channels' regional knowledge capabilities (Owen-Smith and Powell, 2004). These processes interact in complex, non-linear ways displayed graphically in Figure 2.3, to explain regional knowledge asymmetries. Variations in the market value of regional knowledge combinations also contribute significantly to associated regional income disparities (Boschma and Frenken, 2003). Being an evolutionary growth process, successive increasing returns may be triggered from any point within or, of course, beyond the confines of Figure 2.3.

With respect to other sectors, less work has been conducted though research into the analytical base for so doing is presented below. There is of course a large question concerning the validity, reliability and even meaning of such an antediluvian notion as 'sector'. Here is unfortunately not the place to delve into the lethal critique of the notion due to space limitations. However three criticisms can briefly be offered. First, the sector notion is merely a statistician's artefact that is an increasingly misleading representation of reality. Second sector classifications are little changed since their nineteenth-century origins to enable identification of such activities as biotechnology, nanotechnology or new media. Third, as we have seen, technological innovation increasingly progresses by means of the evolution of 'platforms' that combine many technologies that are, in increasing numbers of cases, adaptable across extremely diverse industrial and technological contexts, as a moment's reflection upon the technology platform built around sensors, let alone software or genetics, makes clear.

Nevertheless, perforce much research into economic development must utilise such artifices and the following analysis recognises this fact. In Table 2.4 sectors capturing a wide range of variety are selected for exemplification, and related to our preceding triple categorisation of relevant knowledge types in regard to sectoral knowledge generation, testing and appropriation characteristics. The sectors utilised to illustrate the analysis in Table 2.4 include high-tech (ICT, Bioscience), medium-tech (Automotive, KIBS, New Media) and low-tech (Tourism, Food). Why this classification? Unlike others connected to the knowledge economy interpretations of, for example, OECD (1999) these devolve into whether or not they are strongly science-driven, moderately science-driven or hardly science-driven. For simplicity, each sector's main knowledge base is inserted in appropriate categories. This does not mean they may not be more or less occasionally science-utilising as in the case of food, which buys in testing and trialling examination knowledge for hygiene, packaging, bacterial analysis and so on. Most new media, especially based on Internet service provision, are commoditised rather than science-driven and even the apparently anomalous

Table 2.4 Knowledge: from implicit domains to digital innovation systems

	Implicit (Explore)	*Complicit (Explore–examine)*	*Explicit (Examine–exploit)*
Knowledge domain	(Invention) Bioscience ICT	(Translator) Bioscience ICT	(Appropriation >IPR) Bioscience New Media
Knowledge capability	(Talent) New Media KIBS Automotive	(Research) Food KIBS Automotive	(Technique) Food Automotive KIBS, ICT
Innovation system	(Institutions) Bioscience ICT New Media	(Networks) Tourism New Media	(Digital) Tourism New Media

KIBS (knowledge-intensive business services) is only occasionally science (analytical knowledge)-driven in truth, as with applications of Scholes–Black chaos theorems to hedge fund algorithms and other derivatives and options products. For the purpose of this analysis, software is located in ICT.

To explain these allocations of sectors to categories, it is important to recognise the importance of science-driven, exploration knowledge of an often implicit kind that characterises bioscience and ICT in ways it does not with such cognitive proximity in the other sectors. Also, as we have seen, such sectors rely strongly on innovation systems that mix public and private institutions and funding for research. New media, KIBS and automotive sectors are less institutionally reliant and more 'entrepreneurial' either through large-firm vertical innovation chains or small-firm horizontal co-operation for competition linkages within mainly market settings. In the *complicit* dimension, as we have seen, bioscience and ICT rely on a great many intermediaries in the translation and, particularly, examination (testing/trialling) phases. Food, KIBS and automotive tend to buy research and design capabilities in markets while tourism and new media are rather connected to big institutions for innovation, like broadcaster, content customers, and travel corporations for innovation demand. Food, automotive, ICT and KIBS patent less or in limited ways – software cannot be patented in the US although it can in the EU. Mainly, innovation in these industries relies on technique. Last, when it comes to exploitation, bioscience accounts for some 50 per cent of world patents, ICT is not such a pronounced patenting sector, even excluding software's regulatory anomalies, while new media is widely subject to copyright or trade mark designations. Meanwhile, to the extent that digital networks represent innovative processes available to the market to enhance competitiveness, tourism and new media are major users of such systems directly to the public whereas for the other sectors such system-usage is more private (see Sassen, forthcoming).

Finally, we move to consideration of less the knowledge appropriation chain than kinds of activity within firms in sectors that may be categorised according to their epistemological character in relation to the knowledge value-chain as it moves interactively. Just three sectors are selected for exemplification but readers can easily fit examples from other sectors as appropriate (Table 2.5).

The illustrations from Table 2.5 show very clearly the historical nature of transitions from raw knowledge of a more-or-less implicit 'being articulated' kind through examination for reliability, validity and effectiveness/efficiency testing and trialling, including recombinant innovation such as that of Tim Berners-Lee to enable Internet searching to be facilitated through hypertext links, to eventual marketed innovation outcomes. Hence, while Table 2.4 is somewhat, but by no means entirely, focused on at least potential *proximities* of a spatial kind, but including proximate nodes and their linkages to other distant network nodes, Table 2.5 captures the historic processes by means of which progress of innovations through the iterative

Table 2.5 Knowledge value chains in KIBS, automotive and new media

	Analytical	*Synthetic*	*Symbolic*
Exploration	Scholes–Black	Hybrid fuel	DARPA
Examination	Futures/options	Trialling/testing	Hypertext
Exploitation	Hedge funds	Toyota Prius	Internet

and interactive knowledge value chain actually occurs and may be replicated for all innovations in any sector.[9] Thus a methodology has been presented for historical reconstruction of innovations as 'innovation biographies' linking all kinds of knowledge and innovation actors whether 'fictions' like sectors or realities like firms and their innovation practices as they move ideas from implicit through complicit to explicit commercial outcomes. The analysis allows for both spatial and temporal sensitivities to be emphasised at different stages of analysis.

Conclusions

Thus we have moved analytically from consideration and preliminary evaluation of several distinctive kinds of proximity in regard to better understanding of the manner in which knowledge of various kinds flows under conditions of geographically and non-geographically derived meanings of *proximity*. It seems that geographic proximity is one of the crucial human developmental needs as development brings surges of mobility in quest of geographical proximity to many kinds of social, cultural and economic needs in cities. Cities are the most profound artefact of human civilisation and testimony to the power of proximity in human thought and practice. However, as was shown, not everyone seeks to live in the same city – rather cities are nodes in global knowledge networks and they have specific and differentiated knowledge and other resource assets, developing asymmetrically.

We then moved on to deeper reflection upon the question concerning the difference space makes in relation to localised, globalised, proximate and virtual knowledge exchange. Once again, observation of firm practices in stylised instances of differential utilisation of spatial opportunities tended to underline the mega-observation regarding the salience of proximity needs and demands to the growth of cities. That is, firms can be seen, differentially according to size and specialisation, positioning themselves differently in relation to their spatially asymmetrically distributed needs and demands, notably for knowledge of various kinds. 'Globalisation 2' was proposed (see also Cooke, 2005b) to capture the metamorphosis in knowledge power relations whereby multinational companies are forced to seek out global knowledge and talent pools to access new knowledge for innovation, rather than in 'Globalisation 1' orchestrating production around cheap labour zones, conducting in-house R&D and innovation and adapting products and services marginally to suit globally diverse markets.

This led to theorisation of the nature of knowledge translation, research and directional flows under these new, more complex and externalised conditions typified by 'open innovation' and building upon conventions of 'open science' facilitative of localised knowledge spillovers within these global knowledge and talent pools. The resulting theorisation through conceptual frameworks was then operationalised to show how the dynamics of both clusters and sectors may be more clearly understood utilising the lenses of these frameworks for analytical purposes. Sectoral analysis, though based on the artifice of statistical categories, may deliquesce into platform analysis of the kind more familiar to firms and their knowledge and innovation strategies by taking the proposed 'innovation biographies' approach to industry research. Finally some institutional cases of policy actions to 'tighten' systemic innovation in challenged regional settings were provided as illustrations of key barriers and pathways to securing effective knowledge translation towards innovation, with institutional analysis of actual cases of successful and comparatively less successful integration of innovation systems as means to improve knowledge exploration, examination and exploitation.

3 Bioscientific research and the emergence of knowledge domains

Introduction

The main interest of this chapter is the rapid decline in the capabilities of large pharmaceuticals companies ('big pharma') to develop in-house new therapeutic drug treatments, particularly those deriving from biotechnology, compared to the rapid rise in that precise capability on the part of networks of small, dedicated biotechnology firms (DBFs). This is commented upon in Orsenigo *et al.* (2001), but these authors remain reluctant to see the facts they observe as a weakness of 'big pharma'. Rather, the latter is seen as retaining power through its control over the former through financing R&D contracts with milestone payments and licensing agreements, managing due diligence, and marketing and distributing final treatments or drugs. In this contribution we will present the shift in tacit and exploration knowledge to DBFs as signifying a crisis for multinational drug companies. Such attenuation of the R&D function has not been sought by pharmaceuticals firms, but rather represents a failure to deal with a thoroughgoing paradigm shift. This chapter will examine the nature of that shift and explore ways in which some 'big pharma' are seeking to manage their response.

The chapter has four key sections. The first examines the knowledge management mechanisms by which DBFs use 'embeddedness' to tackle the R&D or 'drug discovery' process and determine the nature of their specific advantage. The second section assesses the adequacy of these mechanisms and how industry and intermediaries judge they need to be strengthened. The third part examines the future of big pharma concerning the cognitive paradigm shift linking 'Mode 2' knowledge production (Gibbons *et al.*, 1994), the demise of 'discovery' methods and rise of 'rational drug design', and fine chemistry versus molecular biology. Finally, the fourth investigates regional development and management control issues arising from clustering of advanced bioscientific knowledge exploration and exploitation in a few globally significant 'megacentres'. Reference will also be made to findings that the 'bioscience megacentre' process is leading to the emergence of a new type of regional policy called *regional science policy* that seeks to overcome the traditional centralising features of nationally formulated science policies (and in the EU, supranational RTD or research and *technology* policy; Cooke, 2003a).

Theoretical approach

In line with the theoretical framework outlined in Chapter 2, the theoretical approach evolves in microeconomic terms from Marshall (1918) and more recent transaction costs theory (Coase, 1937) as developed by Penrose (1959) and Richardson (1972) with their more 'social' 'resource' and 'capabilities' perspective on the firm, rather than the more orthodox neoclassical theorems of, for example, Williamson (1985). This chimes also with Piore and Sabel's (1984) 'flexible specialisation' conception of the advantages of small firm networks in successfully attacking global markets. More recently, authors such as Teece and Pisano (1998) and Best (2001) have explored the 'capabilities' perspective in theoretical and empirical depth.

As noted in Chapter 2, Marshall's initial statement was that small firms gained dynamic externalities from co-location. Proximity gave advantage in terms of specialised skills pools, opportunities for production specialisation and technical or managerial knowledge transfer. These circulated rapidly due to socio-cultural factors like trust, customs, social ties and other institutional characteristics of 'industrial districts'. Porter's (1990, 1998) notion of 'clusters' owes much to these insights, as he readily admits. Nowadays, 'spillovers', including 'knowledge spillovers' (see Audretsch, and Feldman, 1996; Feldman and Audretsch, 1999), are offered in explanation for clustering. These can appear as static rather than dynamic externalities, although in the relevant literature initiative-taking from spillover opportunities arising from R&D outputs and joint patenting is a key research focus.

This chapter explores from a firm-capabilities and, to some extent, an institution-capabilities perspective the modes by which DBFs manage complex types of knowledge ranging from basic scientific to financial, in creating project networks to generate and exploit research to develop therapeutic treatments. The extent, and degree of formality and informality of involvement by (complicit) knowledge intermediaries in this process as compared to direct contact among peers, will be explored. Moreover the process management functions and problems of big pharma in the 'knowledge value chain' from research to financing and final distribution as its control shifts downstream will be of key theoretical interest. This is especially pertinent as a test of the evolving thesis of transnational corporations becoming 'hubs', buying not making key services (Stewart, 2001; Best, 2001). The role of knowledgeable intermediaries as agents in innovation interactions will also be investigated to refine theory.

Theoretically, there are four links connecting the previous, and present, chapter's key interests to the 'design space' of post-genomics (Stankiewicz,

2002). First, following Polanyi (1966) and Nonaka and Takeuchi (1995), there is the interplay of implicit (tacit) and explicit knowledge in the project networks formed among firms with distinctive expertise, like combinatorial chemistry, high throughput, target-based screening, genomics and genomic libraries. We can explore the extent this mimics Nonaka and Takeuchi's (1995) original and Stewart's (2001) recently reviewed 'SECI Process' linking 'Socialisation', 'Externalisation', 'Creation' and back to 'Internalisation' in the eliciting of implicit knowledge, its formulation as explicit or codified knowledge, followed by its re-internalisation as tacit knowledge. In particular, the question as to whether there are important differences related to types of knowledge (e.g. 'exploration' versus 'exploitation') has to be confronted. In this chapter, unlike Chapter 2, the role of *examination* knowledge, the third sub-system of innovation, is not examined in detail.

Second, how adequate are the institutional mechanisms by which such interactions are managed? Are they largely informal and inaccurately accounted? Where is formality strongest, and are 'arm's length' or market exchanges more or less pronounced than 'untraded interdependencies' (Dosi, 1988)? Third, what response does big pharma make, if any, to the shift towards 'rational drug design'? What distinctive strategies are being pursued? Finally, what are key barriers to control for big pharma, and what strategies are pursued to accommodate the current deficits in in-house drug discovery? What lessons have been learned from past experience and what new lessons are being learned currently to adjust to the predominant knowledge value chain relationships and interactions?

Knowledge management mechanisms that are candidates for explaining the specific ways biosciences and biotechnology embed themselves as they do include the following. First, this is the most science-driven sector in the advanced economies. For example, the European Commission (2003) *Third Report on Science and Technology Indicators* (Fig. D6.1.5, p. 419) reveals that 1992–1996 biochemistry and molecular biology were easily the most cited science field in European patents (21%) and, with the exception of 'multidisciplinary' (15%), the next seven fields are varieties of medical subdiscipline like immunology (6%), pharmacology (4%) and genetics (3%). In fact, in the top twenty, only electronic engineering, applied physics, organic chemistry, chemistry and telecommunications (1–3%) appear from outside biosciences and medicine. Second, academic bioscientists control knowledge *exploration*, often establish dedicated biotechnology firms (DBFs) for knowledge *exploitation*, but need specialist investors to assist commercialisation and specialist clinical research organisations to apply *examination* knowledge for clinical research in testing and the various stages of treatment trials. Third, pharmaceuticals firms fund, market and distribute DBF and laboratory research that has been successfully exploited commercially.

Relational embeddedness of the kind advanced by Granovetter (1992) is a reasonable characterisation of the power of deeply institutionalised 'ahead of the curve' science to shift the locational power structure of pharmaceuticals R&D from 'edge city' locations near London, New York, Cologne and Basel to university campus towns like Cambridge, Massachusetts, Cambridge, UK, Heidelberg and Uppsala. We are thus in a Penrosian (1959) universe where organisational and firm capabilities and resources determine the geographical epicentres of location and, accordingly, industry organisation. However, as Woolcock (1998) has argued, embeddedness of any kind takes us only so far, and the concept of relational *autonomy* to innovate is a crucial correlate in dynamic settings where 'lock-in' is to be guarded against (see Cooke and Wills, 1999 for discussion). Nevertheless, the formation of (infra-) structural embedding institutions facilitating relational embeddedness (with autonomy) represents a profound overturning of the normal regional science perspective on industrial location. This holds that geography is an epiphenomenon of industry organisation. In biosciences, industry organisation follows the geography of organisational (universities and other research laboratories) and firm (DBF and 'big pharma') capabilities. Regional *capabilities* thus become crucial to economic accomplishment (Cooke, 2002).

The capabilities of dedicated biotechnology firms

In the 'capabilities' perspective on firms and, it may be added, *regions* that are the knowledge-embedded platforms in which such firms are rooted, it is dynamic capabilities that are the most prized. This is helpful because in the literature on 'knowledge spillovers' it is the dynamic rather than static externalities with which they are associated that are equally highly prized (Feldman and Audretsch, 1999). We noted in Chapter 2 an interesting debate, dating from the work of Jane Jacobs (1969; who also introduced the modern concept of 'social capital') about whether it is the diversified or specialised nature of capabilities in knowledge spillovers that gives the basis for innovatively successful milieux. Jacobs argued in favour of diversification, new combinations of capabilities giving rise to cognitive progress, something with which Feldman and Audretsch broadly agree. But researchers such as Glaeser *et al.* (1992) and Griliches (1992) stressed the superiority of specialisation and the capabilities of fairly narrow 'communities of practice' (Brown and Duguid, 2001), known elsewhere as 'epistemic communities' (Haas, 1992), in delving deeply and reasonably rapidly into a particular scientific sub-field. Empirically both showed how relatively geographically circumscribed knowledge exploitation, for example through patenting activity, actually was. More recently though, Galison (1997), moving well beyond the narrow confines of patenting activity such as that relied upon by Griliches, showed convincingly that new developments in scientific method, broadly consistent with the emergence of 'transdisciplinarity' in Mode 2 knowledge production (Gibbons *et al.*, 1994), are built fundamentally on diversification of knowledge. This debate is ongoing, with Henderson (2003) supporting specialisation and localisation externalities over diversification and urbanisation in producing growth effects for high technology, but not other industry. This feature of contemporary 'complexity' in knowledge management specifically occurring in industrial clusters is the subject of a path-breaking book edited by Curzio and Fortis (2002). We noted the advance on this by Boschma (2005), Cantwell and Iammarino (2003) and Ketels (2006) on the resolution of this by virtue of their discovery of the superiority of 'related variety' in explaining economic accomplishment.

Moving on, in a regional 'megacentre' such as Northern California, to be discussed in more detail in the penultimate section of this chapter, it is clear that the biomedical industry relies crucially on the related variety of information and communication technologies (ICT) to decode and synthesise bioscientific information. This is drawn from the knowledge-intensive ICT base in Silicon Valley, and includes sequencing and screening workstations, photonics and optical networking, low-level electrical energy instrumentation, and software among many 'varieties'. In theoretical terms, this reveals the distinctive structural embeddedness of this megacentre with its legacy of ICT knowledge assets, by contrast with the Massachusetts case, described below, where genomics-based drug research tends to be the capital capability. Convergence between ICT and bioscience has contributed to discoveries in genomics, proteomics, therapeutic cloning and stem cell research, and these in turn enable improved treatments for high-ranking disease targets like cancer, cardiovascular, AIDS, diabetes, and respiratory diseases. But the transdisciplinarity also operates within and between specific sub-fields like molecular biology, combinatorial chemistry, high throughput screening, genomics and bioinformatics. In conducting knowledge *exploration*, multidisciplinary teams of researchers are more prominent than before. In conducting knowledge *exploitation*, DBFs in the distinctive sub-fields form project-based networks interacting also with 'star' scientists and their teams.

Unlike Owen-Smith and Powell (2004), as Zucker *et al.* (1998) see it, such projects involve no or few 'untraded interdependencies'; they are strictly business transactions, with so-called 'pipeline' contracts, confidentiality agreements, time-limits and agreed actions (writing a patent or a paper, for example) and outcomes. Other 'complicit' actors will also enter this *mise-en-scène* at various points, as 'knowledgeable attorneys', consultants or venture capitalists (Suchman, 2000). So we may conclude that *bioscientific megacentres* are realised in the presence of a nurturing 'economic business environment' consisting of: (1) the quality of the inputs available to firms (e.g. human resources, physical infrastructure, availability of information); (2) the availability and sophistication of local suppliers of components, machinery and services, and the presence of platforms of related firms; (3) the sophistication of local demand for advanced products and processes, including the stringency of regulatory environments; and (4) the rules governing the vitality of competition and the incentives for productive modes of rivalry (Porter *et al.*, 2000). The key point to derive from this analysis is that while such 'economic business environments' are not unique to Northern California, they are far from ubiquitous. Because they rely on massive sums of public funding, a moral dilemma arises for policy makers alert to regional disparities. Should they encourage spatial concentration to achieve global excellence or should they also encourage the development of such facilities in less favoured regions? Keep in mind that the latter are where negative health imbalance is often as pronounced as economic weakness in the official statistics. Finally, as health economists and others are coming to realise, health services and their supporting supply firms in pharmaceuticals, biotechnology and research laboratories contribute as much as one-sixth of GDP in some advanced economies (Cassidy, 2002). In the final section of this chapter, regional authorities are shown as increasingly seeking to develop regional capabilities to allow their economies to take a slice of the pie by promoting regional science policies.

As is also shown below, research funding flows freely to 'centres of excellence' in research. Academic and public laboratory scientists evolve 'dynamic capabilities' (Penrose, 1959; Teece and Pisano, 1998) that attract entrepreneurs to engage in tacit knowledge exchange that may also be hemmed in by contractual relations. Thereafter, 'big pharma' establish 'observatory' plants in the cluster, adding to its 'megacentre' strengths. This is a most pronounced tendency in Cambridge, Massachusetts where from 1999 through 2002 Pfizer, Novartis, Amgen, Abbott, Dow and Wyeth all appeared on the scene by acquisition or translocation of R&D facilities. Biogen, one of the earliest Cambridge start-ups, merged with Idec in mid-2003 but Cambridge was chosen as the location for Biogen-Idec, then the world's fourth largest biotechnology firm with a market capitalisation of $12.5 billion, as 'an increasingly popular . . . hotbed for drug research' (Griffith, 2003). However, keep in mind that such firms evolve many non-local trading partners embedded in their wider knowledge value chains. But 'ahead of the curve' exploration knowledge is now the key driver of the co-location process that privileges certain geographical points as knowledge 'megacentres'.

Returning to healthcare as the final market for the products of such a 'hotbed', in Figure 3.1 data are presented on public and private health expenditure in a number of leading OECD countries for 1997. These statistics probably underestimate the whole health economy as described above by a few percentage points. Nevertheless they demonstrate even core health expenditure running at just under 10 per cent for the G7 countries, and much higher, at nearly 14 per cent for the USA.

Statistical data for the USA at regional (state) level reveal just how skewed is the distribution of that element of public expenditure covered by state and federal allocations. It should be noted that about half of US public expenditure on health is explained as follows:

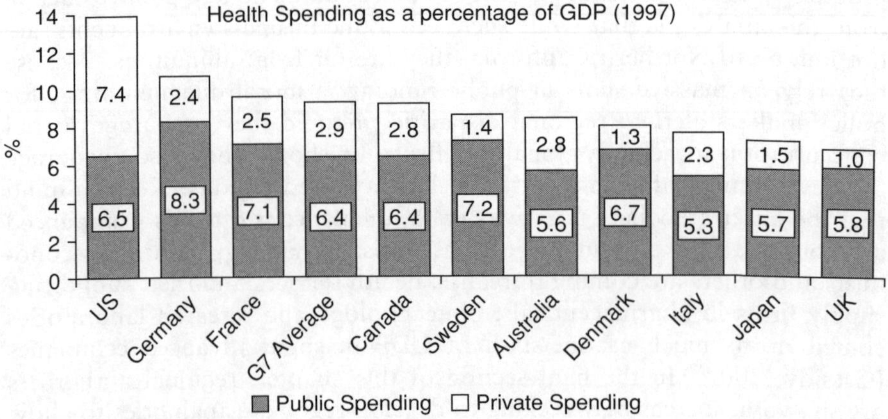

Figure 3.1 Health expenditure in leading OECD countries, 1997

Source: 2000 OECD Health Data

> The system of employer-sponsored coverage emerged to restrain wage inflation during World War 2 and afterward continued when the federal courts ruled that unions could collectively bargain with employers for benefits, including health care coverage. These benefits are considered public sector 'tax expenditures' because they are excluded from workers' wages for purposes of taxation and defined as an untaxed cost of business for employers.
>
> (Milbank Memorial Fund, 2000)

Private spending occurs through private insurance taken out by individuals on top of whatever workplace-related benefits they receive. Since the 1980s there has been a growth in share of the former and a shift from 'fee-for-service' indemnity towards 'managed care' in insurance health plans. Health insurance has thus become more commoditised, suppliers are keen to raise efficiencies in treatment times and use of new technologies, and to reduce casual conduct by physicians of clinical research on patients. Companies and consumers are equally keen to achieve value for money under circumstances where information to enable assessment is at a premium. There has thus been a tendency for geographical concentration of exploration research and, due to insurance industry pressure, clinical research in specific General Clinical Research Centres. For comparable reasons under a differently funded health insurance system in the UK the same kind of concentration into Clinical Research Centres is occurring (Cooke, 2003a).

In Table 3.1 some illustrative data are presented regarding the regional *vis-à-vis* federal *public* expenditure profile for direct health costs, excluding the employers' 'public tax' expenditures. Of particular note are the data pointing to California and New York states accessing 11 per cent and

Table 3.1 State and federal health expenditure, USA top ten, 1999

State	*State funds (\$ million)*	*Federal funds (\$ million)*	*S/F ratio (%)*	*GDP share (%)*
New York	15,009	14,860	99	3.9
California	14,412	11,981	86	2.1
Texas	6,993	8,100	115	2.2
Pennsylvania	6,723	6,110	90	3.3
Florida	5,080	4,469	80	2.2
Illinois	5,536	3,860	70	2.1
Massachusetts	5,125	3,279	63	3.4
Michigan	4,739	4,231	89	2.9
Ohio	6,739	1,620	24	2.3
New Jersey	4,522	2,971	66	3.9

Source: Calculated from Milbank Memorial Fund and US Dept. of Commerce, Bureau of Economic Analysis

12 per cent respectively of the US total for this item. These are, of course, large population centres, California being the larger, though it is noticeable also that while New York spends 3.9 per cent of its GDP in this way, California spends only 2.1 per cent. Thus demographics, politics and per capita income also play their part in explaining differences in level of expenditure. Among other large health expenditure states, but not in the same rank as California and New York, are Texas and Pennsylvania, followed by a group spending around \$8–9 billion in 1999, namely Florida, Illinois, Massachusetts, Michigan and Ohio. Of interest are the ways in which such budgets are composed, such that variance from the US mean GDP share per state (2.6%) is relatively small, including the rest, where West Virginia at 4.8 per cent on the one hand, and Alaska at 0.6 per cent are among the most significant outliers. With a few exceptions, such as Ohio in Table 3.1, federal funding makes a significant contribution. Texas, like other southern Appalachian and south-western states, receives more federal than state disbursements.

However, calculations of such disbursements in relation to state population size show New York, at \$1,572 per capita, and Massachusetts, at \$1,482 to be the most generous spenders; Pennsylvania comes next, at \$1,070, then New Jersey at \$949, Michigan at \$897 ahead of California at \$800, followed by Illinois (\$783), Texas (\$765), Ohio (\$763) and Florida (\$615). Linking back to points made earlier about the increasing 'commoditisation' of health care in the US (and to a growing extent in the UK and perhaps elsewhere), states such as these are becoming active purchasers of higher-quality, more technology-intensive, but also value-for-money health services. It should further be recalled that the statistics under discussion constitute only some 20 per cent of total state health care budgets, but represent baseline funding. Clearly, California is less dependent upon public funding than the East Coast states, that category accounting for only just over half the GDP share it does in New York and New Jersey. As Table 3.2 shows, California's 'mega-budget' for private health care expenditure easily outstrips all others among the entrants in Table 3.1. Hence the relatively low share of GDP allocated to direct public expenditure on health is compensated for by a massive figure of some 10 per cent of GDP being expended on personal health care.

Keeping in mind that Table 3.1 represents one-fifth of each state's health expenditure, requiring an approximate doubling to include the employer's contribution, and that Table 3.2 accounts for some three-fifths, we see the scale of total investment available annually in the leading states' economies. Thus California spends at least $160 billion, New York $145 billion and even modestly sized Massachusetts some $46 billion. Three key points flow from this accounting exercise. First, a few key regions, and in fact cities in those regions have the demographic, financial and scientific scale to afford the whole of the bioscientific and medical knowledge value chain. This moves from the most exploratory, fundamental research into genomics and post-genomics fields like proteomics and molecunomics. This is likely to be conducted at specialist research institutes such as the Whitehead in Cambridge, Massachusetts (partnered with the Sanger Institute, in Cambridge, UK and Washington University at St Louis for the Human Genome project). Knowledge of this kind will likely be applied in specialist medical research institutes at universities, like the Dana-Farber Cancer Institute at Harvard, or the New England Enzyme Centre at Tufts University, Boston. Research in other key fields noted earlier (cancer, cardiovascular, AIDS, diabetes, and respiratory diseases) will be conducted in other independent research institutes and university research centres like Harvard Medical School's pre-clinical research in biochemical and molecular pharmacology,

Table 3.2 Personal health care expenditure, top ten US states, 1998

State	*Personal health care expenditures, 1998 ($ million)*
California	110,057
New York	85,785
Texas	67,750
Florida	59,724
Pennsylvania	51,322
Illinois	44,305
Ohio	42,581
Michigan	35,647
New Jersey	32,695
Massachusetts	30,039

Source: Health Care Financing Administration (2001), Trends in State Health Care Expenditure and Funding, 1980–1998, Washington DC

cell biology, genetics, microbial and molecular genetics, and neurobiology, or affiliates like the Joslin Diabetes Centre. Second, such research institutes and centres both attract and train life sciences talent, giving critical mass to interactive research activity. This, in turn, strongly influences growth in funding through competitive bidding to National Institutes of Health and National Science Foundation programmes. Third, such 'megacentres' interact with the large hospitals, in which clinical research as well as patient treatment occur along with training of physicians. Massachusetts General Hospital and the Brigham and Women's Hospital in Boston are thus important large-scale patient-bases for clinical trialling. There is, accordingly a suitable milieu also for academic entrepreneurship which, combined with Boston's status as a top-three location for venture capital and 'knowledgeable attorneys' (Suchman, 2000), makes it a highly nurturing 'economic business environment' for *exploitation* as well as *exploration* knowledge management in the form of leading biotechnology firms like Immunex (acquired by Amgen), Biogen, Genzyme, Millennium, TransKaryotic Therapies, and others, recently also joined through new openings or acquisitions by the likes of Abbott Laboratories, AstraZeneca, Aventis, Pfizer and Wyeth (merged with American Home Products).

Thus, while scale of expenditure in general health systems clearly matters as an entry ticket to the megacentre 'tournament', it is by no means a sufficient condition. There have also to be world-class science and world-class commercialisation capability. There has to be localised 'social capital' (or relational embeddedness) among the actors present, which can link appropriate partners across epistemic community boundaries. Firms help themselves when they speak with a single voice on matters of common concern, something portrayed in the Boston cluster by the activities of the 280-member Massachusetts Biotechnology Council.

Rational drug design and DBF networks

What are the difficulties faced by DBFs in taking advantage of such nurturing economic business environments, and in what ways are those in the USA better placed to benefit than those in Europe? One key feature that began to differentiate them was the estimated $20 billion per year then available for US biotechnology research from federal investment in, for example, the 1994–98 period studied by Senker and Van Zwanenberg (2001). This compared with the approximately €10 billion spent by European governments over the same period. Further, it is argued, US DBFs can exploit the research findings of National Institutes of Health-funded research more swiftly and efficiently due to the existence of National Science Foundation sponsored Small Business Innovation Research (SBIR) grants enabling DBFs to develop ideas more quickly, thus potentially influencing venture capitalists and 'big pharma' to invest in what elsewhere would appear to be more high-risk ventures. These SBIR grants arise from a requirement that R&D departments spending federal government funds must spend up to 2.5 per cent of their extra-mural research budgets on commissions from SMEs. Moreover, research-minded entrepreneurs requiring continuing interaction with other discovery firms or research institutes have more of these to choose among and thus exploit better networking opportunities.

Jaffe *et al.*'s (1993) finding that knowledge spillovers from universities to firms were relatively regional or even local was also found to be true, but fewer, in the Senker and Van Zwanenberg research on European biopharmaceuticals firms. Owen-Smith *et al.* (2002) also found that US public research organisations (PROs), with which such DBFs are likely to seek to interact, have far more intensive and extensive inter-institutional research networks than European firms. The latter had much smaller, sometimes dyadic, networks and these were often nationally constricted, except for occasional transatlantic contact with a US institute, and consequent potential linkage into wider US knowledge networks (see also Figure 2.1, Chapter 2). However, so few and restricted were the active international linkages that less knowledge exploitation was feasible, and more slowly, than in the case of the better networked US institutes. Hence, it can be argued from our 'capabilities' perspective that the better relational social capital from networking for purposes of knowledge *exploration* and support for knowledge *exploitation* displayed by US research institutes is a sign of superior *institutional capability*. Given the role of key intermediaries in the knowledge management process, this further adds significant value to an already doubly advantaged initial research resource base.

It can be further shown, as Owen-Smith *et al.* (2002) go on to do, that US networks among PROs are hierarchically structured. Over the period from 1988 through 1998, the Boston cluster has remained at the peak of US interorganisational research linkages, and its connections to other clusters has doubled to over 50 per cent of its total contacts over that period. San Diego has moved to second place in the hierarchy over San Francisco, while Seattle and New York have exchanged positions, Seattle rising above New York in terms of the number and strength of its inter-regional PRO linkages. Thus relationships between PROs and firms in their clusters are augmented by the institutional capabilities of both to benefit from spanning therapeutic areas, engaging in multiple stages of the knowledge exploitation value chain, and involving related diversity collaboration. The institutional system takes on certain characteristics that have resulted in the term 'collaboratory' being used to describe such interorganisational networking. As has been suggested, European DBFs and PROs have tended to engage in more attenuated innovation networks, with more specialised than diverse interactions, and a more limited external value chain involvement that is also mostly national in character. Weak structural embeddedness thus makes for weak relational embeddedness.

The weaknesses of European international networks of PROs and DBFs can be understood in terms of the relative immaturity of regional science infrastructures (weak structural embeddedness). This, in turn, echoes the relatively small budgets that have traditionally been available for systemic medical, health and life sciences research integration. This is changing rapidly in some European countries, and has been rather better developed in some smaller EU economies for some time. Thus Germany and France have changed business regulations to encourage innovation and academic entrepreneurship, while a long period of under-funding in the UK health service has been reversed both in service delivery budgets and scientific research budgets. In the UK, the world's largest scientific research charity, the Wellcome Trust, has become highly active in co-funding with the UK government investments in research infrastructure and basic exploration research such as the Human Genome project and new post-genomic research at such centres of excellence as the Sanger Research Institute at Cambridge (UK).

The change from a 'chance discovery' model of scientific research to a 'rational drug design' model based on combinatorial chemistry, molecular biology, high throughput screening, genomics and bioinformatics has meant that those regions and localities with clusters of DBFs of various kinds, linked also to ICT firms and knowledge management intermediaries, are absolutely advantaged, even to the extent of making 'big pharma' dependent on them for key knowledge of both the exploration and exploitation kind. Increasingly such DBFs even manage the due diligence and trial management processes, leaving 'big pharma' to exchange contracts with the DBF networks for the licence to market and distribute the hoped-for biopharmaceutical drug at the end of the pipeline. As noted earlier, this reverses the age-old primacy of industry organisation over spatial organisation. Structural and relational embeddedness in specific megacentres drives industry organisation, even making large corporations 'supplicants' (albeit with deep pockets) to regional centres of exploration knowledge capability.

Which are leading megacentres and what is regional science policy?

We have seen already that Boston is perhaps the leading biosciences megacentre, not because it has the heaviest medical or even bioscientific research budgets, but because it is presently one of, if not the, leading centres for exploration research. So much so that while the Swiss drug company Novartis announced in 2000 a path-breaking agreement to spend $25 million on first access to the results of plant and microbial biology research conducted at the University of California, Berkeley, in the heart of the Northern California biotechnology cluster, in 2002 Novartis announced the establishment of a $250 million Novartis Genomics Research Institute in Cambridge, Massachusetts on the grounds that it was the leading exploration and exploitation centre for genomics and post-genomics knowledge. As noted in Chapter 2, 'megacentres' have large public investment infrastructures, not only competitive and sometimes collaborative firm interactions typical of clusters. Boston's current primacy has not been the product of the operations of the market mechanism alone. In 1999, $770 million of mainly public or charitable research funding was earned for medical and bioscientific research. (Table 7.1, Chapter 7, shows that figure exceeded $1 billion shortly afterwards.) This was marginally less than the amount of National Institutes of Health funding alone passing through the Northern California cluster in 1999, a statistic that increased to $893 million in 2000 (CHI/PWC, 2002). Most of the exploration research conducted in both Cambridge/Boston and Northern California was conducted in institutions dependent on public funding, though private research foundations were also functional in both. In Boston, the Massachusetts Biotechnology Council is an active and successful biotechnology association that lobbies industry and political forums at state and federal levels, pressing for an FDA presence in Boston to offset the advantage enjoyed by emergent firms and research institutes located in Maryland near the head offices of both NIH and FDA.

Let us look more closely at the manner in which the assets of biosciences megacentres like those in Northern and Southern California are now packaged in documentation that promotes the image that is intended to appeal to investors of all kinds in the regional innovation system. The two following examples, from Northern and Southern California, are produced by a non-profit association (The California Healthcare Institute) and a consultancy (Michael Porter's Monitor) respectively. The California Healthcare Institute is a public policy institute for California's 200 leading biotechnology firms and research institutes. It is thus comparable to the Massachusetts Biotechnology Council. Its political brief was expressed clearly by CEO Gollaher who despite noting that 'funding for basic science is strong', bemoans the fact that 'many federal and state lawmakers advocate policies that would impede medical innovation. Our greatest threats include a total ban on human cloning and severe restrictions on stem cell research; a Medicare administration that . . . effectively excludes new products . . . and a leaderless FDA facing the greatest wave of new inventions in history' (Gollaher in CHI/PWC, 2002). The demand was for collaboration among members of the biosciences innovation system to change laws that are perceived as threatening the evolution of the industry in the post-genomic era. California's legislature eventually accommodated the rise of a $3 billion stem cells research fund to circumvent the hostility of the Bush administration in Washington to such studies.

Northern California was presented as the birthplace of biotechnology which, along with biomedical innovatons like cardiac stents, has a strong base of some 819 biomedical/biotechnological firms, employing 86,000 people (28,000 in biotechnology), total R&D of $1.1 billion, NIH grants of $893 million and $4.1 billion in worldwide revenues, including $2.7 billion in exports. New infrastructure projects included University of California San Francisco Medical School's new $1.4 billion Mission Bay bioscience research campus, the new California Institutes for Science and Innovation, and the California State University CSUPERB joint ventures programme where universities and the private sector collaborate in bioscience research, technology transfer, business and even residential development. Much emphasis was placed on survey results showing that significant interaction occurred among firms and the institutional research base in Northern California – a functioning bioregional innovation system supporting a related variety industry platform.

Thus, California's academic research institutions were credited with playing a central role in the growth of nearly one-third of biomedical/biotechnological firms, 42 per cent of firms had at least one research contract with a California research institution, 56 per cent of firms planned to broaden or maintain such agreements and up to 70 per cent of firms had patent licence agreements and planned to maintain or broaden them in future. The key Northern California life sciences and clinical research institutes cited included Stanford University (Biomedical Technology Information Programme), Lawrence Berkeley and Lawrence Livermore National Laboratories, and the University of California, San Francisco Medical School, Berkeley (BioSTAR industry-academic collaboration), Santa Cruz (with Berkeley, the California Institute for Bioengineering, Biotechnology and Quantitative Biomedical Research, QB3) and Davis (Life Sciences Information programme). More than 19,000 were then employed in research in the region, and two of the top ten NIH R&D grant recipients in the US were UCSF and Stanford. A picture of this cluster that is developing the characteristics of a biosciences megacentre is presented at www.biospace.com. But the judgement as to whether it yet is, is occluded by the lobbying points that although some larger firms such as Abbot Laboratories and Genentech were present, most are SMEs, 49 per cent without products on the market, 45 per cent with no revenue in 2000. Finally, of the pharmaceutical pipeline products reported, 53 per cent were in preclinical trials. This is by no means unusual, but nevertheless testifies to the apparent fragility of the *exploitation* aspect of the Northern California cluster, once its strength, but never adequately backed up with strong bioscience *exploration* capabilities and now, belatedly perhaps, seeking to embed them.

In Southern California, the San Diego biotechnology cluster (Figure 3.2) had larger claims to be considered a biosciences megacentre than even that in the North. In Porter's (2002) competitiveness study San Diego's biopharmaceuticals cluster was presented as long established and among the most significant outside Boston, especially for R&D. Cluster employment growth was more than 8,000 from 1988 through 1997 and San Diego had the most rapid growth in patent output compared to the twenty largest US

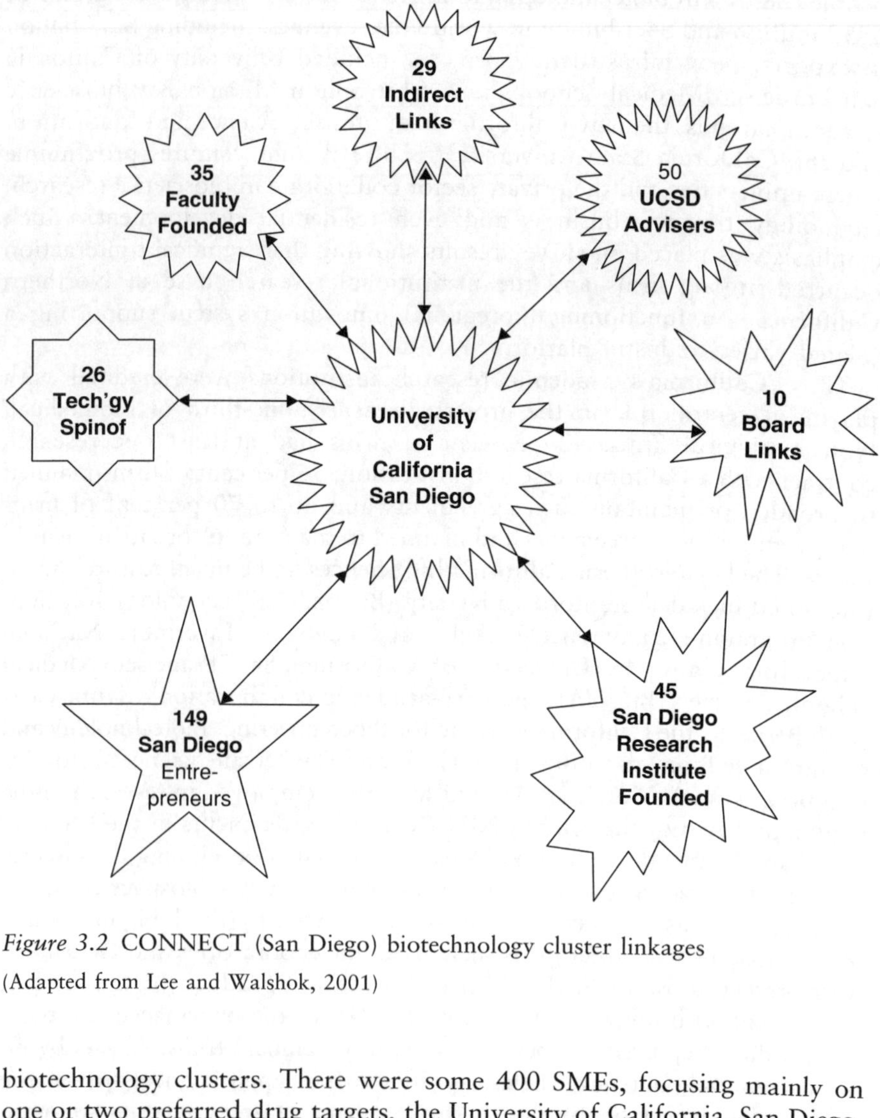

Figure 3.2 CONNECT (San Diego) biotechnology cluster linkages
(Adapted from Lee and Walshok, 2001)

biotechnology clusters. There were some 400 SMEs, focusing mainly on one or two preferred drug targets, the University of California, San Diego, with numerous specialist research centres, and finally, some globally known research institutes, the Salk Institute, the Scripps Research Institute, the Burnham Institute, and the La Jolla Institute for Allergies and Immunology, each focusing upon aspects of life science, medical or clinical research.

The Scripps Institute, since establishment in the 1950s required its researchers to raise their own funds, encouraging collaborative innovation with larger firms (like Dow). By contrast, the Salk Institute does not conduct corporate research but licenses its discoveries and takes equity stakes in companies. UCSD emphasised medical research and academic entrepreneurship. One early fruit of that approach was Hybritech, a 1978 biotechnology start-up, from which more than fifty other local biotechnology DBFs were spun out. In 1986 it was sold to Eli Lilly for $400 million. A further feature of this cluster is its strong and long-established networking propensity, signified by the establishment since 1985 of the UCSD CONNECT network, a model for cluster integration in many other new economy clusters such as Scotland's and Cambridge's (UK) (ICT) networking associations.

In Figure 3.2 the origins of the San Diego biotechnology cluster firms are analysed in relation to the San Diego CONNECT network. It is notable that 'entrepreneurs' (like the first owners of Hybritech) are far more important sources of direct spinout firms than either the university or research institutes. Some such 'entrepreneurs' had no talent for biotechnology but were 'serial entrepreneurs' moving into a promising field in which to realise profitable investments. Overall, the Lee and Walshok (2001) report concludes that the San Diego biotechnology cluster is characterised by the following features:

- 400+ companies
- 248 emergent firms
- 28,000 employees
- UCSD CONNECT – 'a network of professional competencies focused on building shared knowledge . . . for technological companies'
- UCSD, Salk Institute, Scripps Research Institute, Burnham and La Jolla Institutes
- an innovation support infrastructure of investors, consultants and technology intermediaries.

The judgement of Porter's (2002) team was that the close proximity of research centres and firms on the Torrey Pines Mesa was a key advantage in encouraging collaboration and growth. Regarding patenting, San Diego registered 360 patents in biopharmaceuticals in 1997, a rate of 13.17 per 1,000 workers, and the growth rate was the US's fastest, though the intensity was less than nine other bioclusters. Venture capital was invested at a much higher rate than nationally with $421 million having been placed 1995–99, nearly 10 per cent of the national total. But research organisations were the greatest strength, Novartis and Dow having joined the public institutes, making a total of some 16,000 employees in biopharmaceuticals research alone, larger than that specific category in Northern California. However, both Californian clusters had strongly emergent megacentre properties, based especially on their strength in exploration knowledge and an abundance of SME DBFs capable of rapidly forming molecular discovery networks due to geographical proximity and critical mass. In the Orsenigo *et al.* (2001) study of rational drug design research networks, 18 per cent of interacting firms and institutes were in Boston, 16 per cent in Northern

Table 3.3 Comparison of bioregion NIH research funding 2002

Rank (Top 300)	*Institution*	*NIH Research $ million*	*Rank*	*Institution*	*NIH Research $ million*
Massachusetts			**San Francisco**		
12	Harvard U.	248.6	4	UCSF Med.S.	319.7
18	Mass Gen.	232.1	16	Stanford U.	226.3
22	Brig. & W.	192.4	47	UC Davis	103.1
34	Boston U.	132.3	53	UC Berkeley	81.6
51	Dana-Farber	96.3	117	UC Law. Labs	33.3
52	Beth Israel	94.8	144	N. Cal. Inst.	22.5
57	Whitehead I.	91.1	169	Chil. Hos. Oakl.	14.7
60	U. Mass. Med.	87.6	206	UC Santa Cruz	11.5
63	MIT	77.8	247	UC Livermore	8.8
76	Children's H.	62.8	253	N. Cal Cancer C.	5.2
88	Tufts U.	49.9	270	Chiron	2.2
120	N. Eng. M.C.	27.7	San Francisco Total		829.0
133	Boston M.C.	27.2			
148	Joslin Diabetes	20.9	**San Diego**		
166	N. Eng. Rsch. I	14.7	17	UCSD	244.7
200	Mass Eye-Ear	11.9	25	Scripps I	185.6
210	UM Amherst	11.3	96	Salk Inst.	44.9
228	Boston Biomed.	8.5	108	Burnham I	37.2
290	Boston College	6.3	167	SD State U.	16.2
Massachusetts Total		1,494.2	199	Loma Linda U.	13.3
SF/SD Total		1,381.5	227	La Jolla Inst.	10.6
			San Diego Total		552.5

Source: National Institutes of Health

California and 12 per cent in San Diego. The few partners outside the USA were located in Munich (2), Cambridge (2) and Oxford (2).

Changing focus, the Massachusetts NIH budget for 2000 was some $1.09 billion. This contributed significantly to the advantage enjoyed for basic research by the Cambridge–Boston bioregion compared with San Francisco and San Diego. More recent data taken from NIH allocations for 2002 (Table 3.3) confirm the strengthening of the Boston bioregion since 2000. The statistics reveal that the Greater Boston bioregion retains its lead over San Francisco and San Diego combined, with regard to research funding won from the National Institutes of Health. This is because its institutional base is far broader and deeper. This is revealed in the large number of substantial mid-range institutions that are either independent universities, research institutes, research hospitals or specialist centres such as Dana-Farber, Joslin, Beth Israel, Whitehead and so on. Both California bioregions are heavily dependent upon the University of California system, although independent Stanford University makes a significant showing and the Torrey Pines Mesa research institutes in La Jolla, San Diego make a significant research income impact in otherwise institutionally thin San Diego. Nevertheless, as was shown in Figure 3.2 San Diego's institutional base is tightly knit, not least through the activities of UCSD-CONNECT. What Figure 3.2 clearly shows is that while UC San Diego is the catalyst of the bioregion and its cluster, only 26 biotechnology firms were actually university spinouts. Faculty members privately founded 35 firms and 29 firms were indirectly connected. For example, Hybritech was a famous UCSD spinout in 1978 that begat its first spinout, Gen-Probe in 1983. Thereafter, a further 52 firms were spinouts of Hybritech until 2003. Twenty-nine of these firms had founders with some link to UCSD as former students or faculty members. UCSD supplies 50 members of San Diego biotechnology company Advisory Boards, and a further 10 faculty are linked to boards informally but are not members. The San Diego community of research institutes such as Burnham, Salk and Scripps has been responsible for founding 45 spinout firms, while the vast majority of firms (149) were founded by local or in-migrating entrepreneurs with no connection to UCSD, including, for example, some of the Hybritech entrepreneurs originating elsewhere. Emergent firms are those in various stages of pre-incubation, incubation or other early-stage start-up condition without product but possibly with intellectual property.

What does this signify? Once more, the asymmetries in knowledge domains and the associated distribution of bioregional megacentres. An extensive analysis of the response by the large number of non-megacentre, even non-bioscience regions in the USA, Canada and the UK was undertaken (Cooke, 2003a). This showed that 15 US states had undertaken science based analyses and developed science strategies with targets and mechanisms for augmenting their capture of basic science funding in biosciences. Some piecemeal evidence of the outcomes of these are the statistics that show universities like Harvard, Johns Hopkins, UCSF, Stanford and Duke as occupying the highest places in the NIH R&D allocations compared to hitherto. But the USA has also an interesting mechanism for supporting ambitious if underdeveloped universities in lagging states to access national competitive funding. The EPSCOR scheme under the NSF competitive bidding procedure for accessing, first programme, later project funding allowed designated states, e.g. Oklahoma and Mississippi, to bid competitively for such bioscience research funding by lowering the grant aid qualifying bar below that expected for the rest of the USA. This, in EU terms, was a 'structural funds' mechanism within the Framework Funds and it has produced generally positive rather than futile results in the USA. The Bush administration diluted this moderating intervention on such knowledge asymmetries.

What about development of biosciences activities in lagging regions in Europe? Some interesting and instructive cases, not without developmental obstacles, are presented from regions like the Ruhr in Germany, Northeast England, and remote regions of Finland and Sweden in Chapter 9. A particularly interesting case is that representing in a single firm the technological platform shift in pharmaceuticals from synthetic chemistry to 'biologics' and molecular biology more generally is the strategy of DSM. It will be shown in Chapter 9 how this leading producer began its existence as the Netherlands' coal producer Dutch State Mining, the concern responsible for mining. When Limburg coal ran out in the 1960s, the firm and the Limburg region had to embark on what many consider to be a textbook example of 'reinvention' as DSM shifted first into fertilisers, then petrochemicals, performance materials and now life science products. In Chapter 8 I map the progress of developing countries, often against the will of advanced country 'big pharma', to develop biotechnology for domestic healthcare requirements that are simply not met by advanced country producers.

Elsewhere in Europe, some regional bodies began to develop regional science strategies, most notably Scotland that identified its £800 million R&D spend annually and sought to augment it through intra-regional collaboration and the furtherance of existing science-funding mechanisms. Medical and biosciences were key funding areas to be targeted. In Finland a central government policy of seeding regional Centres of Expertise, and later, Centres of Excellence produced regional bioscientific Centres of Excellence modelled on the US method of linking *exploration* to *exploitation*, but through largely public rather than private knowledge exploitation mechanisms. The intervention did not work as well as expected and the promoting agency TEKES was, in 2006, overhauled and given a stronger regional focus as a consequence. Regional governance institutions are somewhat more aware now than they were until only very recently, that health is a large part of their economy, that it links directly to some of the most exciting science being done in the world today, and that as well as making a direct economic contribution through purchasing and employment, it can make an important indirect contribution to economic welfare through possible academic and corporate spinout activities. It is likely that in future regional analysis and policy will be more rather than less influenced by issues such as those discussed in this chapter. This has also happened in another Nordic country, Norway, which in 2006 implemented a formal 'regional innovation systems' approach to bioscientific and other research exploitation. In 2007, Denmark also implemented regional economic governance.

Conclusions

These are brief and orientated towards future implications for regional science and policy rather than being a simple reprise of what has been said. First, of significance to the never-ending debate about the viability of SMEs in a globalising world dominated by multinational firms, events during the past decade or so show just how misplaced arguments prophesying the demise of SME significance can be. It is 'big pharma' that are in crisis as their traditional expertise in fine chemistry is subverted by the molecular biology revolution and the demand for transdisciplinary teams of DBFs to form project-based networks to seek 'rational drug design' solutions based on reagents and their inhibitor compounds at the molecular and even sub-molecular levels. Meanwhile 'big pharma's' drug innovation pipeline dries up as R&D costs escalate, giving a further imperative to externalise projects to the 'knowledge value chain'. DBFs in turn rely effectively upon the deep pockets of 'big pharma' for licensing cash, marketing and distribution. Some DBFs are now quite large and one (Celltech) once bought a mid-size 'pharma' (Medeva) in the UK. Shortly thereafter it was itself purchased by Union Belgique Chemie and its Medeva vaccines facility ended up being acquired by Powderject, then Chiron which itself was acquired by Novartis in 2005. Other global 'pharma' have been merging at pace recently as they seek to reap one-off shareholder value from reducing the competition, as Glaxo, Pfizer, Wyeth, Aventis and Novartis to name a few, testify. The regional science implications of these shifts towards public R&D exploration and exploitation and away from 'big pharma's' traditional metropolitan redoubts require swift and serious analysis. In 2005 numerous acquisitions were made of DBFs by pharma, the largest being Merck of Germany's 2006 acquisition of Serono, Switzerland's 'champion' and third largest DBF in the world for $13 billion. In 2006 also, AstraZeneca bought Cambridge Antibody Technologies, among whose founders was bioscientist Greg Winter, then in late 2006 Glaxo bought another Cambridge DBF, Domantis, one of whose founders was the, by then, Sir Gregory Winter.

In regional policy terms, the new model is more foresight-driven, more collaborative, based on shared vision and leadership than old redistributive policy used to be. This is necessarily policy for securing 'generative growth' (see Cooke, 2003b for fuller discussion of this policy theory). It must take seriously the long-established presence of hospitals and universities and seek to forge links along the biosciences value chain. It must do this in a sometimes hostile environment in which national or federal governments prefer to see a few Centres of Excellence, possibly not too far away from their seats of government rather than in obscure and peripheral regions. Yet policy makers in precisely such regions will surely seize upon the obvious notion that for once in history public sector investments are new sources of innovative development and generative growth, seeking ways of mobilising enterprising coalitions to bid for infrastructure and 'star' scientist recruitment from a variety of funding sources. Or, if this is insufficient, pressing their multi-level governance structures for affirmative research allocation action such as that already pioneered in the USA through EPSCOR.

4 The microbiology revolution and the crisis in pharmaceuticals

Introduction

We are living through a scientific revolution. It relates to bioscientific research and commercialisation and promises to have far-reaching effects in a number of areas of social and economic life, most notably that concerning human healthcare. This is the focus of the present chapter. That is not to say that the impact of the 'molecular biology' revolution on agro-food and environmental treatments arising from biotechnology is not important. It may be more so if it brings major improvements to food, energy and environmental quality and cleanliness. But for the moment, biopharmaceuticals is widely accepted to be the most dynamic application arising from new bioscience, not least because of the possibilities of contributing dramatically to the curing of such highly complex, intellectually challenging and medically exacting diseases as cancer, AIDS, diabetes, and neurological, cardiovascular and respiratory ailments. In relation to the health 'value chain', this represents, according to some estimates, up to one-sixth of GDP in an advanced health economy like the USA (Cassidy, 2002).

Of intellectual interest to social science and particularly *innovation* science, the insights of which have been elaborated by the 'triple helix' concept (Etzkowitz and Leydesdorff, 1997) is the manner in which modern bioscience has caused a massive shift in research firepower away from the 'industry' part of the helix to the 'university' component by virtue of a massive rise in public research investment from the 'government' element. From the dawning of the modern pharmaceuticals industry it has been a textbook case of hierarchical, vertically integrated, in-house R&D firms reaping economies of scale from innovations in synthetic or 'fine' chemistry based on the identification of small molecule compounds and their synthetic 'inhibitors'. Genetic engineering changed all that. But for a long time synthetic chemistry retained hegemony since many of the early compounds like human insulin had a small molecular base that still required chemistry skills for the production of treatments. Genomics and post-genomics now undermine even that role, as large molecules become the disease targets, expressed in proteins consisting of long, interconnected chains of amino acids that cannot be chemically synthesised. The functions of proteins in disease and its inhibition are by no means well understood, hence the massive scope for specific protein research beyond even the knowledge base supplied by the human genome.

This chapter explores the shifting power balances in the 'triple helix' through the lens of modern biosciences, beginning with a discussion of the implications for industry organisation in pharmaceuticals of the rise of 'rational drug design' and the demise of the 'chance discovery' model of scientific research. This will lead into an account in section two of the implications of this shift in epistemology for economic growth and development, particularly focusing on spatial shifts in the epicentres of local and regional development. This will show how clusters of specialist firms now lead the drug discovery process to an increasing extent, occupying a geographical 'design space' (Stankiewicz, 2002) with specialist medical centres and university laboratories that is different from the old pharmaceuticals heartlands. Moreover, traditional 'pharma' are seen to be moving their 'knowledge production' into what are becoming 'bioscience megacentres' rather than simply 'business clusters' by new openings and acquisitions, but mainly by bankrolling 'dedicated biotechnology firms' (DBFs). The term 'megacentre' is French, this author having first seen it in an account of aspirations embodied in the French state's

> Atomic Energy Association, Grenoble's Institut National Polytechnique and the Rhône-Alpes region . . . agreement to establish a new mega-centre for micro- and nanotechnologies. Minatec, as it is to be called, will open in 2004 and aims to become the leading European cluster for students, researchers and entrepreneurs in the field.
>
> (Johnson, 2002)

Informed speculation about the strengthening of 'megacentres' through concentration of new rounds of 'big science' investment by governments and private foundations follows.

Finally, the third section ponders the future for regions that presently seek but do not have great strength in bioscientific, medical and clinical research. So much is biotechnology driving drug development, it will be shown, and so massive are the research and infrastructure funds being concentrated in 'megacentres', that large parts of the population of most countries will not have regional access. Two kinds of response are examined as possible means of moderating this imbalance: scientific affirmative action and regional science policy. Both pose interesting challenges to a traditional 'Enlightenment' view that peer review of scientific excellence alone determines who and where should conduct advanced scientific research.

Rational drug design and the moderation of uncertainty

In a chapter that deserves a wider audience than it has perhaps received, Henderson *et al.* (1999) explored the changing epistemology underlying scientific method in biosciences consequent upon the emergence of biotechnology. They called this the 'molecular biology revolution'. This was, to some extent, predicted by Gibbons *et al.* (1994) as the onset of Mode 2 methods of knowledge production. Mode 2 knowledge production, it will be recalled, is *networked*, *transdisciplinary* and *reflexive* compared to Mode 1 (traditional) which was *hierarchical*, *disciplinary* and determinate. It is cognate to Latour's (1998) interesting distinction between *science* (stable, paternalistic, detached) and *research* (unstable, risky, controversial). Recently, Nowotny *et al.* (2001) corrected what was criticised in Gibbons *et al.*'s (1994) rather generic account of predicted epistemological change by writing about society's and economy's increasing 'contextualisation' of science through interaction with 'research'. True to its time, Gibbons *et al.* (1994) was informed greatly by engineering rather than bioscientific progress. Interestingly, reference is made to Webster and Etzkowitz's (1991) early work on university–industry relations.

What is inappropriate about using an engineering template to characterise knowledge production and exploitation? Schumpeter's theses about entrepreneurship and innovation were thoroughly saturated with an engineering metaphor whereby 'creative destruction' effectively neutralised pre-existing technologies, incumbent firms were out-competed, swarming by imitators occurred as testimony to 'punctuated evolution', and new dominant firms arose from the competitive struggle among innovators and imitators.

Bioscience does not fit the metaphor because, according to Henderson *et al.* (1999), the key shift occurred in *scientific* knowledge – from chemistry to biology. This is far more profound than, say, that from propeller-driven to jet-propelled aircraft engines since the basic science remains aeronautics and knowledge adaptation is less exacting, not to say daunting. The pharmaceuticals industry is, in evolutionary terms, path dependent on synthetic chemistry of the kind by which the Bayer company derived aspirin. As Andermann (1996) puts it, there have been

> attempts to trace the compound back to antiquity through the Ebers papyrus, the Hippocratic writings, and the works of Galen. Such histories tell a simple, linear tale of the numerous 'discoveries' proposed to have led to the use of certain salicylate-containing plants, such as willow bark and wintergreen, or salicylate-related compounds, including salicilin and salicylic acid, as cures for a variety of ailments.

But aspirin was discovered as a by-product of dyestuffs, derived from coal tar and ultimately coal. Remarkably, it was only in the 1970s that aspirin's key mechanism for relieving pain was discovered. This was a result of innovating cascade superfusion bioassay technology (fundamental to genetic screening). The technique was invented by British pharmacologist John Vane. His Nobel prize citation for medicine in 1982 was for the bioassay technique that showed aspirin inhibited the release of a hormone-like substance called *prostaglandin*. This chemical regulates blood vessel elasticity and alteration in functions of blood platelets; it can thus moderate blood clotting, reduce inflammation, and contribute to a healthy heart. This is a good example of the 'chance discovery' nature of synthetic chemistry, underlined when it is understood that Felix Hoffman's re-discovery at Bayer in 1893 of French scientist Charles Gerhardt's earlier, unexploited compound was marketed as a fever suppressant rather than for its later, far more highly valued pain-relief properties.

A second reason the engineering metaphor does not work for bioscience is that firms do not suffer the 'competence destroying' effects of the 'gales of creative destruction' as much as predicted. This is because pharmaceuticals firms contain a wide variety of competences, from research to trialling and due diligence to financing, production, marketing and distribution of commodities sold in bottles, powders or as injectibles. Over the twentieth century they built up global reach with respect to each of these functions, and neither universities nor the majority of DBFs can outperform each of them. However, in the all-important R&D function it looks increasingly to be the case that essential competences are quickly being destroyed. Thus increasingly, as we shall see, 'big pharma', like other kinds of multinational, is becoming a hub that 'systems integrates' the lower order functions where it retains competences and, crucially, funds by means of licence agreements and milestone payments, the high-risk drug discovery process being conducted increasingly by research-intensive DBFs. Of course, drug products derived from synthetic chemistry origins, like Viagra, are not subject to this prognosis, but it is worth noting that in Cooke (2005a) the proportion of biotechnology-derived drugs in the 'pipeline' had risen dramatically by 2000 when there were 90 biotechnology drugs on the US market. Since 1999, 5 to 12 biotechnology drugs have been approved there each year. That number increased sharply from 2002, when 369 biotechnological drugs were in clinical trials, accounting for one-third of all drugs in clinical trials in the United States.

Biotechnology drugs are addressing more than 200 diseases, over one-third devoted to variants of cancer. Preliminary examination of European Medicines Evaluation Agency data suggest that in mid-2002 there were at least 70 such drugs approved in Europe (EMEA, 2002). Despite the 'microbiology revolution' therefore, 'big pharma' have not been swept away but rather operate symbiotically with a necessarily co-operative DBF sector, and with largely public and foundation funded public laboratories, or independently endowed research centres (e.g. Howard Hughes Medical Research Institute, endowed with $37 billion; Cooke, 2003a).

A third point, relating more to Gibbons *et al.* (1994) is that, accordingly, while the biopharmaceuticals industry has developed strong *network* characteristics among DBFs, research laboratories, venture capitalists, specialist lawyers, consultants and accountants, 'big pharma' and hospitals, substantial hierarchy rather than heterarchy alone still characterises all three elements of the 'triple helix'. This point ushers in the essentials of 'megacentres' in biosciences, which are defined in contradistinction to the Porterian (1998) definition of 'clusters'. Whereas the latter is fundamentally concerned with proximate business interactions for competitiveness and commercialisation of innovation, hence with *markets* alone, the megacentre idea captures the knowledge value chain from exploration, through examination to exploitation knowledge (see Cooke, 2005a; 2005b). Thus megacentres are science-driven, public and privately funded institutional complexes that in biosciences have as their ultimate goal the production of patient healthcare. They are hierarchical networks that include industrial hierarchy expressed in the ever-concentrating ownership structure of 'big pharma', government hierarchy regarding basic research funding and regulation of bioscience, and research hierarchy, in that most of them are concentrated in medical schools, hospitals and universities. Their importance lies in the contribution the healthcare-pharmaceuticals-biosciences sector makes to national GDP, probably under-estimated as at least one-sixth for the USA, as we have seen, and 25 per cent for the UK (DTI, 1999).

Because research has became extremely expensive, such that R&D costs for the 28 new drugs introduced worldwide in 2001 coincided with an R&D expenditure by 'big pharma' alone of $28 billion in that year, there has, for a decade, been increasing pressure from shareholders on pharma to reduce the unfavourable ratio of input to output. Simultaneously, pharma found they could not easily adapt from a fine chemistry to a bioscientific epistemology, whereas universities and DBFs could. So the logic of this has been that *transdisciplinarity*, a paradigm characteristic more accurately predicted for Mode 2 knowledge production, has come about. Moreover, small molecular targeting and identification of the 'inhibitor' molecule became the province of an alliance between DBFs and 'big pharma'.

This operates in two main dimensions; the first concerns R&D exploitation, where DBFs and research laboratories in universities, independent institutes or contract research businesses network locally and globally in knowledge transfer through specialisations in distinct sub-professions. These range from genomics, proteomics, and combinatorial chemistry, to bioinformatics, genetic libraries and 'biobanks', bioimaging and target-based high throughput screening. Not only are most of these capabilities far from those of pharmaceuticals firms, but some increasingly require access to 'big science' infrastructure like synchrotrons (for bioimaging) usually funded only by governments or their agencies.

The second concerns such activities as clinical research organisations (CROs) that supply pre-clinical and clinical research, testing and trialling services. These capabilities are supplied by US firms like Quintiles, Covance and Parexel[1] and are sought both by 'big pharma' and by DBFs intent on trialling drug candidates. A second important contract service is biomanufacturing. Such is the range of products and services associated with biotechnology that there is, on the one hand, a dearth of adequate biomanufacturing facilities worldwide, but an exponentially increasing and diversifying level and type of demand. More will be said about this in the next section, but for example UK firm Avecia (formerly Zeneca Specialities, both originally part of ICI) announced in March 2002 a $100 million 'biologics manufacturing investment', to be the largest of its kind in the world. This is part of Avecia's strategy to build a $500 million fine chemicals market share by 2005, for which its biotechnology and pharmaceuticals contract manufacturing businesses are two key drivers. 'Biologics' are the new generation of natural protein-based medicines, accounting for 20 per cent of all new drug approvals, with over 250 in clinical trials. At the other extreme is Cobra BioManufacturing, based on a university science park in Staffordshire, UK, which provides 'good manufacturing practice' (GMP) contract services spanning pre-clinical to early Phase III scale production (clinical trial of new drugs have three phases: animals; small samples of humans; large samples of humans). It manufactures biological products for clinical studies, supplying services in the pharmaceuticals and biotechnology industries. It focuses on the supply of DNA-based therapeutics for pharmaceuticals and biotechnology companies and from discovery through to Phase III scale. Services include DNA manufacture, virus manufacture, protein manufacture, cell banking, bioprocess development, analytical services, quality assurance and regulatory support.

In summary, the 'microbiology revolution' took pharmacology away from chemistry and towards biotechnology and biosciences more broadly. While for a time it appeared that fine chemistry would retain an epistemologically prior position in the new field because the initial goal was the identification of small molecules and their 'inhibitors' that could make mutant genes or cells behave differently, this too has changed. Today the targets are proteins – long strings of amino acids in 'big molecules' – that biotechnology first showed capability of *producing* where they were well understood, like insulin, human growth factor or blood-clotting factor, but today is central to *searching* for new therapies (Henderson *et al.*, 1999). It is the search function that explains the need for proximity of research laboratories and DBFs. This signifies a decline in the attractive power of 'big pharma', rather the reverse, and the further attraction to clusters that are 'megacentres' of much of the rest of the 'healthcare value chain' including CROs and biomanufacturing. Table 4.1 shows indicators of scale in key assets possessed by megacentre candidates such as those in Boston, California, Montreal and Toronto in North America, and Cambridge, Munich and Stockholm in Europe (Cortright and Mayer, 2002; Cooke, 2005a; Niosi and Bas, 2001). The three US megacentres are orders of

Table 4.1 Core biotechnology firms, 2000: comparative US and European performance indicators

Location	*DBFs*	*Life scientists*	*VC*	*Big pharma funding*
Boston	141	4,980	$601.5 m.	$800 m./annum 96-01
San Fran/SV	152	3,090	$1,063.5 m.	$400 m./annum 96-01
San Diego	94	1,430	$432.8 m.	$320 m./annum 96-01
Toronto	73	1,149	$120.0 m.	NA
Montreal	72	822	$60.0 m.	NA
Stockholm-Upp.	60	2,998	$100.0 m.	NA
Munich	56	5,000	$266.0 m.	$54 million (2001)
Cambridge	54	2,650	$250.0 m.	$105 million (2000)

Source: NIH; NRC; BioM, Munich; VINNOVA, Sweden; ERBI, UK

magnitude larger on most indicators than other clusters listed by Cortright and Mayer (2002). Even New York, which scores well on some indicators, scores abysmally at $152 million for venture capital compared to the top three. Clearly, the well-known ten-year advantage of the US is demonstrated in Table 4.1, as is the relative maturity and scale of venture capital and 'big pharma' funding of DBF R&D. But in human capital, the balance shifts somewhat towards, for instance, Munich where firm numbers are respectable. However, even within a European context, Van Reenen (2002) showed German DBFs in total had only 6 drugs in the pipeline compared to 18 for Sweden and 128 for the UK.

To conclude, in a possibly unique contemporary way, changing epistemology has led to a shift in the drivers of the 'triple helix' from the pharmaceuticals industry to capable universities, supported by judicious government plus private foundations. This in turn has led to new regional drivers at the leading edge of the healthcare industry. In the US it is Greater Boston, especially, Cambridge, Massachusetts, home to Harvard University and MIT, and both northern and southern California centred on the University of California Medical School, San Francisco, and the University of California, San Diego rather than the former 'Medicine Chest of the World' in New Jersey; in the UK, it is Cambridge (and rising Oxford) rather than London; and in Sweden it is Uppsala (and rising Lund) rivalling Stockholm. In bioscientific and medical research and commercialisation the university is the epicentre of the 'triple helix', not the seat of government or corporate power.

Bioscience megacentres and the eclipse of 'big pharma'

An eclipse is normally more short-term than a sunset, although both metaphors point to a fascinating reality in an era of globalisation and predominance of multinational capital where small university spinout companies have forced some of the world's largest firms like Pfizer, Glaxo, Merck,

Novartis, Roche, Bayer and Aventis to become supplicants of the knowledge capabilities of a modest number of SMEs based in incubators and science parks. Can the epistemology argument advanced above be the sole cause of this loss of knowledge leadership, or are other key factors at work that may enable a more generic question to be raised about the historic primacy of multinational capital? An analysis by Coe (2002) suggested that there is also a strong market-based reason why 'big pharma' are on the back foot and likely to weaken their hold in biosciences and healthcare in future. Further research by Malerba and Orsenigo (2002) found evidence from the long-run history of the relationship between pharmaceuticals firms and DBFs that was consistent with Coe's (2002) reasoning. They referred to a 'lack of cumulativeness' in innovative activities in the pharmaceuticals industry despite high global concentration of ownership.

This is important in many dimensions. It is important intellectually because it hints that biopharmaceuticals may be a harbinger for other industries dominated by multinationals but faced with uncertainty by technological change. Telecommunications was once the most obvious candidate, especially when UK firm Marconi's share price dropped below one penny in 2002, having been £12 a year previously, and it ended up being acquired a few years later by Sweden's Ericsson. Such was the dot.com shakeout that share options in the likes of Nortel and Lucent become worthless. Closer to bioscience, and as intimated in Chapter 1 and below on numerous instances, the conventional food industry could be a long-term candidate, given the rise in organic food demand after revelations about the lethal effects of economies of scale upon the mass processing of popular foods (Schlosser, 2001).

Second, it is important for the conception of the 'triple helix', its validity and even its ethical basis perceived through the lens of bioscience. For instance, if multinationals can bring influence to bear on the US government to relax regulations intended to assist the eradication of *E. coli* bacteria in meat products, and feed schoolchildren lower grades of meat than even the relaxed regulations allow to be sold in supermarkets, what hope have ethically sounder but financially weaker university researchers of resistance to control of unpalatable research results? The institutional and popular rise in appetite for organic but not functional foods is highlighted in Chapter 9. Less dramatically, what should be the response of academe to the news of a US lawsuit in 2002 between DBFs Genentech and Chiron over the granting of a patent for a piece of research? A substantial portion of biotechnology research is conducted in universities, as we have seen, where openness prevails and information circulates relatively freely. Genentech (60 per cent owned by Roche) was long in dispute with Chiron (then 42 per cent, in 2005 100 per cent owned by Novartis) over the latter's claim that its scientists identified the protein allowing Genentech to develop the $350 million per year breast cancer drug Herceptin.[2] As both biotechnology firms arose from and had traditionally had close links with university research, the broad theories characteristic of academic research lead to broad patents that created potential, though as Chapter 1 showed thus far unrealised, problems for highly specific drug indications. In principle, firms that fund academic research would in future require exact, audited, confidential records from academics, further undermining the Mertonian norms of openness and interaction that had proven the key to the recent primacy of university over corporate research in biosciences and, probably, other advanced technology fields. Simultaneously, the likes of Genentech are under pressure from investors and regulators to be more transparent in disclosure of information about research and trialling results of new drugs, something they and other biotechnology firms vigorously resist.

Alternatively, what might be the implications of Epicyte Pharmaceuticals Inc.'s announcement in July 2002 of having been granted precisely a 'broad patent' covering the production of antibodies in transgenic plants? The patent holder is San Diego's celebrated Scripps Research Institute, home of Epicyte's founders Andrew Hiatt and Mich Hein. This meant that Epicyte, through its patented Plantibodies technology in principle controlled all research into the use of plants. In this case maize was the agent for the broadest spectrum of innovation using monoclonal antibodies for inflammatory, infectious and other diseases, and expected to earn substantial income from licensing by whoever in the US, Japan, Europe and Australia sought to exploit the innovation. Partners in this lucrative deal were, besides Scripps, San Diego-based Dow AgroSciences, UK DBF Biovation, Princeton NJ DBF Medarex, Malvern, PA DBF Centocor, Baltimore, Maryland DBF ReProtect LLC, Johns Hopkins University, Oklahoma Medical Research Foundation, Cornell University and Boyce Thompson Institute (Ithaca, NY). These had, it was then presumed, cornered the future world market for production of biologics from a cheap and universally available GMO source.

Third, understanding of the loss of status of pharmaceuticals corporations in relation to research institutes and DBFs is important for policy both in respect of regulatory and science policy and, related particularly to the latter, economic development policy. For, DBFs today routinely hype the arrival of a possible treatment at 'proof of concept' stage, three or more years before a drug candidate will even have been trialled in order to attract investment from venture capital and 'big pharma' for later stages. That many such candidates do not receive approval means the ethics of hopes raised among incurable patients do not count for much to DBFs pursuing the licensing agreement or first round of venture capital, something that compromises academic integrity among partner institutions where peer review remains a guard against untested claims, but which may itself be transgressed by the image-building imperatives of DBF allies. For science policy, where to direct basic research investment is today more prone to follow market signals, some of which may be of dubious quality, rather than the disinterested pursuit of knowledge for its own sake. And where science policy moves in an era in which, as Gibbons *et al.* (1994) again correctly predicted, there is heightened influence of market value on public investment of 'taxpayers' money' in basic research, while adhering to criteria of peer-reviewed excellence to adjudicate to whom and where basic research funding goes, will inevitably incline investment towards bioscience megacentres. That there are few of these is a feature of 'knowledge economies' (Cooke, 2002), which result in highly disequilibriated locational outcomes. With knowledge economies based in advanced technologies, one of the few economic drivers for regional development authorities, it is inevitable that the 'losers' will seek ways of building up their offer or suffer the consequences. That this has begun to occur is shown empirically in Cooke (2002; 2003a) and the third section of this chapter will briefly reprise some of those findings and explore ways in which regional socio-economic exclusion from the fruits of the 'microbiology revolution' may be, and are being, offset.

Thus market sentiment had already invaded two spheres where it had traditionally been less pronounced than its power in the industrial sphere where it has normally been expected to be strong. The 'triple helix' idea (Etzkowitz and Leydesdorff, 1997) with its tight linkage among these reluctant, if not hostile 'partners', captured this in a general, perhaps rather 'cybernetic' or functionalist way that paid insufficient attention to the implicit and explicit tensions in such 'contracts' as revealed in the litany of litigation that brings academe and enterprise into unwelcome 'face-to-face contact'. Down at the regional and local levels these irksome conventions also express themselves in dissatisfactions from academic researchers with firms that are frequently 'uninterested in blue skies research' but see in university research centres opportunities for 'cheap testing or trialling work'. Equally it masks criticisms from firms of academic 'openness', failure to adhere to 'confidentiality agreements' and 'poor time budgeting' that results in business deadlines having regularly to be revised (these judgements are reported from research into 'academic entrepreneurship' in Cooke *et al.*, 2000). This, of course, is classic 'open science' of the Mertonian kind brought into stark confrontation with the niceties of the private appropriation of knowledge explored in public institutions for exploitation in private ones. It is, once again, at the nub of the 'growth culture' that pervades much science, probably nowhere more so than biopharmaceuticals, in the era of the knowledge economy.

Might this not also be a cause of the retreat by 'big pharma' from initiating the commissioning of academic research towards responding to theirs and their DBF allies' assertions of what it is important to invest research funding in? According to Coe (2002) the pharmaceuticals industry was beginning obviously to display a deepening productivity crisis occasioned by a misbelief that growth in scale by acquisition and merger results in increasing returns to R&D investment. Contrariwise, Coe (2002) showed, there are no, or negligible, economies of scale in pharmaceuticals R&D or sales. The implications of this are profound. Larger companies are no more productive and no more effective at increasing productivity than smaller companies.

(Coe, 2002, p. 2)

So, even though such firms usually explain the rationale for merger of the kind we have seen increasing in major ways around the millennium (Table 4.2) in terms of efficiency improvements in R&D and sales functions, Coe's research showed that there are no significant economies of scale in R&D investment, but that the value : quality of a company's pipeline at time 't' is directly proportional to R&D spend at time 't – 3 years' and the number of R&D staff in the same year. Savings on R&D now will resonate badly three years hence. In a paradoxical way, perhaps because of the 'silo' mentality in hierarchical corporations, 'big pharma' know this because they now routinely spend twice as much on product licensing deals with biotechnology firms. In 1998 it was $4 billion, by 2000 $7 billion and rising. As well as more deals, biotechnology firms that would accept 5 per cent royalties in 1990 were typically commanding 50 per cent of revenues by 2001 (Dyer and Griffith, 2001). According to Dyer (2002) some 500 biotechnology firms were researching new products, and they had 1,300 new compounds in development. Up to 30 per cent of big pharma R&D budgets are now spent on alliances with extramural partners responding, as noted, to the top twenty pharmaceuticals firms by 2001 spending $28 billion on intramural R&D for a yield of only 28 new drug approvals. Pfizer, currently the world's largest pharmaceuticals firm, has over one thousand alliances with DBFs and universities.

Table 4.2 Mergers and acquisitions in global pharmaceuticals companies

Pharmaceutical firm	*Sales 2001*	*Nationality*	*Previous merger*
Sanofi	$72.7 bn.	F	Aventis 2004
Pfizer	$40.8 bn.	US	Pharmacia 2002 Warner Lambert 2000
GlaxoSmithKline	$24.5 bn.	UK	SmithKlineBeecham 2000 Wellcome 1995
Bristol-Myers Squibb	$19.4 bn.	US	Dupont Chemicals 2001
AstraZeneca	$16.5 bn.	UK/SD	Merger
Bayer	$16.3 bn.	G	Acquisition
Aventis	$15.8 bn.	F	Rhone-Poulenc-Hoechst Merger 1998
Johnson & Johnson	$14.8 bn.	US	(45 acquisitions 1989–99)
Novartis	$12.0 bn.	SW	Sandoz-Ciba merger 1996
Wyeth (Amer. Home Prod.)	$11.7 bn.	US	Parke-Davis 1988 American Cyanamid 1994

Source: *Financial Times*, 16 July 2002, p. 31; Network Science

Thus these are testing times for the pharmaceuticals industry, faced with the problem of the 'microbiology revolution', a drying intramural R&D drugs pipeline, and sagging shareholder value. In 2001 eight of the top fifteen pharmaceuticals companies did not win approval for a single new drug, and keeping in mind the ratio overall of $1 billion R&D cost per drug approval (noting Coe's point about a three-year time-lag), it is notable that in 1980 the top twenty companies spent $2 billion for a return in that year of 34 new drug approvals. In other words this evidence also supports Coe's (2002) thesis that merger and acquisition activity at the grand scale noted in Table 4.2 is not visibly associated with more innovation. What happens is that a 'dry pipeline' company of large scale acquires a lesser, more innovative company and boosts its product pipeline substantially at least for a period. Pfizer's acquisition of Swedish firm Pharmacia is a case in point, the detail of which is obscured by the data in Table 4.3. Since 1998 eleven out of Pfizer's eighteen current offerings were developed first by Pharmacia. The same is true to a lesser extent in the contribution Janssen makes to Johnson & Johnson's performance and Parke-Davis' to Wyeth's.

In Table 4.3 the same top firms in the preceding table have been examined to determine the annual number of US Food and Drug Administration approvals of 'original new drugs' occurring 1998 through to 2002 (July). The data are extensive and the table excludes what are called 'efficiency supplement', 'abbreviated' and 'labelling supplement' approvals. It is, accordingly, a rather conservative picture of drug approval trends over the period. Importantly, many 'original' drug approvals are granted for products with the same name and indication as others approved earlier, sometimes in the same year. Hence the pharmaceuticals industry is emphatically one in which incremental innovation is the norm, though 'blockbuster'

Table 4.3 FDA drug approvals of original new drug applications 1998–2002 (July)

Firm	*1998*	*1999*	*2000*	*2001*	*2002 (July)*
Pfizer	3	6	3	2	4
Glaxo	3	5	3	3	1*
Merck	3	2	1	0	1
Bristol-M.S.	1	2	1	1	1
AstraZeneca	2	1	1*	3	0
Aventis	4	2	2	0	0
Johnson & J.	1	2	0	3*	0
Novartis	4	2**	6#	6*#	2
Wyeth	0	4	5***	2*	2
TOTAL	21	26	22	20	11

Source: FDA. NB: * indicates one new drug with previous product name; # indicates three new versions with previous product name

drugs or 'breakthrough' treatments are avidly sought, and sometimes, as with the anti-leukaemia drug Glivec from Novartis via six university research laboratories, innovated.

Clearly, as Table 4.3 indicates, there has been a relatively low (e.g. compared to 1980) level of original new drug approvals and it has been in modest decline among the top, regularly merging for the most part, 'big pharma' firms. What can be said with confidence along lines indicated in Coe's (2002) and Malerba and Orsenigo's (2002) work is that there clearly has been no increase of any significance in recent years in the new drug approvals of the top pharmaceuticals companies.

When to this decline, coinciding with a new round of concentration in the 1995–2002 period when seven major mergers and acquisitions occurred, is added the degree of marginal or incremental innovation represented in nine of the new originals, then the 'drought' is somewhat magnified. So we can conclude this analysis with some confidence. For the foreseeable future and under conditions pertaining to the 'microbiology revolution' and the failure of shareholder-driven demand for stock market returns through mergers and acquisitions to have positive effects on innovation, fewer and fewer new therapeutic treatments can be anticipated to originate in 'big pharma'. Rather, as has been suggested, biotechnology-capable DBFs will increasingly lead the knowledge *exploration* process in tandem with research institutes, DBFs will take increasing responsibility directly or indirectly for services such as clinical trials (knowledge *examination*) and biomanufacturing, with knowledge *exploitation* also increasingly in the hands of DBF innovators funded by 'big pharma' licensing agreements in exchange for their marketing and distribution expertise. Upon cool reflection, that does not leave 'big pharma' with a very big role in the pharmaceuticals industry of the future.

What, it is reasonable to ask, might the future look like given the argument so far? Coe (2002) sees this as a move from 'big pharma' to 'networked pharma', something that is consistent with the line of reasoning presented thus far. This is evident in what has been said so far where from *tactical* issues like sub-contracting production on a temporary basis to *strategic* outsourcing of R&D and other key deliverables, the future of 'networked pharma' as a substantially downsized 'hub' in a network of alliances is the more efficient model to pursue. This is likely to be forced upon pharmaceuticals firms more rapidly as the share of biotechnology protein-based treatments comes to outweigh those derived from fine chemistry. It has been shown that the present ratio is 2 : 1 in favour of fine chemistry, but given current approval rates it is not impossible to imagine that ratio being reversed by 2010. The main drag effects upon the achievement of that vision are the high costs of biotechnology drugs and the high rate of rejection by regulatory authorities, often because of flawed trials arising from DBF inexperience in managing this new function, hitherto the province of 'big pharma'.

Given the vast public research budgets for biosciences, especially in the USA, where a 2003 annual expenditure of $27 billion has been appropriated, there seems to be little preventing new drugs reaching 'proof of concept' stage from where perhaps one in ten reach the market, and FDA estimates of up to twelve new products a year being approved in the early 2000s, mean that 2001's 90 biotechnology drug approvals out of 270 could reach 174 by 2010. If that were to happen the ratio reversal would almost be complete, but it may be that synthetic chemistry drug innovation will decline more quickly towards 2010, in which case it would happen rather sooner. At that point other biotechnology companies than Amgen and Genentech that are long-established and larger than the SME profile of most DBFs, were expected to be joined by others like Genzyme, Chiron and possibly Celltech and Morphosys from Europe. But acquisition was the fate of Chiron and Celltech, leaving only Genzyme to maintain a powerful, independent profile. Thus the dream of some biotechnologists to see the fully integrated pharmaceuticals company (FIPCO) model for biotechnology firms remains, except for the two leaders, unfulfilled. However, that there was in any case some way to go before that FIPCO point was reached is testified to in Table 4.4. These data show that though an historic point was reached in 2000 when the largest biotechnology firm had a larger US market capitalisation than a medium-sized pharmaceuticals firm like Bristol-Myers Squibb, the relative volatility of biotechnology stocks means that the surpassing was short-lived. Amgen bought Immunex in 2002 and the joint $67 billion market capitalisation probably makes it larger than a good many medium-sized pharmaceuticals firms of less than Bristol-Myers Squibb's market capitalisation.

Alternatively, attention should be drawn to the generally upward trend for most biotechnology businesses excepting the 2000-onwards stock market slump, whereas two large pharmaceuticals firms have lost market value over the same period. Moreover, while it should not surprise us that the magnitude of growth in market capitalisation of smaller biotechnology firms is higher than the average for the pharmaceuticals firms in each category, it is still worthwhile remarking that even through the stock market slump, biotechnology firms' market capitalisation grew on average more than six times that for the top pharmaceuticals businesses. This is, of course, consistent with the thesis by Coe (2002) that mergers and acquisitions do not improve business performance much beyond the opportunistic short-term.

Hence, in summary for this section the main argument has been that to the extent that the idea of the 'triple helix' works as a description of the key systems linkages underpinning the knowledge economy, it is not a stable or indeed an equal partnership. It is dynamic and changing quickly. The main shift of interest to innovation studies, industrial economics and regional science is that the 'molecular biology revolution' has seen a weakening of the hold on pharmaceuticals innovation by companies that in the

Table 4.4 Comparison of market capitalisation of biotechnology and pharmaceuticals firms

	Market capitalisation ($ billion)			
Company	*1998*	*2000*	*2001*	*Change 98–01*
Biotechnology firms				
Amgen (US)	49.0	68.7	52.7	+6%
Genentech (US)	18.6	26.8	18.5	0%
Serono (Switz.)	8.6	21.9	17.9	+104%
Immunex (US)	2.0	8.8	14.3	+615%
Genzyme (US)	3.1	10.4	8.4	+170%
IDEC (US)	NA	9.0	8.0	NA
MedImmune (US)	6.2	8.8	7.9	+27%
Chiron (US)	4.1	10.1	7.4	+80%
Gilead (US)	2.2	5.2	6.4	+190%
Biogen (US)	5.1	9.2	6.2	+22%
Celltech (UK)	1.0	9.4	3.7	+270%
Average				+165%
Pharmaceuticals firms				
Pfizer (US)	132.0	264.3	234.1	+77%
Johnson & J. (US)	106.1	135.9	195.6	+84%
Glaxo (UK)	127.2	167.4	148.6	+14%
Merck (US)	164.0	166.5	126.2	–23%
Novartis (Switz.)	137.1	186.7	106.5	–22%
Abbott (US)	62.5	79.6	84.1	+31%
AstraZeneca (UK)	53.9	85.6	81.5	+32%
Wyeth (US)	60.1	80.2	78.6	+31%
Eli Lilly (US)	65.5	94.4	76.4	+17%
Bristol-Myers S. (US)	48.1	50.8	56.9	+18%
Average				+26%

Source: Genetic Engineering News, June 2002

twentieth century effectively made the industry. Table 4.4 reveals no Bayer, Aventis, Hoffman La Roche, or Schering Plough in the higher reaches some of them once dominated. Now, to repeat, the R&D challenge has been accepted by biotechnology firms who are not yet large enough to rival pharmaceuticals firms in market size but, given their current growth rates, may be expected to take the place and possibly acquire some of the currently larger pharmaceuticals businesses. Finally, public budgets for R&D in bioscience and medical sciences in the US are converging. It was noted that 'big pharma' spent some $28 billion on R&D in 2001 and that the US federal research budget for 2003 will be over $27 billion. Once government, research laboratories and DBFs rule the roost, it can be anticipated that pharmaceuticals firms will have to present a more modest disposition regarding therapeutic treatment priorities, and become used to their emerging role as members, albeit 'hubs' with finance, of a new network pharma industry concentrating in megacentres possessed of the fullest extent of the healthcare value chain. This stretches from protein and molecular basic research, applications, medical and clinical research, hospitals for clinical trials, exploitation and commercialisation of therapeutic and diagnostic products, advanced and technical training, pharmaceuticals companies, advanced infrastructure (from PET scanners to a synchrotron), biomanufacturing, clinical services, screening, sequencing and bioimaging, biobanks of cellular and genetic data, investors, specialist lawyers, consultants and management accountants (on proximity of venture capital and DBFs, see Powell *et al.*, 2002). Few of these 'megacentres' exist, as noted (Table 4.1), maybe five in North America and a further three in Europe, But, because of their potential as major contributors to the growth of regional economies, regional development administrations will increasingly target their construction, in whole or in part. What evidence is there that this is happening already, and what barriers stand in the way of a more equal spatial development of the broad healthcare industry as outlined? Finally, what steps are available or have been taken to remove some of the barriers? These are briefly dealt with in the next section.

Regional responses to bioscience megacentre growth opportunities

Thus far, the policy apparatus for seeking 'spread effects' instead of the concentration tendencies of modern bioscience and its advanced knowledge application and exploitation capabilities can be characterised as threefold. These can be summarised as:

- Strategy building, including the establishment of science councils in regions that have 'in principle' delivery mechanisms for implementing science strategies or, earlier, science and technology policies. In some cases, and in smaller countries like Finland, the central government may pursue such strategies for its regions.
- Affirmative Action in national research fund evaluation rules. In this approach a political decision is taken to 'lower the bar' for institutions in lagging regions to build up strength in winning research funds by being funded at lower peer-review assessments than institutions outside designated regions. This is a kind of 'regional policy' within science funding allocation processes, or in EU terms, it would be a Structural Funds mechanism within the Framework Funds.
- 'Capabilities' funding. This is a mechanism with aims similar to 'affirmative action' but it does not interfere with the frequently sacrosanct peer-review rules that are believed to be one of the strengths of Western research funding. Due to the emphasis on the pursuit of academic excellence on the basis of peer review, the best science gets funded. However, it overlooks the concentration effect such a

> mechanism produces. A 'capabilities' fund is separate, 'ring-fenced' allocation to which bids are allowable only from institutions in lagging areas. It can be set at national level or individual regions with appropriate financial authority can decide to establish such a fund.

In Cooke (2003a) an extensive account was given of the emergence of science strategies. In brief, it was argued there are three kinds. The first, and earliest were science and technology policies almost exclusively generated at state or province level in Canada and the USA. They were strongly market facing and set at a level close to the interface between technology and the market rather than basic science. One of the most successful in the long-term was that originated in the 1950s in North Carolina which first resulted in the formulation for Research Triangle Park which attracted many thousands of R&D jobs to a lagging regional economy by diligent marketing and incentive provision aimed at R&D laboratories of firms like IBM, Motorola, Glaxo, Ericsson and Cisco Systems. In a subsequent round research institutes and incubator facilities were successfully targeted in supercomputing, electronics and biotechnology. In 1981, the North Carolina Biotechnology Centre (NTBC) was established, as a commercialisation rather than research facility. In early 2002 NTBC housed some 30 biotechnology businesses, including sites of Aventis, BASF, Bayer, Biogen, Eli Lilly and GlaxoSmithKline among the 90 in the RTP and broader Raleigh–Durham–Chapel Hill area, and 142 in the state.

Fifteen such state-level strategies have been produced in the USA since 1995 and act as guides to state economic development policy, identifying by the use of R&D or innovation indicators which science-based industries are nationally or globally competitive, and thus should be assisted by construction of special infrastructure or otherwise incentivised to grow by endogenous or inward investment means. North Carolina's latest evolution in its S&T policy has been to innovate around investment in its State University of North Carolina system (not to be confused with the University of North Carolina system). This has resulted in the construction near RTC of Centennial Campus, a 'mixed campus' in ICT and biotechnology that juxtaposes departments of the university with R&D laboratories of multinationals like ABB, Ericsson, Eastman Kodak and Bayer along with smaller facilities from the likes of Juniper Networks and Analog Devices.

The second type of science strategy is far more basic science supporting and is something that has begun to develop in the UK since about 2001. The Scottish Science Strategy was the first and targets e-science, bioscience and medical science as described in Cooke (2003a). It was commissioned by the Scottish parliament, advised by the parliament's Science Policy Unit and consistent with an economic development strategy which evolved from the initiative of Scottish Enterprise, the regional development agency, to embark upon a cluster development strategy that included facilitating cluster development in, among other fields, ICT and biotechnology. The strategy identified Scotland's R&D spend from public sources at about \$1.2 billion and identified an above average success (UK comparator) in accessing UK research funds. It seeks to augment these by re-allocations from within the parliamentary budget and by encouraging integration of research strategies among different research laboratories that the parliament funds.

This course has also been followed, despite the absence of a parliament, by the North West region of England. In the aftermath of the debate about the location of the UK's new Diamond Synchrotron (important for bioimaging) which went to Oxford in the heart of southern England's biotechnology belt, the North West used \$40 million of compensation funding to upgrade its nuclear-age synchrotron, build a National Biomanufacturing Centre to conduct and train for biomanufacturing, develop collaborative bioscience projects between Liverpool and Manchester universities, and establish a Science Council. The Science Council reported in summer 2002, recommending special efforts be made to augment basic research funding in six areas, including biosciences, aeronautics, chemicals and textiles. This science strategy also links closely with the regional development strategy of North West England, building upon piecemeal efforts to establish bioscience clusters in Liverpool and Manchester. Because of the existence of delivery mechanisms like the regional development agency, a bid for the UK Biobank was submitted in competition with other regions. This is an indication of growing optimism arising from earlier successes born from the aspiration to move towards 'megacentre' status in the UK. Effort now is directed at securing this \$70 million Biobank facility, a 'hub' with regional linkages to store and make available for research and exploitation the genetic information from DNA samples and the medical records of 500,000 volunteers, aged 45–69, to be funded by the Medical Research Council, the Wellcome Trust and the UK Department of Health. A second English Regional Science and Industry Council was established in North East England in 2002.

Finally, Finland is a good example of a small country that has pursued a regional science strategy from the centre. More is said about this in Chapters 7 and 9 including the implications of the bidding procedure for Centre of Expertise status which was announced in the 1990s. This elicited successful bids in the biotechnology section from four, subsequently six regional science complexes centred on universities, science parks, hospitals, firms and so on. Later, as (Cooke, 2003a) shows, the idea was sufficiently attractive in policy terms that a new Centres of Excellence strategy was over-subscribed yet funds were still found to support such centres in many parts of Finland that showed appetite even in the absence of advanced expertise for modern biotechnology. Despite this Finland has a presence in European biotechnology in biopharmaceuticals and agro-food biotechnology especially in claimed cholesterol-reducing lactobacters. In 2002 without having gone through the process described in Finland, the Netherlands government Ministry of Industry and Ministry of Science and Technology announced a 'megacentre' policy for healthcare, medical science and bioscience sectors by giving incentives to encourage and fund joint research and commercialisation among the universities of Twente, Nijmegen and Wageningen. The lessons of Finnish exuberance were taken on board in the design of an attempt to achieve synergetic scale economies from collaboration among existing institutions.

Affirmative action, something hitherto widely considered to be heretical to the scientific community, was pioneered in the USA. The Experimental Programme to Stimulate Competitive Research (EPSCoR) is a joint programme of the National Science Foundation (NSF) and several US states and territories. It promotes the development of the states' science and technology (S&T) resources through partnerships involving a state's universities, industry, and government, and the federal research and development (R&D) enterprise. EPSCoR operates on the principle that aiding researchers and institutions in securing federal R&D funding will develop a state's research infrastructure and advance economic growth. EPSCoR's goal is to maximise the potential inherent in a state's S&T resources and use those resources as a foundation for economic growth. Nineteen states from Alabama through North Dakota and South Carolina to Puerto Rico have been designated as qualifying EPSCoR recipient states.

Although it has yet to be fully evaluated in terms of scientific impacts or performance improvement, some examples of implemented successes include:

- A University of Alabama optoelectronics and nanotechnology EPSCoR-supported project to build a microfabrication facility, that subsequently has been awarded contracts worth over $1m by local companies for microphotonic device design and fabrication. In addition, an anonymous benefactor in the Huntsville area donated $2m cash for expansion of the cleanroom facility and purchase of additional equipment to expand its nano- and microfabrication capabilities. Moreover, a new $720k DARPA award is a direct result of the availability of the facility.
- The University of Kansas' new Department of Molecular Biosciences built a research infrastructure capable of addressing more complex, inter-disciplinary research across the chemistry/biology interface. The department is hiring faculty and purchasing equipment to support collaborative research and strengthen graduate education. In late 1999, the university hired a protein chemist who collaborated with colleagues to fund, evaluate, purchase and install a new nuclear magnetic resonance (NMR) spectrometer needed to study the structure and chemistry of proteins.
- Creighton University, Nebraska achieved a project for a model study in proteomic analysis. All of the peptide and protein components in the electrically stimulated secretions of the skin of the pickerel frog

> *Rana palustris* have been characterised by mass spectrometry and Edman degradation. The analysis has led to the discovery of 22 peptides with antimicrobial activity and 13 peptides which are toxic to mammalian predators. Many more important new biologically active peptides are similarly expected to be discovered in the secretions and the important research now has the support and interest of a local Nebraska biotechnology company.

Finally, and briefly, because of the likely controversy over introducing any rules that would allow EPSCoR arrangements to prevail in the UK, the government Department of Trade and industry is actively developing the idea of a UK 'Capabilities Fund' that would attempt to stimulate advanced basic research in regions which currently demonstrate an under-performance in attracting research awards. This is connected to a renewal of policy for regional development, particularly in England, where belated attempts to instigate a more innovative culture among firms and policy functionaries followed the setting-up of regional development agencies in 1999. Now, a lesson learned earlier in European regional innovation systems about the inadequacy of the 'learning region' approach has been re-learned (see, most strikingly, Nonaka and Reinmöller, 1998; Hassink and Lagendijk, 2001). That lesson is that it is not enough to rely upon absorbing dated codified knowledge; there is a necessity to build capabilities for integrating up-to-date tacit knowledge into innovation processes to keep or develop a competitive edge. But, of course, the barriers to this lie in inadequacies in the regional science base. Hence the interest in and demand for regional science policy.

Conclusions

In this chapter, an attempt has been made to expose an interesting but seldom analysed phenomenon in the epistemology of science that has had profound initial effects, the repercussions of which will resonate for many decades, upon industry structure and regional development prospects. This is the 'molecular biology revolution' that is sidelining synthetic or fine chemistry in the world of pharmaceuticals, shaking up the cosy assumptions of academic theses regarding equivalence among 'triple helix' institutions, occasioning new kinds of industry organisation in a sector that, altogether, accounts for a large part of national GDP, and forcing far-reaching science and regional policy changes upon administrators such that a new field of 'regional science policy' has emerged.

In section two, the chapter explored some of the key underlying reasons for the shift in the biosciences 'triple helix' away from multinational pharmaceuticals firms – especially in R&D – towards a networked pharma model that this chapter argued is likely to be concentrated in a few biosciences megacentres like California, Boston, Toronto, Montreal,

Munich, Cambridge and Stockholm. Two key reasons are the inadequacies of big pharma in managing the highly complex networked processes of exploration and exploitation that have arisen in post-genomic bioscience and biotechnology, much of which feeds into the burgeoning healthcare industry. The second reason for the coming eclipse of the 'big pharma' is that it pursues a merger and acquisition policy to sustain growth that can be shown to be growth destroying due to failure to improve R&D economies that are not susceptible to scale dynamics.

Finally, the chapter explored a number of policy initiatives aimed at correcting the spatial imbalances that arise from the working through of the 'molecular biology revolution' in geographical space. Three instances of regional science policy were examined and some were found to have brought significant growth opportunities to lagging regions. Two further examples that interfere with basic science funding allocations through 'affirmative action' or construction of a 'capabilities fund' were shown to be testimony to the problem of over-concentration occasioned by megacentre developments and, in the case of EPSCoR, capable of making significant local differences to scientific expertise and capabilities, although on a small scale. The UK was shown to be developing a similar system of affirmative action outside the basic science allocations. This is due to the weight of tradition that stresses the pursuit of excellence based on peer review that is itself a force for spatial concentration of resources that gives rise to the demand for regional science policy in the first place.

5 Academic growth cultures

The rise of bioregional knowledge domains

Introduction

This chapter explores a field, 'globalisation of bioregions', that is of growing interest to social science and policy alike. A number of special journal issues have been published (Cooke, 2003a, 2003b, 2004b; Lawton Smith and Bagchi-Sen, 2004), and others may be anticipated from different stables. Outside the spatial field but well informed by it, an earlier one was published in *Small Business Economics* (2001) subsequently enlarged into a book (Fuchs, 2003). Numerous other books are published in this increasingly dynamic field (e.g. Orsenigo, 1989; McKelvey, 1996; De la Mothe and Niosi, 2000; Carlsson, 2001). From an economic sociology vantage point Powell *et al.* (2002) contribute well in showing close spatial interactions among biotechnology SMEs and venture capitalists from non-spatially designed survey data. Most economists writing about the evolution of the biotechnology sector (many authors refuse to think of a technology as an 'industry') accept that it is a classic case of a science-driven sector, highly active in R&D expenditure and patenting of discoveries, with research increasingly dominated by public (mainly university) laboratories, but exploitation mainly dominated not by large but small firms.[1] This is so for the biggest part of biotechnology, accounting for about 70 per cent of sales, biopharmaceuticals, which is the subject of much of what follows.

Biopharmaceuticals, as a large sub-sector of biotechnology, is fascinating. It belies the prediction that multinational corporations are bound to dominate all aspects of any industry due to economies of scale. While the likes of Pfizer and Glaxo are certainly large global actors by any standards, they are rapidly losing capabilities to lead research. This function has moved – rapidly in the 1990s – to university 'Centres of Excellence' and particularly to networks of dedicated biotechnology firms (DBFs) that often lead important research projects (Orsenigo *et al.*, 2001) relying on 'big pharma' for grants, marketing and distribution. All are agreed that DBFs cluster around leading edge university campuses and that even if they are not the kind of DBFs that depend on immediate interaction with research laboratories, they still agglomerate. A good example of such an agglomerated non-cluster occurs to the south-west of London, centred on Guildford. Why there? Static externalities include good access to Heathrow airport and a suitable talent pool from which to poach. But firms scarcely interact except with global clients. This contrasts with the dynamic externalities from proximate co-presence with many and various significant others in the sector. In the following brief sections, these will be highlighted by reference to key issues of relevance to cluster theory more generally, the specifics of science-driven industry, the post-Schumpeterian symbiotics that characterise the double dependence of small businesses on corporate financial clout and 'big pharma' on DBF and laboratory research excellence. Also posed is the question – raised in the 'spillovers' debate, as to whether most cluster interactions in biotechnology (and yet to be asked for elsewhere) are really forms of market rather than milieu-type transaction, pecuniary rather than non-pecuniary spillovers, and traded rather than untraded interdependencies (Zucker *et al.*, 1998; Owen-Smith and Powell, 2004).

Biotechnology and cluster theory

Biotechnology is easily seen to be an economic activity, technology or platform both of the present but more for the future. Here, as elsewhere, we will call it an enabling technology though, for example, biopharmaceuticals clearly also constitutes a sub-sector of the broader healthcare industry along with research, pharmaceuticals, hospitals, medical schools and a further panoply of support services that together comprise in some countries over 15 per cent and sometimes more of GDP, and in the UK up to 25 per cent of national R&D expenditure (OECD 2000; DTI, 1999). Agricultural bioscience is similarly part of the wider agro-food industry. Biotechnology can be shown to have its own value chain, especially in biopharmaceuticals, although some parts of even that are indistinguishable from agro-food biotechnology or environmental biotechnology, especially regarding enzymes, recombinant DNA and the like. Indeed agricultural 'farming' of DNA antibody sequences for healthcare application was patented, as noted in Chapter 4, by San Diego firm Epicyte through its Plantibodies product. Although objections have been raised by researchers about the scope of such a 'broad patent' for restricting academic research freedom, it does not take extra-sensory insight to see that such technologies can be the saving of some agricultural areas assailed by global competition provided they become more 'knowledge-intensive' in this and other ways. Moreover, as Chapter 1 showed, fears of IPR restrictions tend thus far to be unfounded.

Precisely this sort of Galilean leap seems to be at the heart of the clustering process in this industry. Thus Epicyte's actual patent holder was noted as San Diego's renowned Scripps Research Institute with Dow AgroSciences as the licensing and funding partner. This is the 'holy trinity' of interaction around biotechnology discovery – basic research, 'pharma' or agro-chemicals funding and DBF exploitation for commercial purposes, with 'pharma' earning returns through marketing and distribution. In San Diego, as Chapter 3 showed, are some 400 SMEs engaged in some aspect of biotechnological core or peripheral activity and as Lee and Walshok (2001) showed, 45 were founded by the Scripps, Salk, Burnham or La Jolla research institutes, while 61 were founded directly from the University of California. Many more were founded in the area as a consequence of knowledge flows and investment opportunities for incoming and resident entrepreneurs advertised through activities such as those promoted by the much-emulated UC San Diego CONNECT network.

What differentiates biotechnology clusters like this from others in different sectors is the science-driven nature of the business, and it is worth remembering there are roughly 400 other DBFs in this specific cluster, with many board and advisory cross-links, many of them local but more non-local, each to some extent aping the profile of Epicyte, with strong localised and stronger globalised, or at least North American business linkages. It was argued by Wolter (2002) that this characteristic differentiates biotechnology clusters from Porterian 'market' driven clusters. For the Porterian 'diamond' of factor conditions for competitiveness postulates transparent competitors, clear market opportunities, known production and supply schedules and appropriate market support to enable competing firms to perform effectively. Bioscience lacks this kind of market stability. Most of its incumbents do not make profits, indeed new stock market rules and indeed stock markets like NASDAQ and (in the UK) AIM had to be set up to allow firms not making profits to be publicly invested in. Knowledge is unstable and prone to being swiftly superseded. Skills are in short supply and requirements change rapidly. Deep tacit knowledge of indeterminate but potentially enormous value is in the minds and hands of academics, hence their capacity to generate upfront input returns from licensing income or patent royalties. But many DBFs presently earn little from product or treatment outputs. The key to this unusual kind of economy is proximity to knowledge 'fountainheads' such as those described for Maryland by Feldman and Francis (2003), Munich by Kaiser (2003), Montreal by Niosi and Bas (2003), Cambridge by Casper and Karamanos (2003), and key knowledge centres in Israel by Kaufmann *et al.* (2003). The largest of these may be termed 'megacentres' after the French appellation noted in Chapter 4 and discussed more fully in Chapter 7, for science-led innovation complexes like Grenoble that combine 'ahead of the curve' science with disruptive research, heavy investment in scientific and technological 'talent', and incentives for entrepreneurs to exploit such discoveries commercially (Johnson, 2002). This is a feature of biotechnology that steps beyond the notion of 'clusters' and may presage new developmental opportunities for other science-driven, research-intensive and heavily publicly funded complexes in future, nanotechnology, for example, being the actual trigger for Grenoble's 'megacentre' appellation.

Beyond Schumpeter?

Reference to disruptive research and technologies reminds us of Schumpeter's view of innovation and entrepreneurship as characterised by 'creative destruction' (Schumpeter, 1975). That view has been enormously influential among the neo-Schumpeterian school of innovation theorists and, particularly, innovation systems analysts like Freeman (1987), Lundvall (1988) and Nelson (1993). Inspired by Schumpeter's insights into the heroic efforts of the innovator, whether as the lone entrepreneur or, later, the equally heroic R&D team of the large corporation as the key driver of competition, they developed a heterodox perspective that attracted a large academic and policy following. This was not least because of the school's rejection of the desiccated theorems of neoclassical economics that treated Schumpeter's wellspring of capitalism as a commodity to be acquired 'exogenously' off the shelf. This is no longer believed even by neoclassicals. Although the national innovation systems authors mentioned in this paragraph had no spatial brief – other than for a poorly theorised 'nation state' which in the case of Taiwan and South Korea could look odd, their adherence to Schumpeterian insights implied, by his concept of 'swarming', clusters. These are logical consequences where imitators pile in to mimic the innovator and seek early superprofits before over-competition kills the goose laying the golden eggs. Hence 'creative destruction' as some older technology – sailing ships – gives way to newer – steamships; many shipyards arise only to be competed into a few or none, as competitive advantage moves their epicentre to cheaper regions around the world. It can be argued that the swift destruction of value observable in such processes, and more recently in the bursting of the dot.com bubble (see section five), with its negative upstream implications for the whole ICT industry and world trade, constitutes a problem of 'over-innovation' rather as, in a completely opposite way, German car and machinery makers used to be accused of 'over-engineering'. Mobile telephony and computing may equally be accused of having too many 'bells and whistles' that most people never use.

Biotechnology is not like this, as Chapter 4 suggested. It is in many ways either non-Schumpeterian or post-Schumpeterian, though as we have seen it certainly clusters, even concentrates in 'megacentres' of which there are very few in the world. It does this not to outcompete or otherwise neutralise competitor firms, but rather to maximise access to knowledge at an early stage in the knowledge *exploration* phase that is difficult to *exploit* in the absence of multi-dimensional partnerships of many kinds. In Feldman and Francis' (2003) paper there is a remarkable, even original picture presented of the manner in which the driver of cluster-building in Washington and Maryland's Capitol region was the public sector. Importantly, the National Institutes of Health conduct basic research as well as funding others to do it in university and other research laboratories. Politics mean that sometimes bureaucracies must downsize, something that was frequently accompanied by scientists transferring knowledge of many kinds into the market, not least in DBFs that they subsequently founded. Interactions based on social capital ties with former colleagues often ensued in the market place. Before founding Human Genome decoding firm Celera, Craig Venter was one of the finest exploiters of such networks, not least with the founding CEO of Human Genome Sciences Walter Haseltine, who had been a colleague of Walter Gilbert at Harvard and James Watson, co-discoverer of DNA, and later of Nobel prize-winner David Baltimore (Wickelgren, 2002). Thus, as Feldman and Francis (2003) show, inter-organisational *capabilities* networks from government to firms as distinct from more normally understood laboratory–firm or firm–firm links may also be pronounced in biotechnology.

This is seldom observed in the classical Schumpeterian tradition, which rests to a large degree on engineering metaphors, where entrepreneurship is privileged. Moreover few neo-Schumpeterians devote attention to the propulsive effect of public decisions (for an exception, see Gregersen, 1992) even though regional scientists had long pointed out the centrality of military expenditure to the origins of most ICT clusters (Hall and Markusen, 1985). In Niosi and Bas (2003) it is shown how relations are symbiotic rather than creatively destructive among firms in the Montreal and Toronto clusters. Arguing for the 'technology' definition of biotechnology, they say it is not a generic but a diverse set of activities each with a specificity. Of interest is that in biopharmaceuticals they interact in networks to produce products for different markets. Hence there is cross-sectoral complexity in markets as well as research, although healthcare is the main market. Accordingly, they too point to the 'regional innovation system' nature of biotechnology, spreading wider and integrating more deeply into the science base than is normally captured by the localised, near-market idea of 'cluster'. Thus CROs and biologics suppliers may exist on the orbit rather than at the epicentre of the cluster.

This collaborative aspect of biotechnology is further emphasised by Casper and Karamanos (2003). They show that, for example, regarding joint publication, while there is substantial variation in the propensity to publish with laboratory, other academic or firm partners by Cambridge (UK) biotechnology firms, only 36 per cent of firms are 'sole authors' and the majority of partners are those from the founders or their laboratory. Academic collaborators are equally shared between Cambridge and the rest of the UK with international partners a sizeable minority. Hence, the sector is again rather post-Schumpeterian in its interactive knowledge realisation, at least in respect of the all-important publication of results that firms will likely seek to patent. Moreover, they will, in many cases, either have or anticipate milestone payments from 'pharma' companies with whom they expect to have licensing agreements.

Finally, biotechnolgy is interesting for the manner in which epistemological, methodological and professional shifts in the development of the field created openings for DBFs when it might have been thought that already large pharmaceuticals firms would naturally dominate the new field. This is an important sub-text in Kaiser's (2003) paper on the slow development of biotechnology in Germany, and specifically its main centre at Munich. Pharmaceuticals arose from chemistry not biology and although, as others show, modern biotechnology relies on molecular expertise that chemists pioneered, most of its discoveries derive ultimately from biology. Hence the 'molecular biology revolution' as Henderson *et al.* (1999) call it, somehow superseded the epistemological primacy of chemistry.

Methodologically, chemists retain a hold over such activities as screening for particular molecules that act as inhibitors for those that are disease-causing. But methodologically that process has become infinitely swifter and more automated with the advent of supercomputer-strength high-throughput screening. Such equipment was also developed by specialist ICT companies applying expertise to bioinformatics and the like. A good example of the advantage of the application of such *examination* knowledge is the race to decode the Human Genome. Here, Celera pitted itself against and, for a long time led, an enormous Transatlantic university research centre-based consortium. The reason it could do this was Craig Venter's innovative approach, sequencing segments 'shotgun' fashion rather than whole gene sequences, and Californian firm Perkin-Elmer's decision to produce equipment to enable this to be done rapidly. It was only when the consortium's budget was massively increased so they, too, could afford such equipment that they caught up and the race ended neck-and-neck. This, of course, meant that, professionally-speaking, ICT equipment producers, specialist software engineers and information scientists have become integral to biotechnology.[2]

Clustering: market or milieu?

Strong cases can straightforwardly be made for the characteristic spatial proximity emphasis there is in biotechnology in North America, Europe and, as Chapter 8 shows, Latin America and Asia. But global linkages are also pronounced. Kaufmann *et al.* (2003) show that the academic nature of the background of most Israeli firm founders means they have weak business expertise and cluster accordingly, but near to their laboratory home base in the early stages. The importance of public subsidy in the form of regional support policies is also stressed. However, it is also shown that with maturity, complex business and technological needs cause expansion of networks overseas. In this, the need to access business partners for *examination* knowledge expertise, particularly in testing and trialling for clinical research, is paramount as it is unavailable in such a small country setting. As with early financing requirements that also contribute to a clustering disposition, there is a strong business market impulse for clustering in Israel, one that also drives expansion of networks. In other words, partnering is strategic, but primarily for market reasons. Thus the milieu is in quite pronounced ways one in which business is done with trusted others, rather than, as may sometimes be presumed, through exchanging of much untraded resource. This is echoed in Casper and Karamanos' (2003) study of Cambridge, UK where they conclude that a major factor in clustering is the presence of an 'ideas market' and access to specialised labour markets. This attracts spin-ins as well, a feature shown to be pronounced in Cambridge.[3]

Nevertheless, in Germany, after a number of false starts the BioRegio programme earned praise for rapid stimulation of biotechnology start-up businesses from 1997 through 2002. Of importance in this success was the pre-existence of institutional partnership and a special institution, such as Munich's BioM, capable of promoting start-up activity. But, as Kaiser (2003) notes, this dynamism was equally predicated on the emergence of a venture capital market, itself assisted by the co-funding support available from BioRegio and other regional and local funds. Here, then, public investment was required to stimulate market activity to promote clusters in areas that possessed both organisational and institutional capabilities for clustering to occur.[4] We may say that the institutional dimension constitutes aspects of a milieu effect, but one rather different from the type hypothesised by the GREMI group for whom the public sector seldom plays a leading role (Maillat, 1998). Of course a different but crucial role was, as discussed, shown to have been played out in the Capitol region in the USA in the account of biotechnology growth provided by Feldman and Francis (2003). In Canada, one could say that the clustering developed in classic style for the sector, namely large-scale publicly funded research of high quality created opportunities for entrepreneurship by academics and venture capitalists. Strong links with pharmaceuticals firms strengthened the funding base of DBFs, some of which have been acquired by larger companies, while newer spinouts continue to emerge.

Biotechnology's globalisation: from below?

Today, biotechnology is seen as a *sine qua non* of regional economic development. Paradoxically, as ministers and many scientists bemoan the aspirations of regional agencies everywhere to establish a biotechnology presence, their words are drowned in the rush to implement policies that seek to ensure future regional competitiveness from biotechnology applications in all their myriad forms. This is, of course, reminiscent of the rush to clone Silicon Valley by policy makers world wide during the 1980s and 1990s. Most of these efforts were unsuccessful on a narrow evaluation, in the sense that Silicon Valley remains unique in both the scale of its local and global impact, both during the upswing and more recently the downswing in the ICT industry. It is worth recalling that job losses in Silicon Valley from 2001 through 2003 were 400,000 and San Jose lost 20 per cent of its jobs. The latter statistic, according to a UCLA Anderson business school report, constitutes 'the single largest loss of jobs by any major metropolitan area since at least the second world war' (Bazdarich and Hurd, 2003; Parkes, 2004).

Despite the roller coaster ride that high technology industry clearly provides, these are, on balance, growth sectors over the business cycle, and on a broad evaluation regions that sought to anchor high technology growth during the past two decades have often proved remarkably successful. The Kista science park, north of Stockholm, founded in 1972, still has some 600 companies with 27,000 employees (down from 29,000 in 2000), more than 3,000 students in IT, and numerous university and research institutions active in wireless communication and the mobile Internet. Other such complexes in Helsinki, Ottawa and Cambridge could be mentioned, let alone Austin and Dallas, Texas or Dulles, Virginia and Bangalore, Hyderabad, Shenzen and Shanghai (on the last four, see Chapter 8). None of these is an exact clone of the original, but they are important contributors to the regional economies they inhabit.

Biotechnology has begun to evolve its ecology likewise, albeit at a lesser scale and with different specificities. Thus, first, agro-food biotechnology locales are different from those specialising in biopharmaceuticals – although occasionally they overlap (e.g. San Diego). An interesting case of the latter phenomenon is presented in the Medicon Valley (Sweden–Denmark) study of Coenen *et al.* (2004). The proximate cause of locational variation between biopharmaceutical and agro-food clusters in general is the fundamentally public funding of research in the former and private funding in the latter. Moreover firm structure, organisation and capabilities bring a heavy influence to bear. Thus agro-food is dominated by a few multinational corporations like Monsanto, Bayer, Dow, Unilever, Nestlé, Syngenta and Aventis. Joining them are former chemicals companies transforming themselves into 'biologics' (fermentation and contract DNA manufacturing specialists) firms, the likes of Avecia, DSM, Lonza and rising state-backed firms like BioCon in India and China Biotech. Significant agglomerations of such companies occur in out of the way places like Guelph and Saskatoon in Canada, Wageningen, Netherlands and Basel, Switzerland (Chapter 9). More substantial but somehow still unusual places from a high-tech viewpoint such as St Louis, Missouri also appear on the radar in the form of 'BioBelt', the world's biggest agro-food agglomeration, as Ryan and Phillips (2004) have shown.

Second, unlike the biopharmaceuticals industry these corporates conduct much leading edge research in-house although they are also found in juxtaposition with national research institutes, themselves often located historically in agricultural zones. This explains the concentration in Saskatoon where oilseed rape innovation capabilities have been concentrated since the 1940s, and St Louis with comparable competences in the corn and soybean sectors. Syngenta – the Basel-based agro-food firm formed from the merger of ICI and Novartis' agro-food divisions, established a major cotton research centre in Raleigh-Durham, a traditional centre of expertise in cotton research, now leading in cotton genomics. The same firm was first to decode the rice genome, albeit at its Torrey Mesa Research Institute in San Diego. As many published papers have shown, biopharmaceuticals are unlike this. University research departments and centres of research expertise lead the field in bioscientific research and the exploitation for commercial purposes of biotechnology. Thus the corporate 'pharma' is a supplicant to these leading edge research centres. Recent papers by Lawton Smith (2004) and Bagchi-Sen and Scully (2004) emphasize the centrality of university and medical school research as a magnet for specialist biotechnology firms of small scale that out-license their findings into a supply chain that ends with pharmaceuticals companies marketing and distributing the final product on a global scale. In what follows, three themes of significance to the 'globalisation of bioregions' are highlighted. These are, first, the practices of globalisation in biosciences, second the vehicles for its implantation, and third the evolving, networked global hierarchy in biotechnology.[5]

The practices of globalisation in biosciences

We have already seen that biotechnology underpins a range of industries, and that its broad categorisation into biopharmaceuticals, agro-food and environment/energy biotechnology produces distinctive economic geographies, at least in the first two. Briefly, we can say there is a third distinctive pattern in regard to the main environmental/energy segment, that it is normally found in older industrial regions such as the Ruhr region in Germany or even established chemicals complexes like Grangemouth in Scotland, a 1960s government-planned estuarial 'growth pole' involving the major chemicals companies in the UK in the 1960s. As discussed in more detail in Chapter 6 it currently hosts the aforementioned Avecia and Syngenta British Petroleum (BP), Polyami GE Plastics, Rohm and Haas. The Ruhr region of Germany, discussed in detail in Chapter 9, hosts a substantial bioenvironmental sector spun out in the 1980s and 1990s as the coal and steel industries began to become uncompetitive. Institutionally the co-determination norms that govern industrial management in Germany facilitated such innovations because trade unions and management must 'co-determine' the business restructuring strategy in response to crisis. This meant more sophisticated solutions affecting workers' futures had to be assessed and agreed, adding to the embeddedness effects for the rising bioenvironmental sector of its heavy industrial predecessor. Similar related variety outcomes have been noted for the automotive industry in Detroit and Turin (Klepper, 2002; Giaccaria, 2006).

One segment of biotechnology is based on 'open innovation' (Chesbrough, 2003) among research institutes, pharmaceuticals companies and small, dedicated biotechnology firms (DBFs). The second one in agro-food is organised around direct basic knowledge transfer between large public research institutes and large agro-food biotechnology firms with little or no DBF involvement (although Wageningen has around ten agro-food DBFs and a consortium, the Centre for Biosystems Genomics, a network of four universities, two research institutes and fifteen firms in agro-food biotechnology across the Netherlands in its 'Food Valley' science park). The third model is inter-industry trade by direct spinout, divisional divestiture, or in-house evolution of innovation capabilities from coal, steel and chemicals corporates entering and selling products and services in new environmental or energy markets. The last-named model is more engineering (synthetic) than science-driven (analytical) and thus unaffected in locational terms by the presence or absence of research institutes, but rather remains embedded in its original locational trajectory (Coenen *et al.*, 2004).

The vehicles for implantation of globalised bioscience

Boston's 'open science' does not deter high pharmaceuticals 'pipeline' expenditure among its DBFs.[6] This is further underlined with respect to Figures 5.1 and 5.2 which map collaborative publishing between leading scientists in important or potentially significant bioregions worldwide 1998–2004. Figure 5.1 refers to collaborative publication aimed at five representative European biotechnology journals, Figure 5.2 registers them for the four representative US journals. Figures 5.3 and 5.5 provide comparisons for eight leading (in the SCI top ten) journals, four each from Europe and the US. Three things are of special interest here. First, strong bioregions in Europe and the US collaborate significantly and intensely in collaborative publishing in US journals. Second, intensity of collaboration among European bioregions (and Canadian) is more pronounced in leading European journals than US collaborations. Third, collaboration activity for publication in leading European journals (e.g. *Nature Biotechnology*) is less intense than for US journals (e.g. *Cell*). Interestingly, the co-publication patterns are similar among representative and leading co-publications.

However, in either case the main bioregions listed are the most active collaborative publishing bases, even though in cases like New York and London, they score less highly regarding commercialisation indicators than might be expected. A further point worth noting, which underlines commentary on Japan's weak showing in current bioregion analysis, is that Tokyo is far less active than might be expected, and involved comparably to Uppsala, Zurich or Jerusalem but far less than Cambridge or Oxford. Figure 5.1 has the nodes and networks for five leading European journals.

In Figure 5.2 the network dynamic is to a considerable extent inverted, in that the US collaborative publishing bioregion 'nodes' are much more active, and the European and other 'nodes' are more active towards them

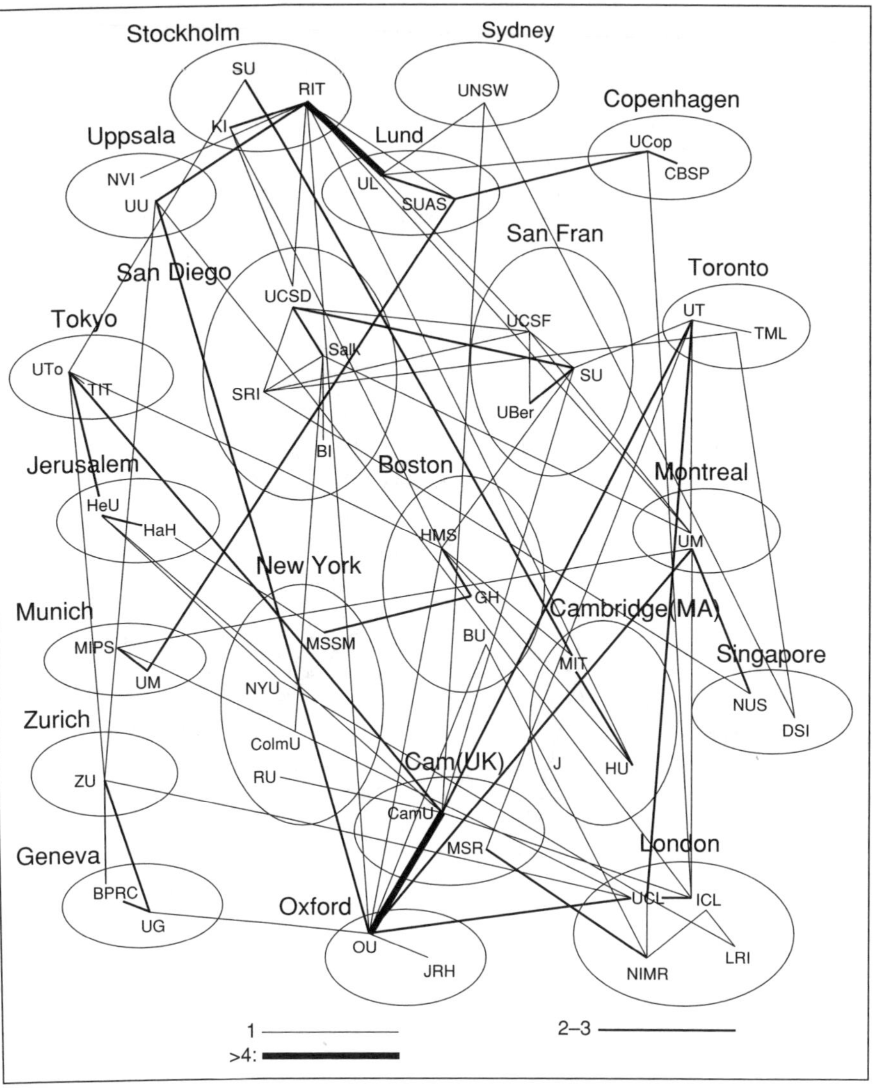

Figure 5.1 Publishing collaborations in five representative European bioscience journals

than the reverse in Figure 5.1. This is thus an excellent way of demonstrating the operation of power in network relationships. This is because Boston and Cambridge, Massachusetts are clearly the most active research publication collaborators, Boston being the location of leading research institutes related to Harvard Medical School. The University of California Scripps Institute and Stanford nodes interact significantly both internally and with

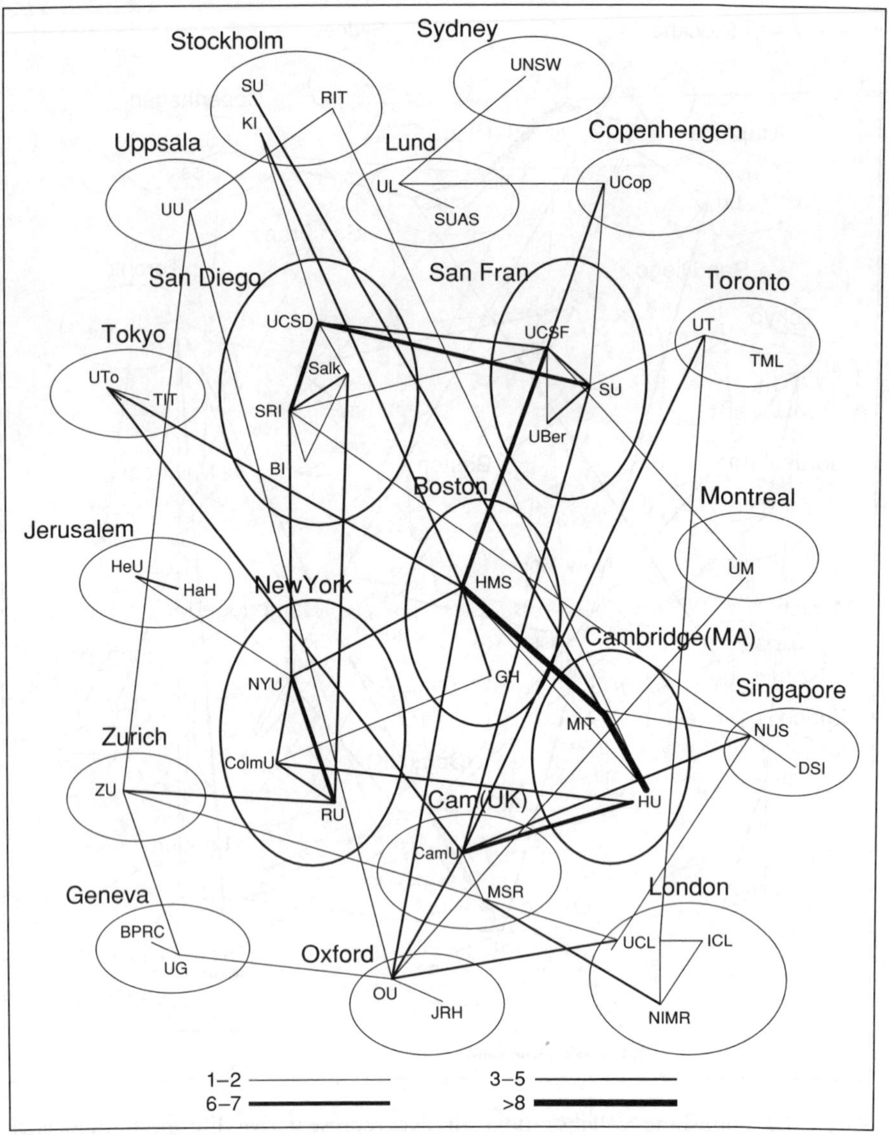

Figure 5.2 Publishing collaborations in four representative US bioscience journals

regard to each other. Inter-nodal collaborations with Harvard Medical School from UC San Francisco Medical School are strong, but so are those from UC San Diego and Scripps with New York University and Rockefeller University, a specialist medical and bioscientific campus once headed by retroviruses Nobel laureate David Baltimore.

To test the extent that these patterns changed when a more tightly structured sample involving eight of the top ten Science Citation Index journals was examined, Figures 5.3 and 5.5 are presented. Note that French involvement, with its leading bioregions of Paris and Grenoble included, appears in these figures. France, like Japan, is somewhat peripheral in the global bioregions co-publication system. Again, it is worth remembering that a key point at issue here is not quantity but quality, as élite scientific network linkage structures on a global scale are the prior category to be anatomised. The four journals from Europe and the four from the US are ranked in the top ten by Science Citation Index criteria. The journals in Figures 5.1 and 5.2 are more mixed, with some well ranked and some less so. There are, of course, criticisms to be made of using scientometrics although the quality versus quantity question has been dealt with by re-asserting that this global bioregions analysis necessarily focuses on the leading clusters. More difficult to identify are concerns that may be justified about the extent to which a journal like *Cell*, for example, favours articles from its Harvard home base The comparisons do not produce significantly different results. Further, as yet unpublished research into publishing in the seven bioscientific fields used in the VINNOVA (2003) study of bioscientific publication shows Harvard Medical School (HMS) to be orders of magnitude more productive in the top three cited journals in many of the seven areas of immunology; molecular biology; microbiology; neuroscience; biotechnology; cell and development biology; and biochemistry and biophysics.

For example, if we examine publication (not co-publication) by institutions in the leading bioregions in the three highest 'impact factor' journals in the seven listed fields, then regarding immunology, HMS published 10 per cent of the articles in the three most cited journals, with Stanford on 4.5 per cent and UCSF on 3.5 per cent. Karolinska Institute led Europe (tied with Salk Institute) at 1.2 per cent, followed by MIT on 0.8 per cent and Cambridge University on 0.5 per cent. HMS topped the share of molecular biology articles in the top three journals in 1998–2004 with 6 per cent of the relevant share, with MIT well behind on 1.7 per cent, UCSF at 1.4 per cent next, then Cambridge and Stanford Universities tied at 0.9 per cent. Of those scoring above 1 per cent Salk Institute, Rockefeller University and UC San Diego follow. Finally, in microbiology HMS is again first on 6 per cent followed by the Scripps Research Institute on 3.4 per cent, Stanford University on 3 per cent and Karolinska Institute on 2.4 per cent. For neuroscience, UCSF is first at 4.5 per cent but HMS is second on 4.1 per cent, Stanford is third on 3.8 per cent, Rockefeller follows on 3.3 per cent while Europe's highest entrant is Cambridge University on 3 per cent. Karolinska Institute is tenth on 1.3 per cent. To conclude, in biotechnology HMS is fifth (4.1%) after University of Toronto (5.5%) and the Scripps Research Institute (5.4%) and Stanford and Cambridge Universities equal on 4.5 per cent. These are followed by Zurich University (3.7%),

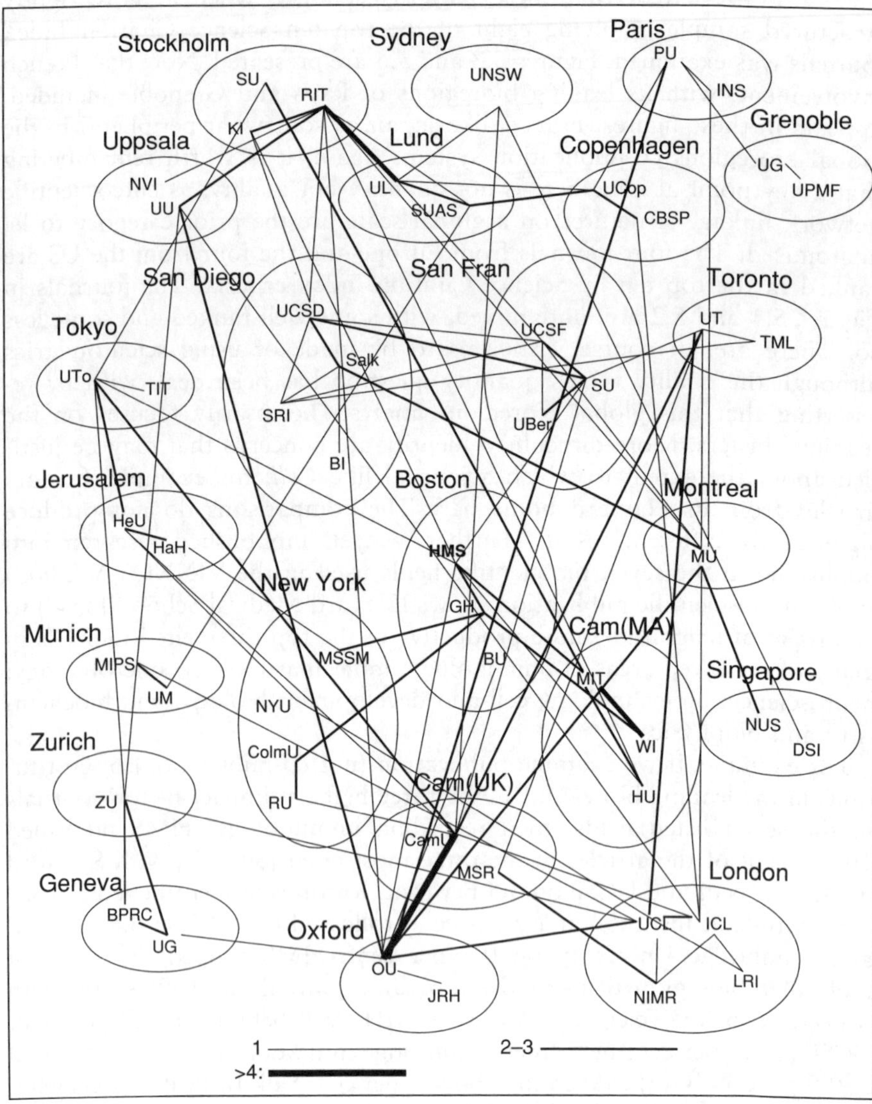

Figure 5.3 Publishing collaborations in four leading European bioscience journals

UCSF and Rockefeller tied on 2.8 per cent and UCSD at 1 per cent. Thus in these five key bioscientific fields alone HMS is first three times, second once and fifth once. Clearly, with or without control of house journals HMS is the leading quality publishing centre for biosciences in the world.

Regarding differences between, for example Figures 5.1 and 5.3, on the one hand, tracking inter-cluster co-publications by 'star' scientists in élite

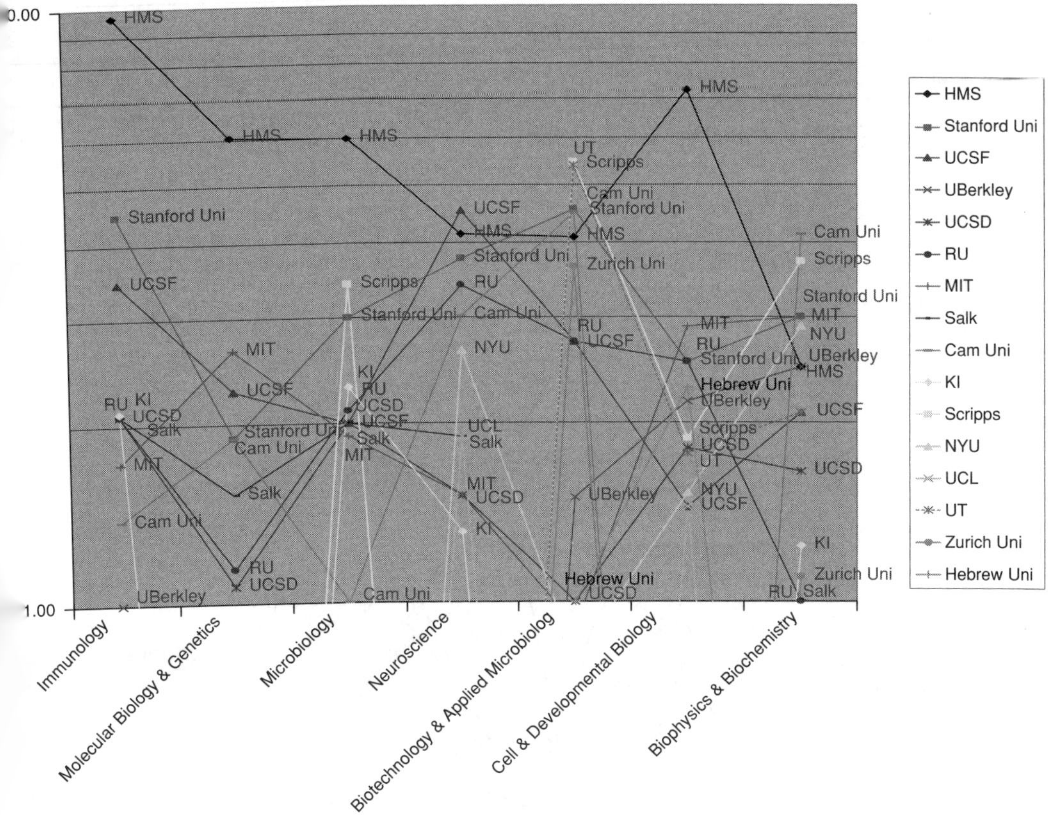

Figure 5.4 Publication shares of bioscience research institutes in the seven main fields of bioscientific research

N.B.: Results (log scale) are based on articles from top two or three journals (with the highest impact factor provided by *Web of Knowledge*) of each sub-field, which are as follows: Immunology: *Annual Review of Immunology*; *Nature Immunology*; *Immunity*; Molecular Biology and Genetics: *Cell Molecular*; *Journal of Cell Biology*; Microbiology: *Microbiology and Molecular Biology Review*; *Annual Review of Microbiology*; *Clinical Microbiology Review*; Neuroscience: *Annual Reviews Neuroscience*; *Trends Neuroscience*; Biotechnology and Applied Microbiology: *Drug Discovery*; *Nature Biotechnology*; Cell and Developmental Biology: *Annual Review of Cell and Developmental Biology*; *Advances in Anatomy, Embryology and Cell Biology*; *Anatomy and Embryology*; Biophysics and Biochemistry: *Annual Review of Biophysics*; *Current Opinions in Biophysics*

European journals and Figures 5.2 and 5.5 tracking the same for US journals, three points are worth making. The first is that between Figures 5.1 and 5.3 Swedish co-publishing declines in intensity between representative and leading European journals. For instance, linkage between Lund and Munich is stronger for representative compared to leading journal co-publication. The

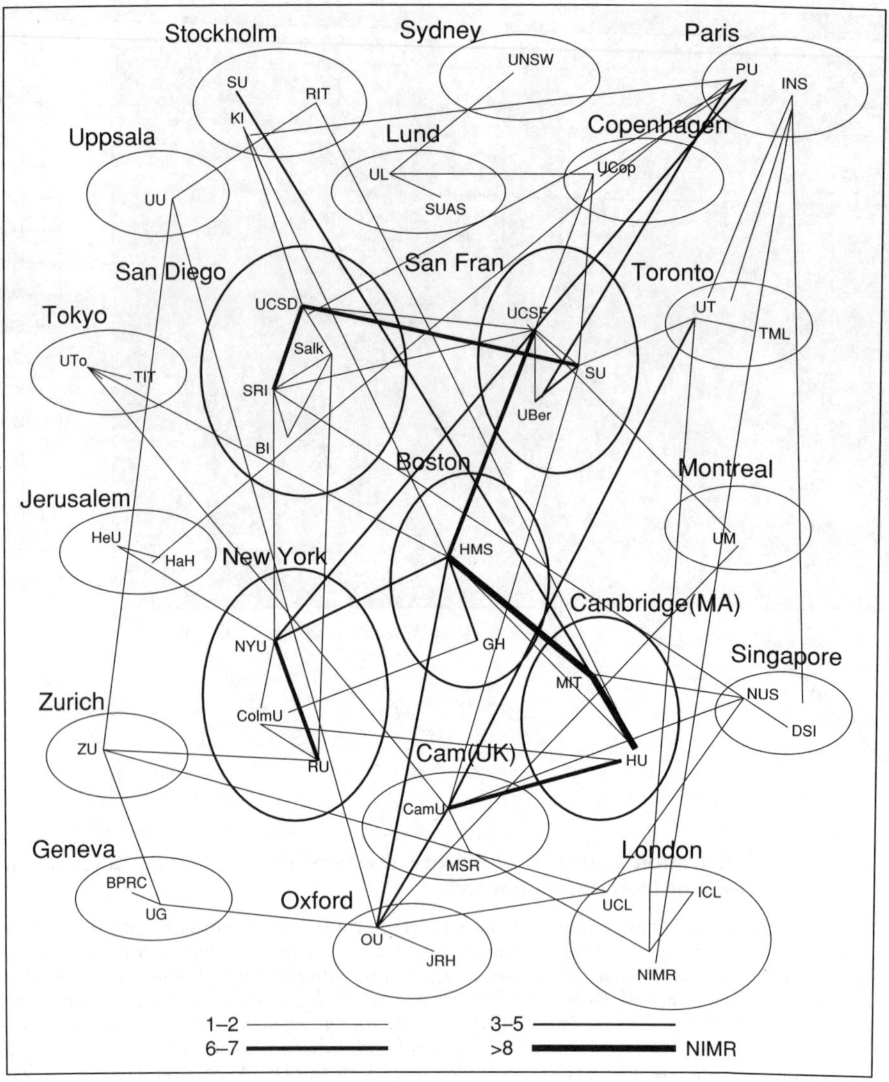

Figure 5.5 Publishing collaborations in four leading US bioscience journals

same is also true for many of the more peripheral bioregions. However intra-Sweden co-publication remains at the same relatively high intensity. Second, while UK links, particularly with Cambridge, Massachusett and Boston remain strong for European journal co-publication, and those involving Californian co-publications in leading European journals diminish somewhat, as do Stockholm's links with Harvard. Finally, regarding the comparison between representative and leading US journal co-publication, there is a concentration in the main routeways or network linkages with the Stockholm–Cambridge (MA)–Boston–New York–Cambridge–Oxford connection strengthening somewhat, but broadly speaking these two figures show less differentiation than the European journal co-publication networks.

The networked hierarchy in biotechnology

Three key things have been shown with implications for the understanding of knowledge management, knowledge spillovers and the roles of collaboration and competition in bioregions. The first is that two kinds of proximity are important to the functioning of knowledge complexes like biosciences in Boston and the northern and southern Californian clusters. These are geographical but also functional proximity (Rallet and Torre, 1998). The first involves, in particular, medical research infrastructure for *exploration* knowledge as well as venture capital for *exploitation* knowledge, i.e. for research on the one hand, and commercialisation on the other. The second point is that where *exploration* knowledge infrastructure is strong, that nexus leads the knowledge management process, pulling more distant 'big pharma' governance elements behind it. Where, by contrast, *exploitation* knowledge institutions are stronger than exploration, they may, either as venture capital or 'big pharma', play a more prominent role. But in either case the key animator is the R&D and exploitation-intensive DBF. DBFs are key 'makers' as well as 'takers' of local and global spillovers; research institutions are more 'makers' than 'takers' locally and globally; while 'big pharma' are today principally 'takers' of localised spillovers from different innovative DBF clusters. They then act as global marketers of these and proprietorial (licensed or acquired) knowledge, generated with a large element of public financing but appropriated privately.

For obvious reasons to do with scale, especially of varieties of financing of DBFs from big pharma on the one hand, and venture capitalists on the other, we conclude that Boston, San Francisco and San Diego are the top US bioregions that also have the greater cluster characterisation of prominent spinout from key knowledge centres, an institutional support set-up like Boston's Massachusetts Biotechnology Council, San Diego's CONNECT network and San Francisco's California Healthcare Institute, and major investment from both main pillars of the private investment sector. We have attempted to access comparable data from many and diverse statistical sources that justify and represent the successful or potentially successful clusters from outside the US, and these are shown in Table 5.1 including the four lesser or unclustered of the seven US bioregions.[7] Global cities like New York, London and even Tokyo have relatively large numbers of DBFs but they are dotted around in isolation and have no established bioregional promotional bodies (such as BioCom in San Diego or the Massachusetts Biotechnology Council in Cambridge) rather than clustered close to key universities as in the listed bioregions.

Table 5.1 Core biotechnology firms, 2000: comparative global performance indicators

Location	*DBFs*	*Life scientists*	*VC*	*Big pharma funding*
Boston	141	4,980	$601.5 m.	$800 m./annum 96-01
San Francisco	152	3,090	$1,063.5 m.	$400 m./annum 96-01
New York	127	4,790	$1,730.0 m.	$151.6 m. (2000)
Lund-Medicon	104	5,950	$80.0 m.	$300 m. (2002)
San Diego	94	1,430	$432.8 m.	$320 m./annum 96-01
Washington DC	83	6,670	$49.5 m.	$360 m. (2000)
Toronto	73	1,149	$120.0 m.	NA
Montreal	72	822	$60.0 m.	NA
Ral-Dur NC	72	910	$192.0 m.	$190 m. (2000)
Zurich	70	1,236	$57.0 m.	NA
Stockholm-Upp.	60	2,998	$90.0 m.	$250 m. (2002)
Munich	56	8,000	$400.0 m.	$54 m. (2001)
Cambridge	54	2,650	$250.0 m.	$105 m. (2000)
Oxford	46	3,250	$120.0 m.	$70 m. (2000)
Singapore	38	1,063	$200.0 m.	$88 m. (2001)
Jerusalem	38	1,015	$300.0 m.	NA
Seattle	30	1,810	$49.5 m.	$580 m. (2000)

Sources: Cortright & Mayer (2002); NIH; NRC; BioM, Munich; VINNOVA, Sweden; Dorey (2003); Kettler & Casper (2000); ERBI, UK, Lawton Smith (2004); Kaufmann *et al.* (2003).

We can see from Table 5.1 the predominance of Boston and San Francisco on all indicators, and the differences between the former (also New York) and the Californian centres are strikingly revealed by these data. Boston's life scientists generate on the order of $285,000 each per annum in National Institutes of Health research funding (New York's generate some $288,000). San Diego's considerably smaller number of life scientists generates $480,000 per capita, substantially more than in Northern California where it is some $226,000. North Carolina, with the second smallest number of life scientists in Table 5.1, scores highest at $510,000 per capita, although Seattle at $276,000 is comparable to Boston, New York and San Francisco. How to interpret these statistics? One way is to note the very large amounts of funding from 'big pharma' going especially to the Boston, and to a lesser extent New York and both Californian centres. Canada's bioregional clusters challenge many elsewhere in the world regarding cluster development. Some of Europe's are rather large on the input side (e.g. life scientists) but less so on the output side (e.g. funding, firms). The process of bioregional cluster evolution has occurred mainly through academic entrepreneurship supported by well-found research infrastructure and local venture capital capabilities. In Israel, there is a highly promising group of bioregions including also Rehovot and Tel Aviv as well as the main concentration in Jerusalem, where patents are highest although firm numbers are of lesser significance.

The most striking feature of the global network of bioscientific knowledge transfer through exploration collaboration as measured by research publication collaborations is, as we have seen, the dominance of US megacentres in such collaborations. However, within that hierarchy, which taken in the round by including examination and exploitation knowledge processes, puts Cambridge-Boston at the peak of the global bioscientific network hierarchy, specific institutional dominance of knowledge production is remarkable. This is shown in Figure 5.4 which ranks specific research institutions from among those benchmarked in Table 5.1 according to their share of publications in the seven main bioscientific fields. The predominance of a single, large and complex institution, Harvard Medical School, is noteworthy to say the least.

As Figure 5.4 showed, HMS published 10 per cent of the articles in the three most cited immunology journals 1998–2004, with Stanford at 4.5 per cent and UCSF at 3.5 per cent. Karolinska Institute led Europe at 1.2 per cent, followed by Cambridge University at 0.5 per cent. HMS topped the share of molecular biology articles in the top three journals with 6 per cent of the relevant share, with MIT well behind at 1.7 per cent, UCSF at 1.4 per cent next, then Cambridge and Stanford Universities tied at 0.9 per cent. HMS is again first in microbiology at 6 per cent followed by the Scripps Research Institute at 3.4 per cent, Stanford University at 3 per cent and Karolinska Institute at 2.4 per cent. HMS is second behind UCSF in neuroscience at 4.1 per cent, Stanford is third at 3.8 per cent, Rockefeller follows at 3.3 per cent, while Europe's highest entrant is Cambridge University at 3 per cent. Karolinska Institute is tenth at 1.3 per cent. In biotechnology HMS is fifth at 4.1 per cent after University of Toronto at 5.5 per cent and the Scripps Research Institute at 5.4 per cent with Stanford and Cambridge Universities equal at 4.5 per cent. These are followed by Zurich University at 3.7 per cent, UCSF and Rockefeller tied at 2.8 per cent and UCSD at 1 per cent. In cell and development biology HMS is again the clear global leader with over 7 per cent leading journal share by citation and MIT trailing second at under 3 per cent. Finally, a relative weakness is biophysics and biochemistry where HMS scores only 2.5 per cent and trails Cambridge University (UK), Scripps, Stanford, MIT, NYU and UC Berkeley. Thus in these seven key bioscientific fields alone HMS is first four times, second once, fifth once and seventh once. Clearly, with or without control of house journals (which, it can be argued, significantly favour 'home' contributions) HMS is the leading quality publishing centre for biosciences in the world. Added to which, for Cambridge-Boston, is the fact that MIT is second for molecular biology and cell biology, fourth in biophysics and biochemistry, and in the top ten for most of the rest.

These data average out performance over the 1998–2004 period. Of significant interest to the analysis presented here are the dynamics involved in the rise of bioscience megacentres over that period. Scientometric data from Web of Science were accessed on specific bioscience publications to test whether or not and in which fields there was concentration of publication by highly cited authors in leading institutes of the kind discussed earlier. Utilising a more conservative methodology than the above, which focused on the highest impact journals by field, in this instance six journals were selected and divided into high, medium and low impact categories, thus giving a more representative canvas. In the fields thus far examined (excluding only microbiology from the seven mentioned above) there is evidence that top research institutes increased their share of articles published 1998–2004 in immunology, molecular biology, cell biology and biophysics with biochemistry. However, in neuroscience and biotechnology the top institutes showed a marginal decline in share. Focusing on the last two momentarily, in neuroscience, the US grew its global share from 63.4 per cent to 70.5 per cent while the EU (excluding UK) declined from 20.8 per cent to 15.2 per cent. The US growth in share thus came from a spread of neuroscience research capabilities to non-élite institutes. However, in biotechnology, the reverse is the case with the EU (and UK) shares rising from 21.3 per cent and 15.2 per cent to 45.4 per cent and 20.10 per cent respectively and the US share declining from 54.9 per cent to 33.2 per cent 1998–2004. Clearly, Europe woke up to the importance of biotechnology research while non-élite US institutions were becoming far more active in neuroscience, indicative perhaps of the 'ahead of the curve' foresight and associated research funding arising from the first-mover biotechnology advantage enjoyed by the US and to a lesser extent the UK since the 1970s.

Moving attention to the fields in which élite bioscientific research institutes have tightened their grip, we focus on the aforementioned Immunology, Molecular Biology, Cell Biology and Biophysics with Biochemistry. The last-named showed the greatest concentration to top institutes, largely explained by the rise in global publication share of the US up from 52.4 per cent in 1998 to 70.5 per cent in 2004. By contrast an 18 per cent concentration to top institutes in cell biology was accompanied by a slight US decline from 55.2 per cent in 1998 to 53.2 per cent in 2004. Growth here came not from Europe (including UK), down from 36.5 per cent to 28.5 per cent, but from other countries. Examination of co-publications in this field shows other countries (mostly in Asia) accounting for the largest rise in co-publication, close to doubling that share from 42.0 per cent to 78.0 per cent in the period. In immunology the US grew from 43.9 per cent to 53.2 per cent largely at Europe's expense, accounted for largely by concentration in top institutes. Finally, in molecular biology a similar change occurred, with the US increasing its share from 42.6 per cent to 53.2 per cent, with other countries rising from 4.6 per cent to 12.6 per cent, again at Europe's expense, down (including UK) from 52.8 per cent to 34.2 per cent. In the space available, only two graphic representations of these trends can be accommodated; these are the contrasting cases of biotechnology revealing globalisation without concentration in top institutes indicated by

Figure 5.6, and the opposite from biophysics showing concentration to US top research institutes in Figure 5.7.

Thus we have seen inside the black box of global research excellence, gaining a detailed understanding of the broader network hierarchy of global knowledge transfer through co-publishing, revealing the nature and extent of the global bioregions network. Recall from Figure 5.1 and Table 5.1, it is clear that a global hierarchy exists in which there are five 'megacentres' (Cooke, 2003a; Niosi and Bas, 2003). Measured by co-publishing activity, these are Boston-Cambridge, New York, San Francisco and San Diego. At the next level are London, Cambridge (UK) and Stockholm. Surrounding these, at lower levels of co-publication activity are global city locations like

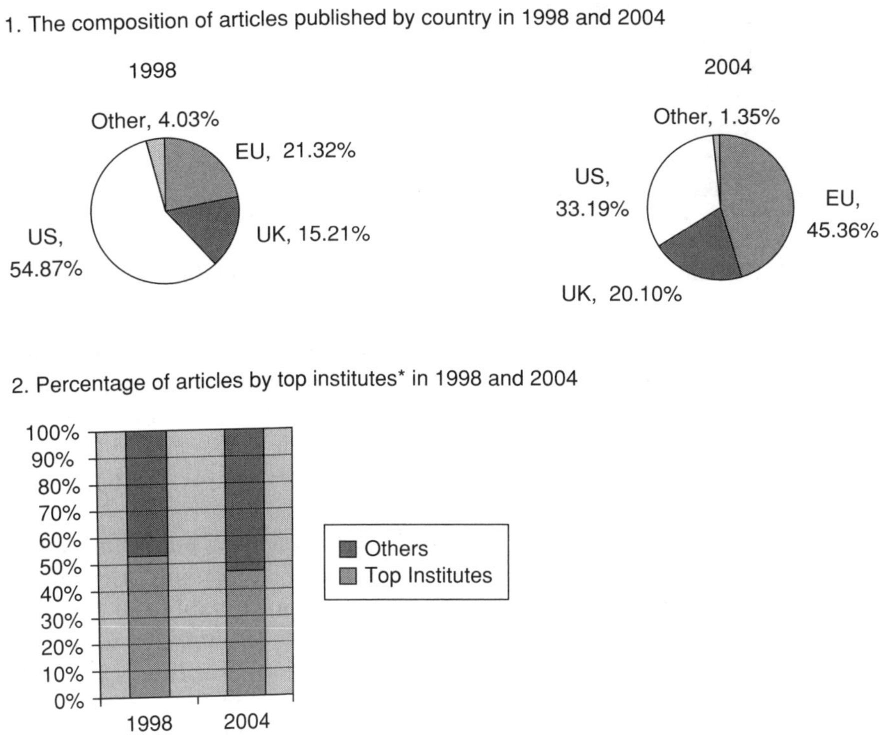

Figure 5.6 Globalisation and deconcentration of biotechnology publishing 1998–2004

* Top bioscience research institutes identified from highly cited returns in high impact factor journals from the Science Citation Index are: Harvard Medical School; Stanford University; University of California Medical School, San Francisco; University of California, Berkeley; University of California, San Diego; Rockefeller University; MIT; the Salk Institute; Cambridge University; Karolinska Institute; the Scripps Institute; New York University; University College, London; University of Toronto; Zurich University; Hebrew University, Jerusalem.

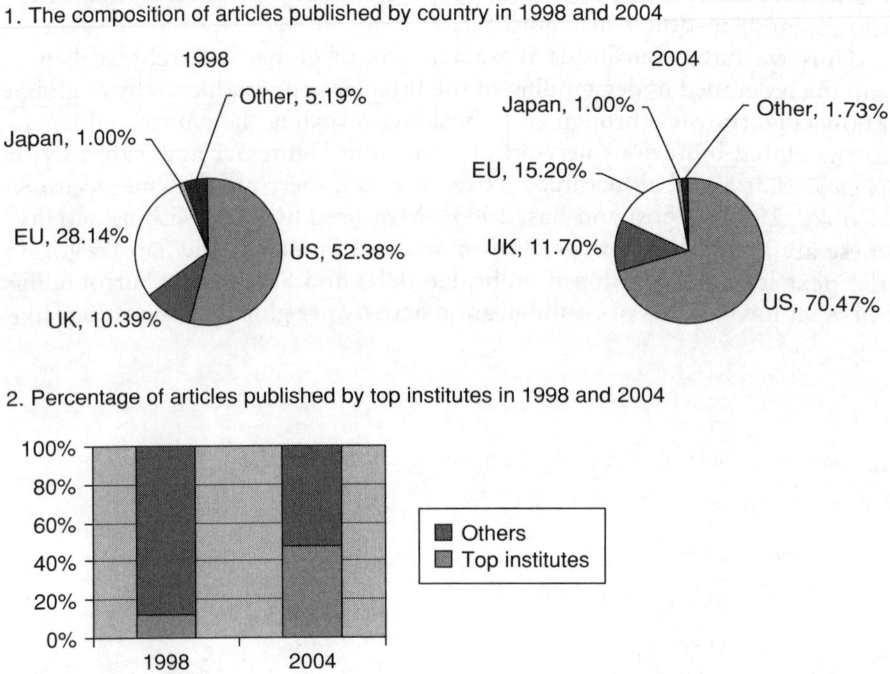

Figure 5.7 Globalisation and concentration of biophysics publishing 1998–2004

Paris, Tokyo and Toronto, but interspersed are other university towns like Lund, Uppsala and Oxford as well as lesser cities like Geneva, Zurich and Grenoble. Bioechnology thus does not fit the 'global cities' thesis (Sassen, 1994) particularly well. Other important but scarcely global cities like Jerusalem, Munich, Montreal and Singapore also enter at this third level of co-publishing intensity.

Notice, though, that in these core fields of bioscientific publishing relatively few of the main publishing centres extend deeply into the lesser network nodes, although this is changing somewhat, notably cell biology and to a lesser extent molecular biology. The key nodes continue to divide asymmetrically between eleven globally key publishing institutes in North America, all bar University of Toronto in the US, four in Europe and one, Hebrew University, Jerusalem in Asia. All the US top institutes are in the five co-publishing megacentres depicted in Figure 5.1. Despite dubious claims that San Diego[8] is the top global cluster, it is clear that with respect to the generation of new exploration knowledge it is Boston (including Cambridge, MA) that in fact scores as the strongest biotechnology cluster.

Moving to the opposite end of the global network hierarchy, as Finegold *et al.* (2004) make clear, Singapore's case is one that runs strongly against the grain of the 'clusters cannot be built' thesis since it is almost entirely the product of state intervention. Although not yet present in the top institutes hierarchy, but a good example of the 'rising-by-co-publishing' strategy, four new research institutes in bioinformatics, genomics, bioprocessing and nanobiotechnology now exist at a cost of $150 million to 2006. Public venture capital of $200 million has been committed to three bioscience investment funds to fund start-ups and attract FDI. A further $100 million is earmarked for attracting up to five globally leading corporate research centres. The Biopolis is Singapore's intended world-class R&D hub for the georegion. The Biopolis is dedicated to biomedical R&D activities and designed to foster a collaborative culture among the institutions present and with the nearby National University of Singapore, the National University Hospital and Singapore's Science Parks. Internationally celebrated scientists have also been attracted, such as Nobel laureate Sidney Brenner; Sir David Lane, discoverer of 'guardian angel' cancer suppressor gene p-53; Alan Colman, leading transgenic animal cloning scientist from Scotland's Roslin Institute; Edison Liu, former head of the US National Cancer Institute; and leading Japanese cancer researcher Yoshaki Ito (see also Chapter 8).

Hence, in exploring both the cluster-formation and globalisation processes they implicate, this analysis throws some light on the complexities of interactions between economic geography and science-based industry. This is so particularly regarding aspects of the networked interactions involved in knitting together the public and private spheres, basic research and commercial exploitation, and tacit and codified knowledge. There are good grounds now for looking upon biotechnology as a pioneer both of 'open innovation' and 'Globalisation 2' in which 'knowledgeable clusters' rather than multinationals drive development more than hitherto. As well as being of intrinsic interest, this analysis may offer hypotheses for other researchers investigating in other related or unrelated fields these challenging processes and the problems and opportunities for regional economic development policy that they entail.

Conclusions

Biotechnology contains much of interest for academic and policy analysts interested in the distinctive ways that clustering specific to biotechnology has occurred in different contexts. It also poses intriguing puzzles for academic specialists interested in biotechnology *per se* or more broadly in the clustering phenomenon. We see that local networks are common but that they may not be as large or as important as distant, even global ones. The industry within which biotechnology of the type mainly discussed here is embedded, pharmaceuticals, does not appear to follow a Schumpeterian innovation or entrepreneurship pathway, being characterised by a double dependence of large corporations on SMEs (for R&D) and the inverse for financing, marketing and distribution. Large corporates merge but do not become out-competed and thus disappear entirely.

Thus Novartis, formed from the merger of Ciba and Sandoz, retains the former as its opthalmics brand, with Sandoz its brand for generic drugs. Moreover, both are heavily dependent on public funding for basic research – the *exploration* knowledge that is the resource to which *exploitation* knowledge is applied. The latter includes knowledge of the kind possessed by venture capitalists to facilitate commercialisation of the few drug candidates that reach the market. Biotechnology thus seems to be more a Penrosian sector in that firm, including small firm, *capabilities* determine industry organisation (Penrose, 1959). Finally, to some extent echoing points made by Zucker *et al.* (1998), biotechnology is not a prime case of milieu effects dominating market effects. As they argue, scientific expertise has meant that since the beginning academics have wielded greater technological influence over firms than was the case with other industries. Firms learned on academics' terms and often, as DBFs, their ownership. As such, untraded interdependencies and even dense local network relationships, at the individual firm level, are less pronounced than might be expected given the obvious cluster concentrations in which biotechnology firms exist.

Noticeable too are the ways in which public funding, policy and expertise can often be much more central to biotechnology clustering than has been recorded for other sectors. But of greatest theoretical interest is the way 'open innovation' hints at 'Globalisation 2' where knowledgeable clusters shift the economic geography of industry organisation towards themselves. Moreover, as the case of Cambridge-Boston and Harvard Medical School showed, some such spaces have begun to exert spatial knowledge monopoly effects that are already bringing 'increasing returns' as multinationals establish R&D facilities to capture 'knowledge spillover' advantages from cluster conventions of both 'open science' and 'open innovation'.

Most interestingly, perhaps, the evidence of intra-field bioscientific research dynamics over relatively recent time is instructive of the importance attached to the field scientifically and industrially. Governments or foundations, through the likes of the US National Institutes of Health or the UK's Wellcome Trust, the world's largest medical research foundation, fund the basic or *exploration* research that is the subject of much of the foregoing analysis. Pharmaceuticals firms spend enormous amounts on in-house research and, increasingly as pioneers of 'open innovation', they conduct R&D outsourcing to research institutes and university centres of excellence as well as dedicated biotechnology firms from whom they license patented knowledge for *exploitation* purposes in healthcare markets. In between are the varieties of *complicit* intermediary organisations and firms assisting in the translation of *implicit* raw research towards *explicit* or codified documentation through the application of *examination* knowledge in clinical trials and various tests conducted by clinical research organisations. So much of this activity is externalised in complex webs of specialist expertise that it is no wonder only a few places in the world can sustain the full value chain of research and commercialisation activities. Despite evidence of the globalisation of bioscientific research, notwithstanding fake claims of cloning success from South Korea that both damage the reputation of the field and yet, strangely, testify to the white-hot nature of the biosciences discovery contest, the field is characterised at its leading edge by interesting instances of global and local collaboration. Open and collaborative science is a necessary accompaniment to innovative progress in dynamic scientific fields such as this.

6 The shifting landscape of bioscience policy

Introduction

We saw earlier that a picture has emerged of a highly asymmetrical landscape of changing science policy priorities. The data on the rise of NIH spending in the recent past is testimony to a major shift in the 2000s towards expenditure on, amongst other things, the renewed 'Crusade Against Cancer'. Mowery and Ziedonis (2000) showed how in the 1980s an earlier shift in priorities towards bioscience expenditure in the 1980s was accompanied by a statistically significant rise in output of biotechnological innovations after a lag of some few years. Other data of more recent provenance show that the rise in the 2000s has been accompanied by a steep rise in approved drugs (new chemical entities or NCEs) for DBFs, the main expertise source for biotechnology exploitation, but not for pharmaceuticals firms. 'Big pharma' are in the midst of a product pipeline 'drought' and have become supplicants to the DBF and university centre of excellence or non-profit research institute community for NCEs.

In this chapter we explore key effects upon science policy of changes in R&D-based clusters given the rise of 'knowledge economies' (Dunning, 2000; Cooke, 2002). To advertise the argument beforehand, it is that the decline in R&D power of large corporations is accompanied by the rise of specialist research firms. The latter include, for example, those referred to as 'discovery companies' in life sciences, along with university and other research labs in proximity to which knowledge-intensive firms increasingly cluster. This is particularly pronounced in biotechnology, but also occurs in other knowledge-intensive sectors like information and communication technologies (ICT), new media and advanced business services. Broadcasters and Bourses are stronger cluster magnets than universities in the last two cases. That life sciences activities cluster geographically along the knowledge value chain from exploration (basic research) through examination (clinical trials) to exploitation (dedicated biotechnology firms – DBFs) is widely known and shown.[1]

Continuing the argument, it will be shown that over the 1990s many regional governance agencies developed interest and capability in formulating policies to network *regional innovation systems*. To some extent multi-level governance hierarchies have evolved, as suggested in Lundvall and Borrás (1997) and Cooke *et al.* (2000), where national governments are mainly responsible for delivering science policy and basic research funding, while regional governance systems (involving public and private actors) deliver innovation programmes. These are usually near-market incentives to firms to build innovation networks, access co-funding and engage in joint marketing to enhance innovative potential and competitiveness. In Europe, the European Union was less directly involved than member states in basic research funding, more in research and technology development (RTD) and, while co-funding *innovation* initiatives, leaving these to regions to deliver along with their own and any national programmes on the ground. Towards the end we briefly address the question of whether the incessant rise in resources for bioscience and healthcare more generally may cease, given rising and widespread governmental concern at the apparent effects of climate change upon living conditions on the planet.

In what follows, the first main section will show how and why knowledge economies create regionalised innovation networks and clusters in life sciences more generally and biotechnology specifically. In section two, the implications of this are explored for the large-scale national innovation system model of 'big science' and, for example, 'big pharma' following an expensive 'chance discovery' methodology, also involving big departmental research laboratories in universities. Then, finally, evidence is mobilised to show the emergence of regional science policy and its funding mechanisms. A model that responds to regional science funding requirements within the national basic science funding remit is also discussed.

Knowledge economies and their regionalisation

A common misconception among non-regional scientists is that when regional analysis is done it inevitably means somehow ignoring other spatial, economic or political scales. As will be shown, the contrary is the actual position, particularly where science, technology and innovation are in focus. Indeed, what has been shown elsewhere (Lagendijk, 2001; Cooke, 2001b) is that by excluding the regional level from analysis, major innovation interactions between key knowledge generation and exploitation actors are likely to be overlooked. As Dicken (2001) sees it, from the TNC perspective, regionalisation enables faster delivery, more customisation and smaller inventories than globalisation. But this does not mean that TNCs become less global, rather they use whatever advantages may be available to them in seeking value chain efficiencies. So it is with regard to what might be termed the 'knowledge value chain' that this exploration is directed.[2] What is this and how might it be changing? We know of the changed emphases in knowledge production proposed by Gibbons *et al.* (1994). Key differences involved are the move from Mode 1 to Mode 2 conventions like disciplinary purity to *transdisciplinarity*, organisational hierarchy to *flexibility and diversity*, and value freedom to *reflexivity*. Related to reflexivity, the authors argue, are *quality*-related questions of a new kind concerning the competitiveness of knowledge outcomes in the market, or cost effectiveness and social acceptability.

Critique of this view focused on its technocratic introspection and failure to show science having to engage with social forces. As noted in Chapter 4, in a subsequent book (Nowotny *et al.*, 2001) some of these authors respond with an acceptance of Latour's (1998) identification of a struggle between 'science' and 'research'. The former is stable, the latter unstable, and the rise of research expresses a 'contextualisation' of science by society, politics and economy despite its claims to objectivity. Thus more multidisciplinary concerns and network interactions typify modern life sciences, as Powell *et al.* (1996) and Orsenigo *et al.* (2001) among many others show. Indeed the former go as far as to assert that 'knowledge is in the networks', a revision of the traditional primacy of codified over tacit knowledge. Orsenigo and colleagues explain this in terms of the heterogeneous nature of the cognitive skills demanded in bioscientific research. The economic and ethical demands of 'research' give rise to collaborative learning through transdisciplinary network relationships. Science's engagement with society initially occurred on *its* terms, as 'public understanding of science', now rightly criticised for its patronising disposition. It is, accordingly, difficult to see voluntary reflexivity in knowledge production in life sciences and others. Rather, ethical regulatory powers and company protocols to constrain 'value free' excesses in the field have had to be imposed externally. Contrariwise, as we shall see, the inclination for stock market imperatives to interfere with peer review norms of scientific reporting is increasingly driving announcements of scientific discovery and raising further ethical issues.

Thus in 2001 Millennium Pharmaceuticals and Human Genome Sciences both made press announcements when experiments reached Phase 1[3] of clinical trials, at least three years away from possible approval. In the past such announcements would be made on applying for approval. Also dedicated biotechnology firms (DBFs) have recently made announcements on experiments still at basic research stage, such as Advanced Cell Technology's claim to have cloned a human embryo and PPL's to have cloned pigs, both in advance of peer review and publication of results. This is doubly problematic when, at approval stage, large numbers of candidate treatments are rejected, as 30 were by the US Food and Drug Administration during 2001. These included Chiron's anti-sepsis drug, Immunex's cardiac infarction treatment and Maxim's melanoma product. A US head of bioethics was quoted that 'these companies must raise enormous amounts of money and the only way to do that is to put a hard spin on any good news' (Griffith, 2002).

An effect of the rise of research is that the mode of knowledge production has shifted, as revealed in life sciences R&D databases such as that of Pammolli *et al.* (2000) at the University of Siena. For example of the fifty DBFs in the database pursuing 'designer molecule' rather than 'discovery' methodologies 74 per cent operated in global, hierarchical networks with big pharma developers. From 1992 onwards the incidence of R&D projects involving combinatorial chemistry, target-based screening, genomics and genomic libraries doubled. All but one of the 26 per cent of specialists followed the leaders after 1992, mostly in bioinformatics. Strikingly, 54 per cent of the DBFs responsible for originating all these R&D projects were in four key US clusters: Cambridge, MA (18%), San Francisco–San Jose, CA (16%), San Diego, CA (12%), and Maryland (8%). Hence we see a highly globalised, hierarchical knowledge generation model in which leading edge research is initiated by multidisciplinary DBFs in clusters linking with (often many) large pharmaceuticals firms, research institutes and other DBFs as developers. It is plain that the clusters are increasingly the locus of knowledge generation.

Confirmation of the relativities of basic research funding in biosciences is provided in Table 6.1, which nevertheless throws up some puzzles. The leading position of Johns Hopkins University is not well reflected in the performance of the Baltimore and Washington cluster.

This is probably because the number of life scientists is boosted enormously by the National Institutes of Health that, while conducting intramural as well as extramural research, clearly are engaged in other activities that depress the ratio of NIH funding per life scientist (Feldman and Francis, 2003).

Table 6.1 Top ten National Institutes of Health funded research institutions, 2000–2003

Rank (2000)	*Institution*	*Funding 2000*	*Funding 2003*	*Rank (2003)*
1	Johns Hopkins University	$419.3 million	$555.9 million	1
2	University of Pennsylvania	$321.2 million	$434.5 million	3
3	University of Washington	$302.5 million	$440.9 million	2
4	University of California, San Francisco	$295.2 million	$420.7 million	4
5	Washington U., St Louis	$279.5 million	$383.2 million	5
6	University of Michigan	$260.4 million	$362.1 million	6
7	Harvard University	$250.4 million	$301.6 million	11
8	UCLA	$243.5 million	$347.0 million	8
9	Yale University	$242.7 million	$303.5 million	10
10	Columbia University	$226.6 million	$291.3 million	13

Source: National Institutes of Health

NB: New entrants to Top Ten 2003: University of Pittsburgh $348.2 (7th); Duke University $345.8 million (9th)

Other centres, like the University of Washington, UCSF, Harvard and Columbia are key contributory factors to the cluster strengths of the places listed. Moreover, North Carolina's campus at Chapel Hill is listed at position 13 with $207.1 million and Duke University is at 15 with $197.2 million. San Diego's outstanding contract research performance is explained by the presence of the Science Applications International Corporation[4] at 14 with $198.9 million and UCSD at 16 with $190.5 million. The Scripps Institute enters at 26 with $138.8 million and the Salk and Burnham Institutes lower but with a combined contribution of $72 million. In the case of Boston, Table 6.2 shows the other key research institutes that contribute to what it is justifiable to call its top bioregion status.

Thus it is clear that from the perspective of medical and bioscientific research the Greater Boston area has in reasonable proximity at least a further ten substantially funded specialist research institutions that, with nationally seventh placed Harvard University brought in for the year 2000 funding to the value of $1.09 billion. Examples of research and exploitation partnerships between these and DBFs in the region include: Curis and Harvard University (genetic signalling); Genzyme and Massachusetts General Hospital (HIV/AIDS), Dana-Farber Cancer Institute and Beth Israel Medical Centre (melanoma clinical trials), and Ariad with the Whitehead Institute, Massachusetts Institute of Technology and Harvard University (cell sequencing research).

The rise of research over science explains the rise of DBFs over big pharma in new knowledge generation. But DBFs still need large drugs firms to fund their discovery programmes. Thus despite most of the major companies

Table 6.2 Principal NIH-funded research institutions in Massachusetts 'after Harvard' 2000–2003

US Rank	*Institution*	*NIH Funding 2000*	*2003*	*Rank*
17	Massachusetts General Hospital	$180.5 million	$292.5 m.	17
22	Brigham and Women's Hospital	$162.5 million	$220.3 m.	25
38	Boston University	$108.2 million	$292.5 m.	13
47	Dana-Farber Cancer Institute	$87.2 million	$122.2 m.	51
53	Beth Israel Deaconess Medical Centre	$82.1 million	$103.3 m.	56
54	Whitehead Institute for Biomedical Research	$81.3 million	$100.9 m.	57
60	University of Massachusetts Medical School	$73.9 million	$96.1 m.	58
58	Massachusetts Institute of Technology	$75.0 million	$94.2 m.	64
74	Children's Hospital	$52.9 million	$85.4 m.	70
86	Tufts University	$37.5 million	$68.6 m.	82

Source: National Institutes of Health

experiencing drought in the product pipeline, they accommodate to new realities. For example, Swiss company Novartis announced in early 2002 that Glivec, its already successful chronic myeloid leukaemia (CML) drug also works for stomach tumours (Table 6.3). The US Food and Drug Administration (FDA) hastened approval for Glivec to save the lives of leukaemia sufferers and the product was granted orphan drug status in the US and the UK for gastrointestinal stromal tumours.[5] Novartis' methodology for developing Glivec was a prototype for the rational drug design or 'silver bullet' rather than 'chance' mode of drug discovery. This entailed genomic research to target the precise molecule giving rise to the mutation causing CML.

Actually, Glivec is rather unusual in that DBFs were not directly involved in the progress towards production of the therapeutic treatment. Rather, as may be seen in Table 6.3 the elapsed time from initial discovery to final approval was forty years. This is too long for DBF and venture capitalist survival of the normal ten years or so from proof of concept to hoped-for initial public offering of the DBF to the stock market. Today, combinatorial chemistry that allows vast numbers of compounds to be rapidly and systematically screened through high throughput screening (HTS), applied

Table 6.3 Institutional and corporate history of Novartis CLM treatment 'Glivec'

Date	*Institution*	*Name*	*Indication*
1960	U. of Pennsylvania	Nowell/ Hungerford	Blood Chromosome 22 'Philadelphia Chromosome'
1973	U. of Chicago	Rowley	C22 translocated to C9 discovery
1986–7	Whitehead Institute, Cambridge, MA	Baltimore	Bcr-Abl Protein: Tyrosine Kinase (Cell Regulator)
1992	Dana-Farber Cancer Institute, Boston	Druker	Bcr-Abl > CM leukaemia; mutant enzyme jams cell-signals discovery
1993	Oregon Health Sciences University	Druker/ Ciba-Geigy	Reagent and inhibitor for Tyrosine Kinase activities
1993	Ciba-Geigy	Leyden/Matter	ST1571 inhibitor compound (Glivec) selected
1998–2000	Novartis	Druker	Clinical Trials and FDA approval
1998		Nowell/Rowley	Lasker Medical Research Award

Source: *Journal of the National Cancer Institute*, 5 January 2000, http://www.nci.nih.gov/clinical_trials

also to methods for sequencing genes, speeds up the process. Research DBFs have advantages in swiftly bringing together networks of distinctively skilled researchers and technologists to target specific molecules.

It was university and private research institute scientists who conducted the fundamental research that resulted in Novartis releasing the world's first approved drug directly to turn off the signal of a protein known to cause cancer. In other words, the 'rational drug design' approach pioneered in cancer treatment by Novartis was really the culmination of university and research institute processes of origination, and final development through the company. Though it started as a 'chance' discovery of the Philadelphia chromosome, it evolved into a process in which precise molecular targeting became possible. This is expected to become an important, possibly the paradigmatic methodology in the post-genomic era.

In the Glivec case the importance of particular centres of research excellence in a strong biotechnology region like Greater Boston is evident. Random discoveries were made elsewhere early on, but key milestones were reached in the 1980s and early 1990s, first at the Whitehead Institute (Cambridge), subsequently co-leader of the Human Genome project, and the Dana-Farber Cancer Institute (Harvard), in identifying and then understanding the mechanism causing a mutant enzyme to jam the signal that normally prevents massive over-production of white blood cells, hence CML. This built on the prize-winning research elsewhere that first identified and then found where the key piece of missing DNA had translocated, which was a valuable research by-product. Thereafter, the main development technology moved with Brian Druker, the holder of the reagent that matched Ciba-Geigy's inhibitor compounds for Tyrosine Kinase activities, from Harvard to Oregon and Basel, Switzerland. This meant that Ciba had only to check its bioinformatics library. Such is the esteem in which Greater Boston is held by Novartis that in 2002 the company announced the location of its new $250 million Genomics Research Institute there. Comparable research clustering occurred in San Diego around the Scripps, Salk and Burnham Institutes, San Francisco in relation to the University of California Medical School and continues in, for example, Seattle in relation to the Fred Hutchinson Cancer Institute and Cambridge, UK in relation to many but particularly the MRC Molecular Biology Research laboratory which has hosted eleven Nobel prize winners in its time.

A key interim conclusion here is *not* that only a few places in the world (mostly in the USA) can ever pretend to global excellence in life sciences and their biotechnology exploitation capabilities. Healthcare, let alone agro-food and environmental applications are enormous, constituting some 20 per cent of GDP (DTI, 1999) in advanced economies. That biotechnology is far more thoroughgoing than an esoteric toolkit is testified to by the political scramble to find ways at regional level to facilitate development of relevant capabilities worldwide (see Cooke, 2005b). In the UK, for example, research performance assessments have resulted in the majority of life sciences research funding being allocated within the so-called 'Golden Triangle' among London, Oxford and Cambridge.

Clearly healthcare R&D expenditure generally and that focusing on biopharmaceuticals specifically has been on the rise, if US data are any guide to the trend. However, two factors may be bringing this apparent bonanza to an end, though neither signifies more than a reduction in the rate of increase or a possible stabilisation at approximately current levels. The first concerns the practices of 'big pharma' regarding R&D expenditures in the light of drying pipelines and the effectiveness of R&D productivity by DBFs. By contrast, inspection of the distinctive levels of R&D expenditure in Figure 6.1 show two things. First, the productivity advantage of DBFs is remarkable with over twenty new drugs approved per year by 2004 for an expenditure of about one-third the amount of R&D expenditure by pharmaceuticals firms. And second, by 2004 pharma R&D expenditure had begun to stabilise, in all probability preceding the beginnings of a decline, based upon recognition that the results of their massive expenditure were beginning to look nugatory at best, and far less effective than their smaller biotechnology counterparts.

These R&D sums being invested are of sufficient scale that more DBFs must today be conducting R&D, which firms of such scale might not have done to the same extent in earlier times. Meanwhile, although biotechnology R&D rose modestly from under $10 billion to about $15 billion, its rate of new drugs approvals doubled from 11 to 20 per year 1998–2004, while pharma's rate fell from 24 to 11 over the same period. Moreover many of pharma's 11 were modifications to existing products rather than totally new drugs.

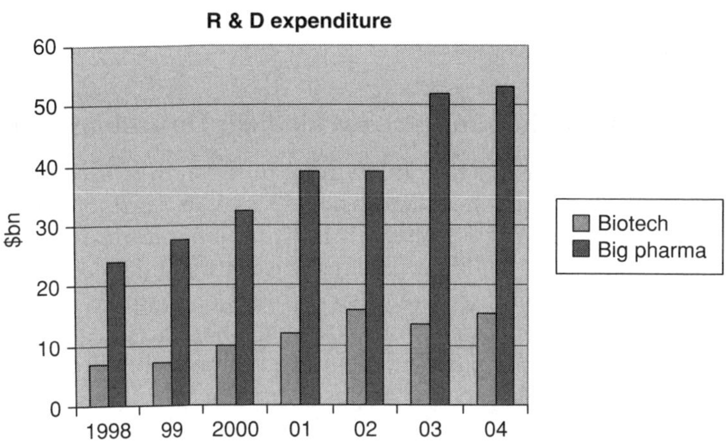

Figure 6.1 United States pharma–biotech R&D split

Source: *The Economist*, June 2005

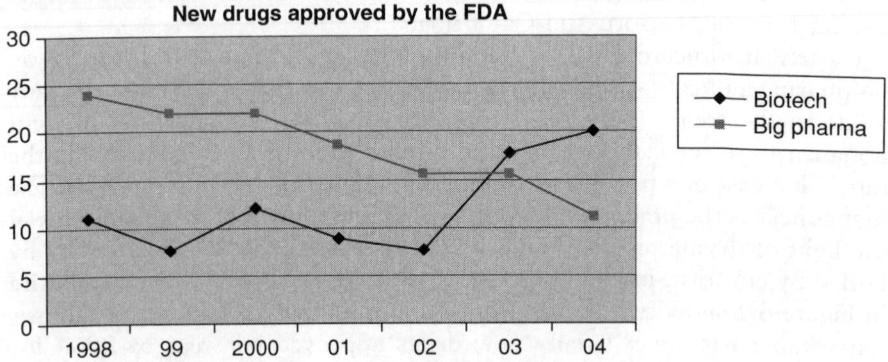

Figure 6.2 US pharma–biotech innovation performance

Source: *The Economist*, June 2005

The other straw in the wind of change in government expenditure may also be heralded first by big business. Thus oil companies like BP have begun to invest large sums of R&D funding in climate change research – $8 million at the new centre at Imperial College, London. They and even food multinationals are beginning to invest in serious R&D into biofuels such as corn and soya, while governments like Sweden have set exacting targets to eliminate oil dependency by the 2020s. Already substantial amounts of energy are derived there from birch trees either as oil substitutes or high calorific wooden pellets. Expectations are that governments will support such efforts by business by, to some extent, redirecting science policy expenditure towards publicly funded research into non-oil, renewable and especially biofuel research.

Strategic science policy and science funding: Do regions matter?

It has been argued thus far that knowledge production is becoming rather strongly regionalised in particular clusters. This is because of the growing importance of university and research institute laboratories to clusters of DBFs that exploit and commercialise basic scientific knowledge, with the support of venture capitalists and other business or legal services. Simultaneously, possibly more distant multinational pharmaceuticals companies are investors in milestone payments that fund the research in exchange for future expectations of licences or acquisitions. Thus the innovation system in this sector is both highly regionalised, for research and early exploitation, and highly globalised for development and, later, distribution and marketing. Is there evidence of a growth in *regional* science policy and consequent allocations? Moreover, if there is, does it take a different *form* from the state or pooled public (EU) one described above? And if globally networked regional clusters predominate in biotechnology, as there is abundant evidence they do (see Swann *et al.*, 1998 and papers in *Small Business Economics,* Special Issue, 2001) is this leading to new policy thinking? Is 'ground-up' strategic science policy and funding within specialist 'knowledge economy' regions occurring for other sectors too? In the space available this can only be explored for biopharmaceuticals, but the resulting evidence makes the hypothesis worth exploring further. The key point, tackled later, is that while regional technology and innovation policy are not new, regional *science* policy is, in the sense of regional administrations pressing for augmented basic research budgets or for central governments, where policy supremacy for science allocations lies, to find ways to assist less research-favoured regions *à la* EPSCoR (see below). Despite Ashcroft *et al.*'s (1995) critique of, for example, UK innovation policy, there is no evidence of any UK regional science strategies such as that promulgated by the Scottish Executive in 2001 because there weren't any at that time.

Strategic research (as defined by Irvine and Martin, 1984) has become less military and more civilian since the end of the Cold War (despite the rise of funding against bioterrorism). It is arguable that it has been forced to become more 'relevant' in the sense of more market-facing and ethically sensitive, as we have seen (Nowotny *et al.*, 2001). Importantly, it has allowed a repositioning of major science policy priorities towards the health rather than the defence of civilian populations. Even in the 1980s, the US spent nearly 50 per cent of its national academic and related research budget on life sciences. In 1987 US public healthcare R&D (funded by the National Institutes of Health) was $7.4 billion; by 1999 it was $14.8 billion and for 2003 it was over $27.3 billion, causing the following comment from the director of science policy at the American Association for the Advancement of Science:

> As a result, NIH is now the 800-pound gorilla of the research community, accounting for 42 per cent of all non-defense R&D, more than half of all federally-funded basic research, and nearly two-thirds of all federal support for R&D in colleges and universities.
>
> (Teich, 1999)

The 2003 appropriation was 51 per cent of federal basic research, a $3.7 billion increase over 2002 (Burnell, 2002). To this must be added portions of National Science Foundation, National Aeronautics and Space Administration and Department of Energy (human genome) research budgets. It is abundantly clear that healthcare is driving the US basic research-funding portfolio as never before. In the UK and Germany totals for biosciences research are of the order respectively of $2 billion and $1 billion annually. Senker and Van Zwanenberg (2001) estimated annual public EU life sciences R&D expenditure at $10 billion, about half that in the US.

The very large sums of research funding now going to regional biosciences/biotechnology clusters in the US and their younger equivalents in European countries give these locations both the resources and expertise to develop as implicit if not explicit 'centres of excellence'. Because of the perceived relevance and political virtue of life sciences research, health research budgets have mushroomed. Correspondingly, financial pressure on hospitals to treat, not conduct research on, patients (see below) has undermined clinical scientific opportunity to take advantage of the molecular biology revolution. Hence, such Centres of Excellence attract further funding, from their regional governments, from bilateral industry research investments. Thus Novartis and University of California Berkeley announced a $25 million licensing agreement for life sciences patents, but a year later in 2002 Novartis announced a $250 million Genomics Research Institute investment in Cambridge, Massachusetts to tap into the knowledge and talent arising from Human Genome research. There are also great resources from endowed institutes and medical foundations, which in the shape of those such as the Howard Hughes Medical Institute ($13 billion endowment) and in the UK the Wellcome Trust with a $21 billion endowment and $1 billion annual expenditure are of major significance. The latter explicitly operates a Centres of Excellence programme, which in the UK involves the universities of Glasgow (parasitology) and Edinburgh (cell biology) in Scotland, Manchester (cell matrix), Cambridge (cancer jointly with Cancer Research Campaign), Oxford (human genetics) and London (history of medicine). The Howard Hughes Medical Institute primarily funds researchers rather than centres but has laboratories at the universities of Maryland, California–Los Angeles, Washington–St Louis, Rockefeller and internationally.

In the US there are, of course, numerous other charitable and corporate medical foundations, the largest of which are shown in Table 6.4. Grants

Table 6.4 Top ten US medical research foundations and corporate foundations

Foundations	*Grants 1999*	*Endowment 2000*	*Corporate foundations*	*Grants 2000*
1. RW Johnson	$372 m.	$8.8 bn	Aventis Foundation	$41.6 m.
2. D&L Packard	$114 m.	$9.8 bn	Proctor & Gamble Fndn.	$30.4 m.
3. California Endow.	$91 m.	$3.5 bn	Merck Fndn.	$28.8 m.
4. Whitaker Fndn.	$50 m.	NA	Pfizer Fndn.	$25.5 m.
5. B&M Gates Fndn.	$48 m.	$21.2 bn	Eli Lilly Fndn.	$17.1 m.
6. Burroughs Wellcm.	$37 m.	$0.8 bn	Bristol-Myers Squibb	$15.8 m.
7. Rockefeller	$36 m.	$3.6 bn	Monsanto Fndn.	$14.0 m.
8. D. Reynolds	$35 m.	$1.3 bn	Medtronic Fndn.	$12.0 m.
9. Starr	$34 m.	$4.5 bn	Abbott Fndn.	$10.0 m.
10. WM Keck	$32 m.	$1.5 bn	Glaxo Wellcome	$7.0 m.

Source: Lawrence (2001)

from these augment the large-scale NIH awards and add further to the resource base of Centres of Excellence. Indeed the more a regional centre is designated as such, the more likely it is to attract further funding. The UK is unique in having a single charitable health research trust that spends per year an amount equivalent to the sum the top ten US charitable foundations spend together. This makes the Wellcome Trust a strategic science funder and policy maker in its own right in the UK. It is as, if not more, important as the UK government in determining bioscientific and health research expenditure flows, as a glance at selected highlights of its funding portfolio during 2000–2001 demonstrates (Table 6.5).

The key point to note from Table 6.5 and below is the trend towards regional Centres of Excellence and the manner in which in the UK the Wellcome Trust increasingly sees its role as regenerating parts of the national innovation system for health which has been damaged by a lack of policy or underfunding crises that have had negative effects on important parts of the national system, especially the National Health Service. Thus although, as Table 6.5 shows for the May 2001 entry, Wellcome grants under the Strategic Research Investment Fund scheme were diffused, a third of the £125 million, some £40 million was awarded to Leicester, Edinburgh, Leeds and Manchester (UMIST) universities, seven awards went to London, three to Oxford, two each to Sheffield, Cambridge and Cardiff, and a further two to Edinburgh. This reflects an emergent picture underlined in the UK government report on biotechnology clusters (DTI, 1999). That proposed regional centres in the above-named places plus regions with collaborating regional universities, as in Yorkshire ('White Rose' partnership) with Sheffield (2 awards), York (1 award) and Leeds (1 award) or in Scotland Glasgow (1), Dundee (1) along with Edinburgh and, in Wales, one centre based in Cardiff, could expect to be candidates for development outside the 'golden triangle' of Cambridge–Oxford–London.

Table 6.5 Wellcome Trust grant announcements 2000–2001

Date	*Headline*	*Funding*	*Recipients*
April 2000	Joint Infrastructure	£129 million	Ulster, Dundee Birmingham JIF
July 2000	Joint Infrastructure	£225 million	New (SRIF) Prog.
July 2000	Genome Bioinformatics	£8 million	Cambridge (Sanger)
October 2000	Genome Sequencing JV	£8 million	Cambridge (Sanger)
October 2000	C. for Molec. Medicine	£7 m. annually	Cambridge (Addenb.)
April 2001	Science Centres/ Infrastr.	£76 million	Scottish Universities
May 2001	Scientific Rsch. Facilities	£125 million	34 SRIF Grants
July 2001	Synchrotron	£110 million	Oxford
October 2001	Post-Genomic Research	£300 million	Cambridge (Sanger)
November 2001	Clinical Research Facility	£3.8 million	Southampton

Source: Wellcome Trust

This is predominantly where, given appropriate quality bids, Centres of Expertise may be expected to become Centres of Excellence,[6] and the move, first successfully demonstrated with the establishment of numerous Centres of Research in Stanford University (Gibbons, 2000), of specialist research away from large university teaching departments gets under way. This also applies to another big casualty of the changing research mode, clinical research in hospitals. In the UK, the latter is underlined in Wellcome's policy of funding clinical research facilities (CRFs) because:

> while the UK has the inestimable advantage of a National Health Service . . . the financial pressures on the NHS and healthcare reforms have created many obstacles to patient-oriented research. Not least of these is the enormous pressure on beds; patients requiring treatment obviously take priority, leaving no spare capacity that could be used by researchers.
>
> (Wellcome Trust, 2002)

The Southampton facility (Table 6.5, November 2001 entry) is one of five CRFs, the first opened earlier that year in Edinburgh; the others will be at Manchester, Birmingham and Cambridge. Thus regionalisation of special clinical research as well as medical and bioscientific Centres of Excellence is occurring as a policy initiative being implemented by the Wellcome Trust in response to changes induced in the traditional model by government health policy. Specifically it is seeking to maximise patient treatment capability on internationally low public expenditure, at the expense of clinical research. Thus the government's Culyer Task Force found ways of segmenting costs of NHS research and paying for it with a levy on healthcare purchasing. These funds are only to be used for recurrent costs, while foundations and research councils cover direct research costs. Absence of funding for fixed capital developments led to Wellcome Trust policy to invest in CRFs. They are modelled on US General Clinical Research Centres (GCRCs) of which there are 78. Their existence was initiated by US health insurance companies' refusal to pay for research in hospital beds, and now even outpatient clinics to which research moved are too busy for research. GCRCs cost $170 million per year to run and are funded through grant applications from the centres to the National Institutes of Health. The Wellcome Trust programme is funded at £20.5 million.

As government funding constraints have placed the NHS under ever greater financial pressure, clinical research capability is facing diminished capacity. Providers are thus becoming more entrepreneurial in their response and seeking significant funding from non-public sources, notably foundations like the Wellcome Trust, co-funded for CRCs thus far by university hospitals and local health service administrations (the NHS Trusts). Having large patient databases for research is a necessity in the new world of molecular medicine and rational drug design. It is not difficult to see the evolution of regionalised 'knowledge value chains' from basic university or research institute Centres of Excellence such as the Sanger Institute for genomic and post-genomic research, through to medical and clinical research at Centres of Excellence in university hospitals or schools of biosciences, to biotechnology institutes or centres, and gene centres or gene parks where exploitation and commercialisation are conducted by academic entrepreneurs interacting with clusters of DBFs.

The model can be observed in Greater Boston where all these facilities are in place, where over $1 billion in research funding alone is spent annually, much of it in collaborative partnerships among universities, special research Centres of Excellence (e.g. the Whitehead or Dana-Farber Institutes at MIT and Harvard), large hospitals like Massachusetts General or Brigham and Women's, GCRGs, incubation and successful start-up and more mature biotechnology firms like Ariad (founded by Nobel laureate David Baltimore, discoverer of the key cell regulator enabling Glivec to be produced by Novartis), AlphaGene, Dyax, Genetics Institute, Genome Therapeutics, Genzyme and Progenitor. Reference to the University of Siena database (Orsenigo *et al.*, 2001) shows that non-regional research partners are Aventis, Bayer, Bristol-Myers Squibb, Chugai Pharma, DuPont, Merck, and Pharmacia. These are increasingly engaged as investors, developers (though see below, section five), distributors and marketers for the products and services of the regional biopharmaceuticals innovation system in Massachusetts, focused on Greater Boston and with its epicentre at Cambridge (Cooke, 2001a). Massachusetts has for many years had a regional science policy to support with tax-breaks and other incentives high technology industry, once seen as responsible for the 'Massachusetts Miracle' in mid-sized computers, and now experiencing a resurgence through the promotion of biotechnology, biomedical and venture capital 'clusters' (Best, 2000).

Regional science policy: the basic model and some national variants

It has been argued thus far that bioscience underwent a cognitive, methodological and technological evolution that appears to have been expressed as an empirical punctuation point around 1992, though much of that change had been in the pipeline well before that. Some move into Mode 2 knowledge production was enforced as transdisciplinary research networks among research Centres of Excellence, academic entrepreneurs and successful start-up DBFs began accessing dynamic externalities in the form of knowledge spillovers from co-location in geographical proximity to exploit opportunities for rational drug design. However, such networks remained hierarchical both because of élite science (the 'star' system; Zucker *et al.*, 1998) and the continuing involvement of big pharma companies, less as originators than as developers of therapeutic solutions coming from biotechnology. The argument then evolved to discussion of a 'knowledge value chain' in life sciences spanning the arc from basic post-genomic and proteomic research through clinical research and treatment to innovation and commercial exploitation by clustered DBFs. It was then argued that this exists in only a few regions of the world, that Centres of Excellence are competitive and attract or possess large financial resources, and that their regional and technological innovation system governances have explicit or implicit science policies. Other governances will seek to emulate these leaders and have, indeed, begun doing so.

In the US and Europe the regional 'Clusters of Excellence' include Southern California, centred on San Diego, Northern California centred on Silicon Valley, and Massachusetts centred on Boston. In Europe such clusters are found, on a smaller scale than in the US, in the UK at Cambridge in the Eastern England regional development agency (RDA) area, possibly also Oxford (South East RDA) and Scotland (a triangle including Dundee, Edinburgh and Glasgow); in Sweden, Stockholm-Uppsala; and in Germany, Munich in Bavaria although two other BioRegios also exist (Dohse, 2001; Cooke, 2001a, 2002). In these innovative 'biotech cities', it is vital to recognise the regional innovation systems in which they operate. These supply finance (e.g. Bavaria sold its state energy company and established a high-tech fund, which invests in biotechnology [and ICT] research and commercialisation activity), and in Scotland, as we shall see, 'regional' funds for implementing its science policy for biotechnology are pooled among its RDA, parliament and university funding body the Scottish Higher Education Funding Council (SHEFC). As part of its modest move to regionalise administratively, the Swedish national government in 2001 established VINNOVA, the Swedish Development Agency for Innovation Systems with responsibilities to invest in *regional*, technological systems in biotechnology and other advanced technology sectors. There is also a unique cross-border policy and R&D body (Øforsk) to exploit the new bridge, by building an Öresund regional biosciences innovation system between Denmark's 'Medicon Valley' near Copenhagen and the Ideon Science Park bioscience cluster at Lund near Malmö, where AstraZeneca has a large R&D facility. We shall explore a further Nordic case in depth below, which is the case of Finland.

Observing these developments, regions with aspirations and some perceived or actual potential to emulate the élite can relatively easily be identified. Two in the US are worth exploring briefly. These are North Carolina and Maryland, both of which have emergent regional biotechnology innovation systems. In the case of North Carolina, there has been an effective science policy since the 1950s when the Governor got approval for the Research Triangle Park (RTP). This had the limited objectives of attracting R&D jobs with no presumption that synergies would flow among the facilities locating there. Subsequently major support was given to boosting the research capabilities of the three universities, Duke, UNC Chapel Hill and NC State, but especially the last two public ones. Duke's private endowment has ensured that its medical school and bioscientific research profile have prospered, while in the 1990s NC State was the recipient of major state funding to develop it as a Technology Campus with industrial R&D laboratories co-located with science and technology departments. In between, in 1981, the North Carolina Biotechnology Centre (NTBC) was established on RTP (as, at approximately the same time, were the NC Supercomputer and Electronics Centres). NTBC was not a research but a commercialisation facility. In early 2002 NTBC housed some thirty biotechnology businesses, including sites of Aventis, BASF, Bayer, Biogen, Eli Lilly and GlaxoSmithKline among the 90 in the RTP and broader Raleigh–Durham–Chapel Hill area, and 142 in the state.

Duke University Medical Centre is prestigious in basic and clinical biomedical research, with cancer and urology being leading fields for which the centre is ranked sixth in the US. Basic scientific research is wide ranging and operates in 38 laboratories including biochemistry, cell biology, genetics, immunology, microbiology neurobiology, pharmacology and cancer biology. The Duke Comprehensive Cancer Centre is accredited by the National Cancer Institute (NCI) and conducts clinical research, patient care and teaching in cancer immunobiology, prevention, detection and control, cell regulation and transmembrane signalling, cellular and structural biology, experimental therapeutics, molecular oncology, and cancer genetics. UNC School of Medicine is unofficially ranked twenty-second in the US and its strongest research field is biomedical engineering in which expertise is found in medical imaging, biomedical computer communication, medical informatics, neuroscience engineering, bioelectronics and sensors, physiological system modelling, biomaterials and real-time computer systems. The Lineberger Comprehensive Cancer Centre is one of the NCI national network of Cancer Centre Programme facilities specialising in biomedicine.

Maryland is also an important US centre for bioscience, hosting 210 bioscience businesses, half in research services, testing and contract manufacturing, two strong university systems, organisations like FDA and NSF, and a large number of federal research laboratories, notably the NIH system. Much of this activity is clustered along the I-270 'Technology Corridor' (Bethesda–Rockville–Frederick) and around Johns Hopkins University in Baltimore. The Howard Hughes Medical Institute research laboratories are nearby in Chevy Chase. NIH has 25 institutes and centres, including the US National Human Genome Research Institute at Bethesda and the National Cancer Institute at Rockville. The Johns Hopkins University is ranked first among US universities in receipt of federal R&D funds, the School of Medicine is first in receipt of NIH extramural funding, and unofficially ranked second nationally after Harvard Medical School. Its research expertise is focused on AIDS, biomedical engineering, cancer, clinical immunology, genetics, molecular biology, neuroscience, organ transplantation, and urology. The University of Maryland, Baltimore is a rapidly expanding biomedical research centre in partnership with the

University of Maryland Medical School System, the Veterans Administration Medical Centre, and the Medical Biotechnology Centre, specialising in molecular genetics and human molecular biology. There is also a UM Biotechnology Institute specialising in basic science applications to health, marine environmental and agricultural biotechnology, protein engineering and structural biotechnology.

Science and technology policies: market facilitating

Both states inherited buoyant technology markets from past public investment decisions. On the basis of these strengths and to a high degree influential upon them, both states are among the 13 in the US to have adopted statewide strategic science and technology policies, from between 1991 and 1995 (American Association for the Advancement of Science, 1999). The main goal of each policy has been enhancement of economic growth and improving standards of living by capitalising on the state's research base. Policies recognised the importance of sustaining and strengthening the R&D capacity of university research and training. In North Carolina, building on the success of Research Triangle Park, strategy focused on further stimulating exploitation of biotechnology and other technologies, and continuing to strengthen R&D capacity. New strategies emphasised stimulating indigenous entrepreneurship and promoting generative rather than redistributive growth. Maryland's strategy included recommendations for exploiting commercialisation potential of technology from its strong universities and federal research laboratories. Both Maryland's and North Carolina's policies were initiated by their governors, but others were the result of private initiative. Usually they began by analysing the strengths and weaknesses of the state economy and research infrastructure. In many cases they then went on to identify knowledge-based industry clusters, arguing that the state's economic base was passing from an old to a new economy character. Strategic policies were proposed to meet the challenge. In North Carolina's case this involved seeking input from six task forces and nine focus groups, using the North Carolina Alliance for Competitive Technologies as the governance body for the process. Both Maryland and North Carolina included specific outcome measures, such as quantifiable growth rate of technology businesses, industry support for university R&D, and new start-up companies. However the AAAS assessment of these policies was that they were insufficiently detailed and mostly failed to address issues of social exclusion.

In 2000, North Carolina published *Vision 2030: Science and Technology Driving the New Economy* based on a new approach emphasising *visioning* based on a statewide foresight options process. It will be shown later that this is becoming a more widely adopted approach to regional science policy. It involves cluster identification and regional stakeholding to attempt to commit industry and university administrations to invest in co-funding actual initiatives intended to be implemented. UNC Chapel Hill organised regional conferences, focus groups, cluster analyses, and global benchmarking, and produced the North Carolina Innovation Index. Recommendations included evolving a knowledge economy through supporting venture capital, public funding and tax incentives, marketing North Carolina globally as a knowledge economy, and designing a globally competitive R&D tax credit. Maryland's newest policy statement *The Maryland Technology and Innovation Index* was launched in late 1999 with similar style and content, using comparative benchmarking indicators addressing performance, dynamics and resources, and using the Maryland Technology Alliance of private sector, academic, federal and state government organisations as the catalyst.

Science strategy: science-led growth from below

Devolution in the UK has opened up a responsibility for democratically elected executives in Scotland, Northern Ireland and Wales to formulate science policies. Wales developed the EU's first Regional Technology Plan in 1994 and relies on an updated version under the Regional Innovation Strategy 2 programme from Brussels. The Welsh strategy has guided the establishment of expenditure patterns on technology and innovation under the Structural Funds Objective 1 action lines. This includes establishing a Knowledge Exploitation Fund, technology counsellors in universities and other infrastructures in support of *innovation* rather than basic science strategy. Northern Ireland was in a better position because of the existence of the Industrial Research and Technology Unit (IRTU) which, through its annual corporate planning process, designs technology and innovation, if not science policy in the region. It has now become a division of Invest Northern Ireland, the new integrated regional development agency. It is noticeable that, despite its peripherality and political troubles, Northern Ireland has developed a discernible science and technology policy not unlike but more piecemeal than Scotland's. Thus biosciences and ICT (especially telecommunications and Internet software) have been supported with research funding from IRTU, contest-successes for UK grants to enhance academic entrepreneurship, and the construction of nine incubators for the two target sectors. The necessity for regional science policy in both Wales and Northern Ireland is demonstrated by the evidence that at £34 and £24 per student respectively, the UK government's low investment in science funding there compares unfavourably with the £44 per head in England and £58 in Scotland. Research performance, measured since 1986 in the UK Research Assessment Exercise, explains the disparities to some extent. In this context, it is noteworthy that Northern Ireland's most significant biomedical research initiative, the University of Ulster's £14.5 million Centre for Molecular Biosciences, was equally co-funded by the Northern Ireland 'Support Programme for University Research' fund and Atlantic Philanthropies, an Irish-American foundation which, since 1982, has invested

$1.3 billion in higher education worldwide, 28 per cent of which was in Northern Ireland and the Republic of Ireland. The donor, Charles Feeney, also funded the Sinn Fein office in Washington. The university's vice-chancellor, bemoaning a 20 per cent decline in the region's funding for academic research through the UK system said: 'With devolution, we have found greater awareness of the importance of the research base than when we had direct rule' (Farrar, 2002).

Scotland was first in the UK to seize the opportunity to develop a regional science policy, its Minister of Science publishing in January 2001 *A Science Strategy for Scotland*. It was preceded by a report in 2000 from the Science Strategy Review Group and informed by Scotland's Science Policy Unit. The report shows that about £800 million is spent on scientific research in Scotland annually, and that Scottish universities won £141 million or 11 per cent of the UK Research Councils budget in 2000, a percentage point or so above the country's share of the UK population or GDP. The *Science Strategy* makes it clear that although Scotland's economy performs at about the UK norm, market forces alone cannot be relied on for economic growth to occur, but that Scotland's basic science advantage and government activity more generally have to be directed increasingly at sustaining world scientific leadership in a few feasible areas and raising commercialisation and entrepreneurship opportunities arising from science. The report prioritises bioscience and genomics, medical research, and e-science as the three areas of world leadership in basic science that the executive will support in particular. This means maximising targeted science research expenditure for these areas, including improving relationships between university and biological research institute research facilities in Scotland. To assist this the executive commits itself to investment in Scotland's joint Science Research Investment Fund.

Making an important commitment towards science funding in the UK as a whole, it aims to assist in setting in place a more transparent research funding methodology to ensure that underfunding of the kind widely perceived to have bedevilled UK science for decades cannot happen again. Scotland's problems of low industrial R&D and a high proportion of small businesses are to be moderated by connecting to economic growth initiatives such as the Scottish Executive's *The Way Forward – a Framework for Economic Development*; *The Knowledge Economy Cross-cutting Initiative*; and the *Digital Scotland Task Force*. Accordingly, it commits to keeping the 'Proof of Concept Fund' (see below), setting up a National Health Service Technology Transfer Office, revitalising UK-originated small business research and technology awards, assisting academic entrepreneurship, using Foresight to identify future scientific challenges and opportunities, and recruiting investment and scientists from overseas.

This was clearly a more interventionist set of commitments than were discernible in the more 'market-following' policies described previously. As in Northern Ireland and Wales, government has to do more because of market arrest, in a context of greater reliance on market forces where they are strong, which in the UK means, in effect, the aforementioned 'Golden Triangle'. Scotland was advantaged in bringing forward its fairly robust commitments to science support by preceding work done by Scottish Enterprise. The £40 million 'Proof of Concept Fund' established in 1999 is a good illustration. It allows scientists in the prioritised sectors, among which biotechnology and ICT were the first to benefit, to buy-out teaching and administration time to conduct research leading to academic entrepreneurship. The fund was formed from contributions by Scottish Enterprise, the Scottish Executive and SHEFC. Notably, no private co-funding was committed to the fund. Scottish Enterprise estimated in late 2002 that 14 biotechnology projects had been funded and that, since March 1999, 28 new biotechnology companies have been created, equivalent to a growth rate of 30 per cent per annum. This compares favourably with the European average of 17 per cent per annum over the same period. Scotland is home to 20 per cent of the biotech companies in the UK and is recognised as one of the fastest growing regions for start-ups. Thus far, policy to support commercialisation of bioscience has been successful; also we have seen that Scotland has received major funding, including both Research Council and Wellcome Trust grants, in support of its leading university Centres of Excellence and their research. Scotland now has a strategic science policy and it remains to be seen if the effectiveness shown without one can be enhanced consequentially.

Science strategy: science-led strategy from above

The last case to be explored is that of Finland, a small country that has emphasised the importance of developing Centres of Expertise in its regions, supporting university-centred basic research, commercial exploitation and cluster-building in biotechnology as it did with global success in relation to ICT and the rise of Nokia to global prominence in mobile telephony. The model is also one in which foresight and envisioning play a role in bringing about a consensus among business, academia and industry to invest in Centres of Expertise in locations that already show some comparative advantage. Centres of Expertise in Biotechnology arose from a Ministry of Education national research programme on biotechnology in 1987.

The aim was to develop four regional centres of biotechnological expertise by 1992, planned to be affiliated to those Finnish universities assessed to have the appropriate potential. The selected centres were at Helsinki, Turku, Kuopio and Oulu. The programme was evaluated and continued in 1996 then extended to 2000. Financing came from the Ministries of Trade and Industry, Agriculture and Forestry, and Social Affairs and Health as well as Education. Other centres were added such as Tampere and Seinajöki. The arrangement for enlargement of the network is one whereby if a municipality is sufficiently committed to serious long-term investment in biotechnology, by funding a number of chairs in universities, for example, and provided they pass exacting tests of expertise, they can become eligible for designation and funding as a Centre of Expertise. This has led to an excess of demand for such designation and the programme has been terminated in consequence. Centres specialise within biotechnology such that Oulu, Turku, Tampere and Kuopio focus on medical research and co-operation with the pharmaceutical industry. Helsinki and Seinajöki specialise in agro-food biotechnology and some agro-food R&D is also performed in Kuopio, Oulu and Turku (Academy of Finland, 2002).

Tekes, the state technology agency, has invested some $90 million in biotechnology, some 27 per cent of its total investment portfolio. The Centres of Expertise programme receives $4.1 million annually from Tekes and the Academy of Finland. Thus some 40 per cent of these two agencies' budgets is in support of biotechnology. Also, the National Programme for Research on Biotechnology, begun in 1988, invests an annual amount of some $13.5 million in biotechnology. Further expenditure on the Genome Research Programme and the Cell Biology Research Programme attracts $4.5 million and $1.8 million annually over six- and three-year programme periods respectively. In 1993, the Ministry of Education set a new Centres of Excellence standard, seeking to identify ten 'top units'. By 2000 26 had actually been established of which nine are in biosciences and biotechnology. In March 2001 a further $102 million rising to $151 million by 2006 was committed by Tekes, the Academy of Finland, Sitra (Finnish national R&D Fund), Finnish Bioindustries, and a substantial group of pharmaceuticals companies to 'Medicine 2000' addressing biomedicine, medicine development and pharmaceutical development research and technology.

Finland's commitment to evolve a strong biosciences and biotechnology capability is remarkable, with proportionately comparable shares of total national R&D budgets (some 40%) as the US. The fact that its agro-food firms are responsible for nutraceuticals innovations like anti-cholesterol product Benecol (Raisio Ltd), lactobacter drinks and UHT infant food (Valio Ltd), and xylite sweeteners (Danisco-Cultor Ltd) suggests where current strength lies. Orion Pharma and Orion Diagnostica are the two leading biotechnology players, the former having the leading Parkinson's treatment Comtess newly released, the latter targeting the global point-of-care market for *in vitro* diagnostic products. Orion Pharma collaborates with all the Finnish Centres of Excellence, but particularly the regional centres at the Universities of Helsinki and Kuopio, the Helsinki Biotechnology Institute, and increasingly with the Universities of Tampere and Oulu – the latter also being Orion Diagnostica's main research partner. The Finnish national innovation system is highly integrated but state-led with a knowledge value chain involving the Finnish Academy funding basic research, Sitra funding R&D, VTT conducting research and technology transfer, Tekes funding technology development, and Centres of Excellence in universities working directly with large firms, start-ups and spin-offs in clusters on state and locally funded Science and Medical Technology Parks like Hermia at Tampere, Oulu Technopolis, and Medipark or DataCity and BioCity at Turku. In the report on Finnish Life Sciences by Tulkki *et al.* (2001) these regional innovation systems are presented as worked models of the Finnish view of the functioning of Silicon Valley. The key difference is the involvement of large firms and public investment in the commercialisation process, substituting for an arrested market for key innovation support services. In this respect, it has been influential upon the German regional biotechnology clusters commercialisation programme BioRegio that similarly sought a 'corporatist' version of the 'basic research–academic entrepreneurship–venture capital' model that was pioneered in California (Dohse, 2001).

Conclusions

This chapter started with a question about the existence and observability of a new phenomenon, regional science policy. Its likelihood was implied by a number of important changes in global politics (ending of the Cold War), scientific research funding (major transfers from defence to healthcare), knowledge production (Mode 1 to Mode 2), bioscientific research approach (molecular biology), drug research methodologies (chance discovery to rational drug design), R&D leadership ('big pharma' laboratories to university 'Centres of Excellence') and innovation leadership ('big pharma' to DBF clusters). So developed are these relationships that inevitable ethical clashes have arisen. Thus troubled US firm Bristol-Myers Squibb, in trying to forge closer DBF links, invested $2 billion in 20 per cent of ImClone to access Erbitux, a colon cancer drug. One of ImClone's key shareholders was Martha Stewart, doyenne of US home-making who has been under investigation for insider dealing. The problem arose when FDA approval was withheld due to faulty clinical trialling by ImClone. This case has caused big pharma to question its absorptive capacity as it has become mainly marketer-distributor to DBFs like ImClone, or Celltech with a similar deal with Pharmacia, and Isis Pharmaceuticals with Eli Lilly. Attempts by Bristol-Myers to take over the development due diligence function with ImClone failed.

Clearly, with some 500 DBFs worldwide researching 1,300 compounds for new biotechnology products, it is no surprise that up to 30 per cent of big pharma R&D budgets are now spent on alliances with extramural partners when the top twenty pharmaceuticals firms in 2001 spent $28 billion on intramural R&D for a yield of only 28 new drug approvals. Pfizer, currently the world's largest pharmaceuticals firm has over one thousand alliances with DBFs and universities in response to the drought. So the knowledge-based clusters and the university or research institute Centres of Excellence at their hearts continue to be the pacemakers in molecular bioscience research and rational drug design. The chapter showed also how changes in funding régimes for healthcare, diminishing the traditional 'free-rider' system of clinical research in hospitals in favour of development of clinical Centres of Excellence, was hastening this process. Moreover, the vast amounts of Research Council and foundation research funding for Centres of Excellence accelerates it even further. Of course, such regional clusters, drawing on national funding to meet global market demand, are by no means ubiquitous. This is because, abundant though funding is, it is increasingly excellence-driven when it comes to funding allocations. Under such circumstances, alliance and partnership-based cluster governance has been shown to be an asset, and the functional presence of regional innovation systems with the full knowledge value chain in place and the lobbying and grantsmanship expertise that comes with a sophisticated science and innovation support system are invaluable.

While regions became familiar with the importance of regional innovation systems and strategies in the 1990s (Cooke, 2001c), the current evolutionary position in life sciences research requires learning to apply those skills to creation of the infrastructure of excellence that provides the foundation for regional technological systems, namely strong and varied basic and applied research capabilities. The logic of this points to the future rise of the formulation and implementation of regional science policy. This is evolving, as in the UK two Northern English regions established Regional Science Councils, and one, the North West, published its Regional Science Strategy in 2002. Finland decentralised development through its science strategy, Scotland built from below. Both are peripheral, with relatively weak market mechanisms but a strong science base. Each has developed focused science policies with strong public funding targeted at a few world-class life sciences sectors. North Carolina moved to the kind of foresight-led, envisioning, stakeholder plus action leadership process pioneered in Massachusetts, then adopted in Scotland and later in Germany's regional cluster solution to its biotechnology innovation deficit, BioRegio. National funding bodies have to respond by making more transparent the allocation of research funding, as demanded in Scotland's science strategy. They can devolve regional funds or designate annual tranches for regional science development. Most scientists react in horror to citing criteria of excellence and equivalence in regard to research grants, infrastructure funding or investment in the ever developing Centres of Excellence located in regional clusters. Safeguards would be needed to prevent the target inflation and excessive spread of investment revealed in the Finnish programmes. But equally, if regions show enterprise in mobilising strategic capabilities, they merit appropriate reward for so doing. This exists in the US Experimental Program to Stimulate Competitive Research (EPSCoR) that assists 22 less favoured states by accepting marginally lower-scoring science grant applications than elsewhere (www.her.nsf.gov/epscor). Regional science policy is beginning to prove a key precondition for the fulfilment of regional development visions in the knowledge economy.

7 The cluster model in biotechnology

Nodes in global networks

Introduction

It is clear from research conducted by Orsenigo *et al.* (2001) that many of the above-mentioned knowledge spillovers among post-genomics research and business partners involve proximity, but not exclusively so. Biotechnology and biosciences research are particularly global in reach even if the US is overwhelmingly the dominant product market, and indeed producer. They studied the network structure of projects to develop post-genomic therapeutic treatments. Although they failed to examine the geography of these networks, it can be reconstructed and Figures 7.1 and 7.2 do so. We shall see some differences in the first analysis, which represents the network linkages for advanced genomic and post-genomic research projects among Californian DBFs. Figure 7.1 shows more intra-regional networking among Californian DBFs than does Figure 7.2 and the Cambridge–Boston–Massachusetts circuitry. However, it is worth noting that a substantial portion of this occurs between members of California's two biosciences clusters as well as within each of them. Massachusetts firms are the largest regional group of partners outside California. The largest network is *Aurora* (San Diego), *Cytovia* (San Diego), *Axys* (San Francisco) and *Protein Design* (Silicon Valley) for fluorescent genomic screening. *Incyte*, the leading bioinformatics DBF, from Palo Alto, Silicon Valley partnered with *Scriptgen* (San Diego – also in the ArQule, Massachusetts network), *Genentech* of San Francisco, the University of California, San Francisco medical school (where *Genentech* began), and *Ariad* of Cambridge, Massachusetts also on innovative genomic screening technologies. Three issues regarding the proximate clustering and distant networking propensities firms and other organisations of consequence show.

First, it is noteworthy that while there are some localised cluster linkages among DBFs, it is likely that those seeking to evolve their research towards commercialisation particularly swiftly, building on knowledge spillovers from proximate interactions of a pecuniary or non-pecuniary kind (see below) will seek, in all probability, distant commercialisation partners mainly because of proprietorial considerations. Moreover, the talent or expertise necessary to progress a drug or diagnostics candidate along the

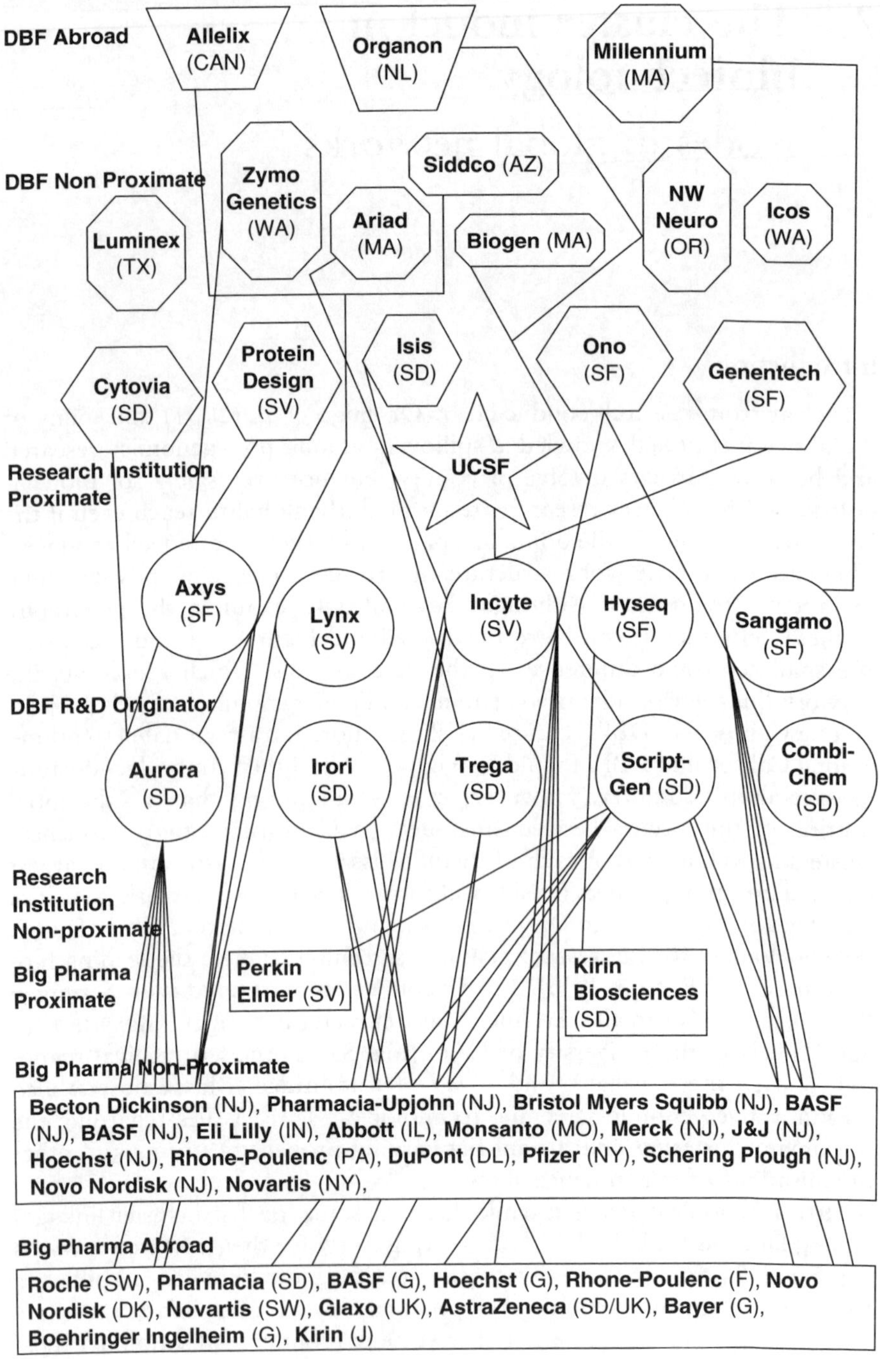

Figure 7.1 Biotechnology linkages: Californian DBF R&D organisations (based on Orsenigo *et al.*, 2001)

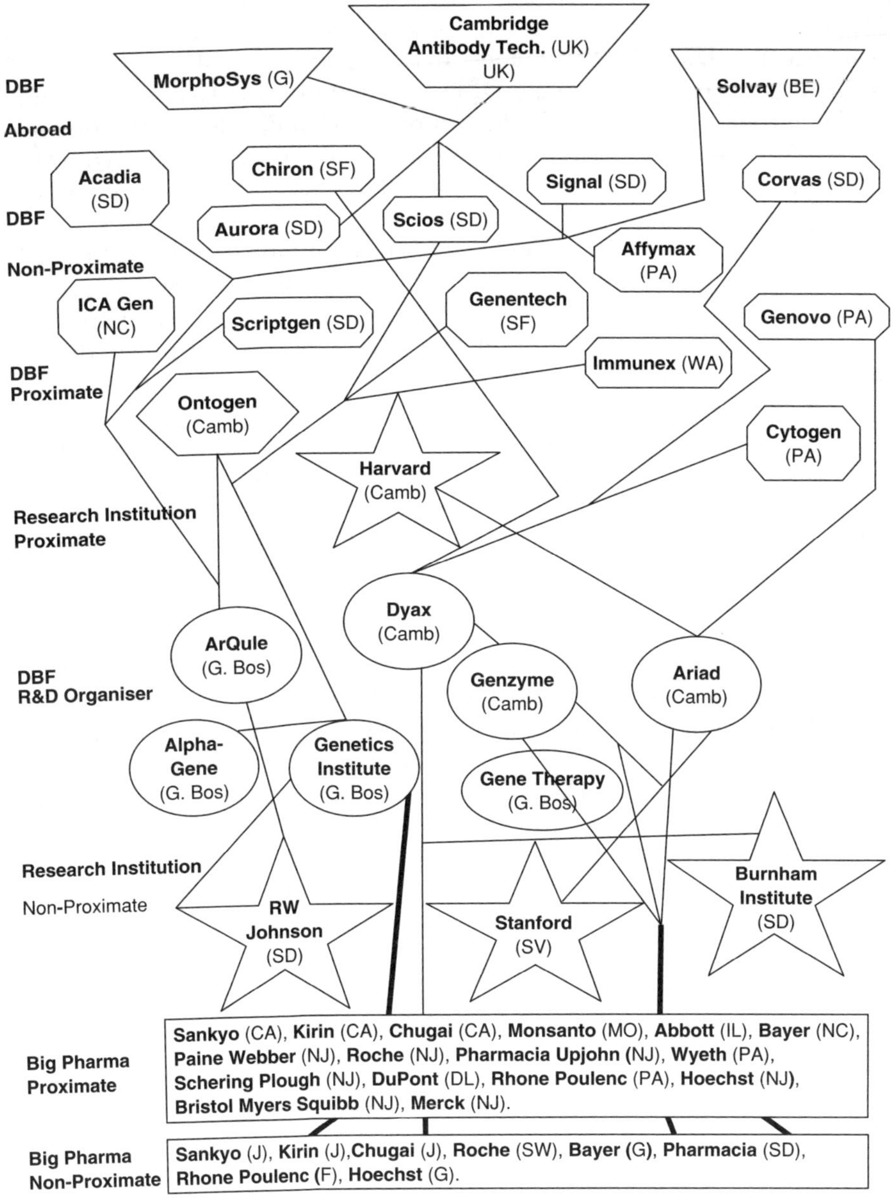

Figure 7.2 Massachusetts DBF-biotechnology research linkages

NB: Thin line 1 link; thick line 5 links

pipeline may, if our 'knowledge domains' concept in theory developed in Chapter 2 is correct, be asymmetrically distributed. Thus networking from Boston to San Diego or San Francisco would be hastened by knowing of the great strength of Californian DBFs in required fields like, for example, combinatorial chemistry, high throughput screening and bioinformatics. Indeed DBFs and many larger specialists globally dealt with Incyte (now Pfizer) at Palo Alto, in Silicon Valley, near San Francisco because it had the most comprehensive bioinformatics libraries.

A second feature differentiating the Californian biotechnological knowledge exploration set-up is the relative dearth of institutional connections to pure or applied research laboratories (see this in Table 3.3, Chapter 3). Unlike Massachusetts, there are relatively few specialist institutes, though the Scripps, Salk and Burnham Institutes are actively focused in bioscientific research in San Diego. Further, as can be seen in Figure 7.2, the last-named was involved in the Dyax project led from Cambridge, Massachusetts. In California it seems that technological advance is highly focused in DBFs and their networks of DBF scientific and technological partners.

Finally, regarding the commercialisation dimension, that is the advantages of proximity to firms that 'make it happen' and help to turn a scientific finding into a firm that commercialises a drug, treatment or diagnostic test, namely venture capitalists, specialist lawyers and consultants, there is econometric and case study evidence that these knowledge demands cause them to locate their investment a mean distance of one hour's driving time from their office base for the most part.[1] These are 'pipeline' type relationships, sealed from prying eyes and ears. This 'market' perspective focuses specifically on those contractual relationships where exacting transactions involve potentially large returns to partners from academe and enterprise (Zucker *et al.*, 1999). But the other 'social' position observes, albeit with social anthropological data, a different characterisation of the successful or potentially successful bioregion (Owen-Smith and Powell, 2004). That success is based on the practice of 'open science' transformed into a cluster convention of knowledge sharing rather than secreting.

Cluster analyses

The research by Zucker *et al.* (1998, 1999) tended to see only market relationships between biotechnology academics and entrepreneurs. These were contractually bound, exclusive and project focused, whether writing a paper or a patent. In particular they stressed how in the early days, academics ruled the roost in terms of expertise at understanding gene technologies. However, in light of the remnants of adhesion to Mertonian norms, this seems, on the face of it a somewhat one-dimensional, possibly *economistic* view of academe. Owen-Smith and Powell (2004) examined the Boston biotechnology cluster anthropologically, recognising there to be certain contractual norms of confidentiality but highlighting also those of open science and sociability among scientists. The following are key processes by which dynamic, embedded, place-based capabilities are expressed in research, knowledge transfer, and commercialisation of bioscience.

- The difference between 'channels' (open) and 'pipelines' (closed). The former offer more opportunity for knowledge capability enhancement since they are more 'leaky' and 'irrigate' more, albeit proximate, incumbents. Pipelines offer more capable means of proprietary knowledge transfer over great geographical distances based on contractual agreements, which are less 'leaky' because they are closed rather than open.
- Public research organisations (PROs) are a primary magnet for profit-seeking DBFs and large pharmaceuticals firms because they operate an 'open science' policy, which in the knowledge economy era promises innovation opportunities. These are widely considered to be the source of productivity improvement, greater firm competitiveness, and accordingly economic growth.

Over time the PRO 'conventions' of 'open science' influence DBFs in their network interactions with other DBFs. Although PROs may not remain the main intermediaries among DBFs, as the latter grow in number and engage in commercialisation of exploration knowledge and exploitation of such knowledge through patenting, they experience greater gains through the combination of proximity and conventions, than through either proximity alone or conventions alone. This is dynamic knowledge networking capability transformed into a regional capability, which in turn attracts large pharma firms seeking membership in the 'community'.

However, it can be argued that California has less of the fundamental genomic and post-genomic research and more strength in technology development and applications. This makes the region attractive to 'big pharma' as the long list of domestic and foreign pharmaceuticals listed as partners to Californian DBFs in Figure 7.1 testifies.

Of importance are three features of the Massachusetts system of genomic and post-genomic R&D projects aimed at exploring and exploiting advanced platform technologies for future drug development. First, while there are some localised 'cluster' linkages among DBFs, it is likely that those 'in the hunt' are in a competitive position locally because of the region's strong basic research, and seek distant DBF exploitation partners both because of proprietorial considerations, and especially because of the great strength among Californian DBFs, notably in San Diego, in the required combinatorial chemistry, high throughput screening and bioinformatics technologies. Locally, in Massachusetts, the main network was that linking ArQule, Ontogeny, Genetics Institute[2] and AlphaGene. But this is less than the Dyax-led phage display project network, the DBF partners in which include three Californian firms (Chiron, Scios – both also in the Genetics Institute network – and Corvas), along with three others, Cytogen in Princeton, NJ,

MorphoSys in Munich, Germany, and Cambridge Antibody Technologies in the UK. Both involve combinatorial chemistry, the first for molecular array mapping, the second for bioimaging of bacterial viruses (phage display).

Second, it is notable how common links are to be found between Massachusetts DBFs operating in this advanced field and the plethora of 'big pharma' partners that intend to license or acquire the successful biotechnologies arising from the research projects in exchange for milestone payments and other contractual relations with DBF project originators. In Massachusetts, links with foreign, including Japanese, pharmaceuticals firms is pronounced. Finally, testifying to the presence, to be itemised below, of leading research institutes in abundance in the Boston area, there are some links to such organisations too, although once more it seems that for technological applications in the field, links with Californian institutes like the Burnham and R. W. Johnson in San Diego are again favoured.

Some 'big pharma' have research institutes, like Abbott, Kirin, and Roche Biosciences in California, but headquarters are normally on the east coast, particularly in New Jersey, the state that once advertised itself as the 'Medicine Chest of the World'. Thus, as in Massachusetts, proximity is relatively unimportant for client relations between DBFs and 'big pharma'; it is – on a project-by-project basis, relatively important for a few key DBF originators; but for many *exploitation* research activities in the sample discussed, the greatest importance lies in combining strengths derived in the 'home-base' cluster with those found in other 'world-class' clusters. Also, in a competitive arena, it makes sense to link with firms outside the home base where they have complementary assets. However, it is possible to infer that there is a distinction between the Massachusetts and Californian clusters in biosciences, where the former may be stronger in basic medical and life sciences knowledge while the latter is stronger in clinical and biotechnological applications knowledge. These *knowledge domain* differences are predictable from the theoretical framework advanced in Chapter 2 in terms of path dependent knowledge asymmetries depending on initial knowledge endowments. Massachusetts' possibilities for strong bioelectronics disappeared with the eclipse of its mini-computer industry while California's thrived, but Massachusetts had far greater depth in its bioscientific research platform, giving it advantage in the genomics and post-genomics era.

Thus we see, beyond the information society and into the knowledge economy, the importance of Massachusetts' varied scientific research base. This makes it more of a biosciences 'megacentre' covering substantial parts of the public-sector biosciences 'knowledge value chain' than is the case in technology-intensive California. This in turn may help us a little in understanding why a leading drug discovery pharmaceuticals firm such as Novartis first invested $25 million to access biotechnology research results from University of California, Berkeley in the late 1990s, but in 2002 announced it was investing $250 million in the Novartis Institutes of Biomedical Research in Cambridge, Massachusetts. It is clear that the post-genomic research found in a centre such as the Whitehead Institute in Cambridge,

Massachusetts (partnered with Cambridge, UK's Sanger Institute in successfully completing the decoding of the Human Genome) gave Boston an edge over California in basic bioscience. This has also brought a wave of new entries into the Boston biotechnology cluster by the likes of Abbott Labs (formerly BASF Bioresearch), Aventis (formerly Hoechst-Ariad Genomics Centre), and Wyeth (formerly Genetics Institute), and openings such as AstraZeneca's Biosciences R&D Centre, Pfizer's Discovery Technology Centre, and Californian firm Amgen's R&D facility (formerly Immunex), also in Cambridge, Massachusetts. We have already proposed that Greater Boston is the leading biosciences megacentre, not because it has the heaviest medical or even bioscientific research budgets, but because it is presently the leading centre for exploration research.[3] This was conclusively shown to have happened since 2002 in Table 3.3, Chapter 3.

Switzerland's big pharma penetration of the San Francisco, San Diego and Boston bioregions

Roche and Novartis have been the world's most active acquirers and partners of Californian DBFs, though they are not alone in having such links. In San Diego alone Eli Lilly entered collaboration with the Scripps Institute, gaining rights of first refusal on discoveries in exchange for $50 million. Then, in 1986 Lilly acquired Hybritech, one of the earliest DBFs, only to dispose of it subsequently, thereafter investing in ownership of a leading diabetes therapeutic from Ligand Pharma. In 1996 Lilly entered a collaboration with Neurocrine Biosciences, and in 2001 did the same with Isis Pharma ($200 million). In 1996 Schering-Plough acquired Canji, a gene therapy firm with late-stage clinical trials. Johnson & Johnson also, like Lilly, began interaction early, entering a collaboration agreement with Scripps Institute, then in 1995 taking an 11 per cent stake in Amylin, enlarging this in later years. From 1995 to 1999 it also collaborated with Neurocrine Biosciences and in 1996 Johnson & Johnson created an integrated Genomics Research Institute in La Jolla, extending it in 2002 after signing a partnership deal with Maxia Pharma in 2001. Warner-Lambert (now Pfizer) acquired Agouron Pharma, the most successful San Diego biotechnology firm, employing 1,000, for $2.1 billion to access its HIV treatment. Pfizer began with a research collaboration in 1991 with Ligand Pharma, integrated Agouron into its worldwide operations in 2000 with the acquisition of Warner-Lambert, and in 2002, opened the first stage of a new research centre ($155 million) in La Jolla on the Agouron site. In 1999 Merck acquired Sibia Neurosciences, and invested in its research laboratories, expanding them substantially. In 1998 Ireland's Elan Corporation entered partnership with Ligand Pharma and in 2000 acquired Dura Pharma for $1.5 billion, centralising its biopharmaceuticals operations in La Jolla, then entering eight further biotechnology collaborations. Finally, Japanese pharmas Chugai and Sankyo established research facilities in San Diego in 1995 and 1998 respectively.

However, these activities pale into insignificance in comparison with Novartis' systemic network formation in both San Francisco and San Diego from 1977 in the former case and 1990 in the latter. These are summarised for San Francisco in Table 7.1. The symbiotic integration of Novartis with the San Diego bioregion warrants deeper exploration to show how small-country big pharma gain advantage from *embeddedness*[4] in the cluster. The strongest of these, agreed by Sandoz in 1992 and effective from 1997, is the ten-year research collaboration with the Scripps Research Institute (founded in 1955). It complemented in-house research by Novartis in immunology, neurological science, and cardiovascular diseases by giving first access to Scripps research results in these fields, and the right to commercialise 47 per cent of Scripps discoveries. Controversial because of infringing 'open science' conventions, the initially agreed financing had to be reduced from $300 to $200 million but Scripps researchers gained the right to submit research proposals to Novartis, effectively restoring the cut. A comparable agreement, continued until commercialisation by Novartis, was initiated by Ciba-Geigy with Isis Pharma. Accordingly, the AIDS-induced retinitis drug Vitravene was introduced in 1998.

Table 7.1 Novartis collaborations with San Francisco biotechnology DBFs and institutes (after Zeller, 2004)

Year	*Partner*	*Deal*
1977–1988	ALZA	Controlling stake – transdermal kits; divestment and collaboration, mktg.
1986–1995	Chiron	Joint venture, vaccines; research agt.in growth factors. Acquires 48% of Chiron in 1994; accesses gene sequencing
1990	Protein Design Labs	Research, anticancer antibodies
1991	Athena Neurosciences	Anti-spasm drug in-licensing
1991–1997	SyStemix	Collaboration immunology; 60% acquisition; stem cells and immunology collaboration; completes acquisition
1992	Affymax	Collaboration catalytic antibodies
1997–2001	Affymetrix	Acquires Gene Chip technology
1997–1998	Incyte Pharma	Bioinformatics software agreement
1997	Titan Pharma	Iloperidone global marketing rights acq.
1998	UC Berkeley	Licensing ag-bio discoveries, first refusal
1998	Stanford U.	Transplantation technology collaboration
1999	Versicor	Research collaboration, antibacterial NAS
1999	Rigel	Five drug target collaborations
2000	Axys Pharma	Combinatorial chemistry library access

By 1997, Novartis had implemented a new functional genomics strategy. In 2002 a new $250 million genomics research institute in San Diego was announced, named the Genomics Institute of the Novartis Research Foundation (GNF). The 200 staff complemented in-house research teams at institutes in Basel and New Jersey (later also Cambridge, Massachusetts for an equivalent investment[5]). Several GNF scientists also have faculty appointments at Scripps, and 17 per cent of post-doctoral researchers work with GNF scientists. GNF also gave rise to the Joint Centre for Structural Genomics (JCSG) and the Institute for Childhood and Neglected Diseases (ICND) funded as consortia by the US National Institutes of Health. Scripps and these other institutes are more entrepreneurial than universities. Intellectual property that Novartis is disinterested in can lead to spinout firms being formed with GNF board members.[6] Venture capital comes from the Novartis BioVenture Fund.[7] Established in 2000 with $100 million available, by 2002 the fund had invested in eleven Californian firms, four of which are in San Diego, by which time Novartis had moved the BioVenture Fund's headquarters from Basel to GNF in La Jolla, San Diego. Novartis collaborations with DBFs and institutes in San Diego are listed in Table 7.2. It is worth noting that many academic as well as DBF partnerships are also made by GNF in San Diego, including such firms as LifeSpan Biosciences, Molsoft, Syrxx, Sequenom, Xenogen and Immusol covering bioinformatics, genetic and proteomic mapping and oncology. In addition GNF partners the Salk Institute, UC-San Diego and, of course, the Scripps Institute. Hence, it can be seen that the extended Novartis knowledge chain is deeply embedded in the two Californian bioregions, but more in the form of a corporate, regional-technological innovation system in San Diego (and more particularly La Jolla), whereas in San Francisco it is more market than system-focused. We shall see shortly what the relationships

Table 7.2 Novartis collaborations with San Diego biotechnology DBFs and institutes

Year	*Partner*	*Deal*
1989	Cytel	Immunosuppression
1990–1995	Isis Pharma	Antisense technology
1992	Sibia	Amino acids receptors
1992–	Scripps Research I.	R&D
1995	IDUN Pharma	Neurological
1995–1997	Neurocrine	Multiple Sclerosis
1997	BioSite	Immunosuppression
1998	MolSim	Simulation technology
1998	Trega	Combinatorial chemistry
1998	CombiChem	Combinatorial chemistry
1999	Diversa	Seed research
1999	Invitrogen	Functional genomics

Source: Zeller (2004)

are in Massachusetts where, to repeat, a new Novartis Institutes of Biomedical Research opened in 2004. But before that, it is worth displaying equivalent information regarding the other Swiss big pharma representative, Hoffmann-La Roche, known commercially as Roche (Table 7.3). The approach taken by Roche involves partnering agreements with innovative, often relatively immature but specialist DBFs and these located rather more in San Francisco, although some are in San Diego. Many of these agreements are relatively recent, marking a change in Roche strategy from its acquisition of 60 per cent (today 58.2%) of the highly successful biotechnology pioneer firm Genentech in 1990 to a more flexible, short-term acquisition of technology and knowledge to fit the Roche product portfolio. Results for both Californian bioregions are given in Table 7.3.

Finally, note has been taken of the approaches of Novartis and, to a lesser extent, Roche as multinational big pharma companies from small-country Switzerland extending knowledge capabilities by integrating within leading global bioregions. What is the evidence from Cambridge, MA regarding 'embedding' – in this case in the world's leading genomics research and exploitation knowledge base? As we have seen, Boston outstrips both San Francisco and San Diego on many but not all biotechnology benchmarking indicators used here. In Table 7.4 the collaborative links of Novartis in the Cambridge–Boston bioregion are shown. The Roche website revealed none in this bioregion.

We have seen how well-provided the Greater Boston bioregion is with a diverse set of knowledge exploration, examination and exploitation institutions and firms. Thus it is clear that Novartis gains distinctive expertise, with the opening in 2003 and a further commitment of $4 billion investment over ten years, of the Novartis Institutes for Biomedical Research (NIBR)

Table 7.3 Roche collaborations with San Francisco and San Diego DBFs

Year	*Location*	*Partner*	*Deal*
1990	San Francisco	Genentech	60% stake in firm
1996	San Francisco	Gilead	Hepatitis C; Influenza
1996–1998	San Francisco	PDL	Inflammation drug
2000	San Francisco	Valeant	Anti-viral treatment
2001	San Francisco	Telik	Proteomics
2001	San Francisco	Gryphon	Anaemia
2001	San Francisco	Tularik	Therapeutic antibodies
2002	San Francisco	Lipomics	Metabolomics
2002	San Francisco	Kosan Biosciences	Polyketides (organic)
2003	San Francisco	Maxygen	Interferon; HIV
1998	San Diego	Agouron	HIV treatment
2000	San Diego	Pharmingen	Immunology license
2001	San Diego	Anadys	Anaemia
2002	San Diego	Syrrxx	Proteomics

Source: Roche and DBF Websites

Table 7.4 Novartis collaborations with Greater Boston biotechnology DBFs and institutes

Year	*Partner*	*Deal*
1982–1984	Genetics Institute	Immunology; interleukin-2; growth factor
1984–1986	Collaborative Rsch.	Cardiac infarction enzyme
1985–1989	Corning Glass	Diagnostics
1986	Biogen	Vaccine tissue
1989	Repligen	Retroviruses
1991	Dana Farber Inst.	Oncology and signal transduction R&D agt
1993	Procept	Auto-immune substances
1993–2000	BioTransplant	Xenotransplantation
1996–1998	Focal	Surgery materials
1997	Alexion	Viral vectors gene therapy
1997	Avant Immuno	Immunotherapeutics transplantation
1999	Cubist Pharma	Anti-infection technology
2000	Vertex Pharma	Protein kinases research

Source: Zeller (2002)

in Cambridge compared to San Diego. NIBR constitutes the primary pharmaceutical research arm in the company's strategy of post-genomic drug discovery, concentrating on the key therapeutic areas of cardiovascular disease, diabetes, infectious diseases, functional genomics, and oncology. With the aim to gain a better understanding of the molecular mechanisms of disease, the company is integrating previously segregated scientific disciplines, fostering interaction among scientists from both within and outside of Novartis and developing partnerships with academic research institutions and DBFs. This was underlined politically by the cities of Boston and Basel signing a cultural partnership in 2000. Novartis activities in San Diego are more to do with technologies like databases, combinatorial chemistry, simulation technologies and cloning technologies and even agricultural research. In San Francisco, the emphasis is more on stem cells and gene therapy, especially also technological applications in bioinformatics and GeneChips. This confirms the earlier suggestion that Cambridge–Boston is the world-leading post-genomics research bioregion while California's bioregions have strength in platform, diagnostic bioinformatics and gene therapy technologies linked to California's global excellence in computing and software, and that Novartis takes advantage from those distinctive knowledge categories in its global drive to redefine the process of drug discovery.

It is evident how Swiss big pharma conduct their research. This can be summarised for Novartis[8] – one of the leading organisational models for the modern pharmaceuticals industry – according to the following five points.

- Novartis innovation strategy is founded on three supports: internal research; linkage of internal and external research (e.g. GNF); and collaborations with external partners.

- Collaborations are varied but, first, they support in-house efforts to acquire therapeutic lead substances, drug targets and disease models, new discovery technologies, and entry into new fields. Second, partners supply active substances for integration into in-house development pipelines for clinical trials. A further activity is developing new drug discovery processes, as in the collaboration with Vertex.
- Collaborations with the Scripps Research Institute and Dana Farber Cancer Institute are to get new hypotheses of diseases and drug targets.
- From such inputs Novartis gets (e.g. with Vertex) exclusive worldwide development, manufacturing and marketing rights to clinically and commercially relevant drug candidates it develops with Vertex, earning royalties on collaborative products thus marketed.
- It is important to organise the optimal R&D mix, concentrating certain research activities in-house or externally according to corporate competences in the main global locations of Novartis Centres of Excellence. These are mapped onto globally excellent research bioregions.

If we compare that strategy with other big pharma, it is possible to see that Novartis has thought about and developed the 'organic' model to a fuller extent than any other big pharma. Most of the others are more market-oriented and opportunistic. For example, Glaxo makes partnership arrangements with DBFs but still considers in-house R&D to be a priority even though its pipeline is drying as in-house R&D fails to deliver significant returns on investment. Hence, in 2002, seeing biotechnology capabilities being increasingly monopolised by DBFs and universities, *Glaxo* restructured its global R&D effort into six main divisions with scientists liberated from corporate bureaucracy to some extent and incentivised with stock options and equivalents to emulate DBF research performance. Merck is also of the belief that in-house research is the preferable form by which corporate R&D should be organised. Alternatively, as we have seen, Merck engages in short term collaborations and will continue to do so, but meantime intends to control as much of the R&D chain internally to the corporation. Pfizer has mainly kept its pipeline from drying by acquisition. Accordingly, eight of its thirteen NASs were derived from its acquisition of Pharmacia. There is strong evidence that such mega-mergers as Pfizer and Glaxo have engaged in do not in the long term create greater R&D productivity. Hence Pfizer also has over 1,000 collaborations with DBFs and research institutes globally to attempt to cover the field for unexpected new hypotheses or discoveries.

One reason why pharma cannot reduce their collaborations with DBFs and institutes is that not only do they have declining capability in R&D but also in the kind of competence that it might be thought economies of scale would actually benefit. Sadly the following suggests that capabilities in exploiting new technologies, notably high throughput screening, are no better than research capabilities. The change from synthetic chemistry to biology as the epistemological basis for pharmaceuticals companies brought to the fore genetics, database, screening and bioinformatics technologies, allowing pharmaceuticals firms to utilise their natural economies of scale in experimentation if not R&D. The automated, mass-production analysis of patients and 'in silico' simulations using large databases to mobilise the combinatorial chemistry required to identify compounds that act as molecular disease-inhibitors gave them an optimistic future. The main reason concerned time-economies and 'throughput' capabilities that only the administrative capabilities of scale could satisfactorily master according to industry research conducted by the author noted below.[9]

Intriguing, therefore, to read the following analysis that re-directs our attention from the failings of 'big pharma' in basic research, resulting on their heavily increased reliance upon DBF capabilities, to newly emergent failures in precisely the capabilities identified as those of 'scale' that would triumph in the face of the DBF challenge. Traditionally many pharmaceutical companies had large industrial-chemical or consumer-products businesses to smooth out the cycles of drug research. The industry shed those businesses over the years to focus on prescription medicines, which brought higher stock-market valuations. But the rise of generic copying once patents expired meant that companies needed consistently to bring new products to market or lose market-share and profits. Hence the attraction of 'combinatorial chemistry', the new technology of the early to mid-1990s specifically identified in Nightingale's (2000) influential paper. The technology involved selection of chemical molecules, and their multiple combinations. Machines created thousands of chemicals almost overnight compared to the weeks humans took to do combinatorial chemistry. Robots connected elements of each chemical into small vials containing samples of a bodily substance involved in a disease – for example, the protein triggering production of cholesterol. If the two reacted in the desired way, a 'hit' was registered, latterly using luminescence technologies in some cases. This testing process is known as high-throughput screening (HTS). Most large pharmaceuticals firms install the new machines in sites formerly housing their laboratories, and many invested large sums on contracts with small companies specialising in HTS. For example, GlaxoSmithKline spent more than $500 million to buy a combinatorial chemistry company.

However, the automation processes failed to work as anticipated. The head of discovery chemistry at Bristol-Myers Squibb has referred to the first five or six years of the new technology as a 'nightmare', observing that many chemists became fixated on creating thousands or millions of chemicals for testing without thinking about whether any of them could turn into a usable treatment. Test tube compounds were broken down too easily in the human stomach, issues that 'craft-based' capabilities embedded in traditional chemists were usually assessed beforehand. Some results meant that scientists were forced to wrestle intellectually with chemicals that were almost impossible to deliver in humans. The struggle often led to serious delays in development schedules when, for example, a drug to prevent infection might work *in vitro* but fail to dissolve in water, the medium used in intravenous drips.[10] Since this débâcle, some firms have begun subcontracting this function to specialist screening firms, others have reduced scale. Bristol-Myers has tried to create a better mix of high technology and old-fashioned lab work with a screening machine capable of testing a million chemicals at once, taking as little as 10 microlitres of each chemical. Scores of mice are available nearby, allowing scientists quickly to test the 'hits' in animals. This avoids a problem whereby scientists spend months or years refining a compound in a test tube only to find it doesn't work in living things. Schering-Plough's main research laboratory has a machine that allows making several dozen variations of a chemical, all in the liquid state. The number is much smaller than older machines produce and the scientist can hand-pick the combinations and get higher quality. Hence screening is returning towards craft-skill and computational hybrids, having demonstrated total automation to be unworkable.

Bioregions in Germany

An important point in the recent history of biotechnology clustering in Germany was represented by BioRegio, the German federal initiative started in 1995 with funding from 1997 through 2002 to support new firm formation in biotechnology clusters. As a contest, it favoured well networked regions (actually cities or groups of cities) yet by 2003 BioRegio had considerably assisted the formation of new biotechnology businesses in Germany. BioRegio fits into a lengthy history of German federal and *land* schemes to support biotechnology, but differed from its predecessors by its success in giving a stimulus to the commercialisation aim which had often been the ambition of previous programmes, but never satisfactorily fulfilled. Other programmes, e.g. BioProfile and BioChance, complemented the firm formation emphasis of BioRegio later on. There was some debate indicating that BioRegio was unfair to innovative firms outside BioRegio areas, but it was clearly recognised that the geographical focus and development of firms in proximity to research institutes and local venture capital meant that whatever policy measure was adopted to boost start-ups, clusters would remain the distinctive mode of business organisation in biotechnology, as indeed they have worldwide (Dohse, 2001). This is because they offer vital external economies that promote productivity, innovation and new business formation, hence the competitiveness of biotechnology firms compared to slower-moving big pharma which, in Germany, had been heavily dependent on foreign biotechnology firms to enter the market.

Despite a growth in private venture capital associated with BioRegio, public funding remains important to the German venture capital industry and it is inconceivable that government intervention will not remain a central part of the future of German biotechnology industry development. German biotechnology firms and independent observers are in consensus that BioRegio helped close Germany's technology and, particularly, commercialisation gap in biotechnology, and the view was widely expressed that government support should be continued after 2002. BioProfile and BioChance continued this trajectory from 1999 onwards. Weaknesses remain in Germany's biopharmaceuticals industry, and Bayer, for example, is far less significant globally than it once was, while Hoechst was swallowed up in what is now the French firm Sanofi that acquired Aventis, the merged Hoechst-Rhone Poulenc entity, while BASF has withdrawn from medical biotechnology. Hence, for health-related biotechnology Germany's small business segment is relatively even more important than in the UK, where Glaxo is the second-ranked pharmaceuticals firm in the world and AstraZeneca is also a significant player. Thus the winning BioRegios of Rhineland, Rhein–Neckar–Dreieck and Munich are of particular interest and focus as Germany's leading biotechnology clusters. The strong biotechnology agglomeration in Berlin is profiled for comparative purposes.

Rhineland cluster

Given its history as a heavy industrial region undergoing major restructuring in coal, steel, chemicals and heavy engineering, particularly in the Ruhr, towards newer growth industries, the *land* of North Rhine Westphalia launched numerous technology-orientated initiatives such as new technological institutes with near-market research functions, technology parks and innovation networks, set up under the TPZ (Future Technology Programme) initiative. Amongst these was the *land*'s first biotechnology programme (*Landesinitiative Bio- und Gentechnik e.V.*) to support small and medium-sized enterprises. The biotechnology initiative was superseded in 1994 by the establishment of BioGenTech. This agency was a non-profit organisation with representation from industry, academia, trade unions and government. It acts as an intermediary body linking biotechnology start-ups, an expert network of 200 members, venture capitalists and partners from industry. A wide range of medical biotechnology areas was prioritised under its programme of support, and environmental and agro-food technologies were also supported.

Some 15 different networks were initially established, including a venture capital network of local but also internationally operational firms and groups, a competence and training network, and a management and coaching network. BioGenTech was seeking to become a commercial company and to sell services to the industry on that basis. In December 2002 it merged with two other life science agencies to form the Life Science Agency GmbH to that end. The research strengths of the *land* include the Max Planck Institute for Plant Breeding research at Cologne, around which larger (e.g. Monsanto, DSV and Agrevo) and smaller firms are clustered. In

1998 a letter of intent was signed between the governments of NRW and Saskatchewan, Canada to improve collaboration in the field of agro-food biotechnology. Also in Cologne is the Max Delbrück Laboratory (also part of the Max Planck Society) specialising in plant genetics. The Max Planck Institute for Neurological research, specialising in (photo)receptors, signal transduction and recombinant proteins is at Mülheim an der Ruhr. A Helmholtz Institute exists at Aachen (biomedicine and cryobiology), and a Fraunhofer Institute for Environmental Chemistry and Ecotoxicology at Schmallenberg. Altogether the *land* has some 167 research institutes, many employing relatively small numbers of researchers, but with representation across different elements of the biotechnology spectrum. In the early days of BioRegio funding most federal funding went to expanding companies while *land* funding went directly into start-ups. As Table 7.5 shows, by 2005 Rhineland occupied a lagging position similar to Rhein–Neckar (mainly Heidelberg) compared to Munich and Berlin regarding both firm and drug candidate numbers. As of 2000 and the peak of the biotechnology boom, the sectoral distribution of *companies* working in the field of biotechnology meant that 22 per cent were in diagnostics, 12 per cent pharmarceuticals, 7 per cent in agro-food biotechnology, 18 per cent in environmental protection, 9 per cent in filtration engineering and 10 per cent in bioanalysis. Within biotechnology the emphasis within Rhineland firms lies clearly on platform and diagnostic technologies rather than drug production, as the meagre pipeline shown in Table 7.5 indicates.

Finally, to what extent can clustering be said to be a feature of the Rhineland BioRegio or areas adjacent to it? Key institutional conditions present include: a strong science base, expanding numbers of firms, qualified staff, available physical infrastructure (e.g. the Rechtsrheinisches Technologie Zentrum, RTZ in Cologne, a 5,000-metre-square biotechnology incubator opened in 1999 with space for 50 firms and 450 employees and a high-quality central laboratory), availability of finance, business support

Table 7.5 Performance in core pharmaceutical biotechnology cluster firms 2002–2005

	Rhineland	*Heidelberg*	*Munich*	*Berlin*
Number of firms				
2002	29	31	63	55
2003	28	27	63	50
2005	26	27	59	54
Drug candidates 2005				
Pre-clinical	8	8	47	44
Phase 1	4	2	19	5
Phase 2/3	4	3	17	12

Source: Ernst and Young (2006)

services, a skilled workforce, effective networks and a supportive policy environment. However, the number of biotechnology start-ups was at a peak of 26 in 1990–1991, and in decline to a figure of 12 in 1994–1995, recovering over 1996–1997 after the announcement and including the first operational year of BioRegio (BioGenTech, 1998). The ITS Expert Mission report (DTI/FCO), 1998 reported 11 start-ups and eight company expansions since 1996 in biotechnology in the Rhineland BioRegio area. Because of the unrelated variety in the cluster it can be deduced that cluster-style networking through collaboration and partnership is not pronounced (this is discussed further in Chapter 9 and Omland and Ernst, 2004). As Table 7.5 shows, the number of core biotechnology firms in Rhineland remains 26 by 2005. Hence while the Rhineland BioRegio has many appropriate infrastructural conditions for stimulating the development of new biotechnology firms, whether a significant cluster of growing biotechnology firms will appear remains doubtful on present evidence.

Rhine–Neckar–Dreieck

Heidelberg is Germany's oldest university and has one of the best science bases for biotechnology. Two Max Planck Institutes, for cell biology and medical research, are in the region, as is the German (Helmholtz) Cancer Research Centre (DKFZ). The European Molecular Biology Laboratory, and European Molecular Biology Organisation are there, along with one of the four Gene Centres, the Resource Centre of the German Human Genome Project, two further medical genetics institutes and two plant genetics centres. Three other universities, Mannheim, Ludwigshafen and Kaiserslautern and three polytechnics complete the research and training spectrum. Two of Germany's leading big pharma firms are nearby, namely Boehringer Mannheim Roche Diagnostics (Mannheim), and Merck (Darmstadt). But the heart of the BioRegio is the Heidelberg-based commercialisation organisation, the Biotechnology Centre Heidelberg (BTH). This is a three-tired organisation consisting of a commercial business consultancy, a seed capital fund and a non-profit biotechnology liaison and advisory service. Central to BTH's functioning is Heidelberg Innovation GmbH (HI) a commercial consultancy that takes company equity in exchange for drawing up market analyses, business and financing plans, assisting in capital acquisition and providing early phase business support for start-ups. It is a network organisation, relaying information, partnering organisations seeking contact with local biotechnology companies and linking to research institutes and local authorities.

The key initial financing element of BTH is BioScience Venture. This was established by local big pharma and banks, managed by HI, and acts as a seed fund and lead investor in early start-ups. It also seeks international venture capital to finance second round developments. Assessments of project viability were made with advice from HI and BioRegio Rhine-Neckar e.V., the third element of BTH. The last-named seeks out commercial projects and recommended the most promising for BioRegio public funding support. Business proposals ran at some 50 per year from 1996, yet between 1996 and 1998, for example, only nine start-ups had been established, a figure that had risen to seventeen (including biochip and biosoftware firms) by 1999.

The total number of biotechnology SMEs (excluding start-ups) was 20 in July 1998. As Table 7.5 shows, the number of core pharmaceutical biotechnology firms in this Bioregion was 31 in 2002 and 27 in 2005. In Table 7.6, the core pharmaceutical biotechnology and diagnostics/platform technologies firms in the heart of the Bioregion at Heidelberg are shown. These data reveal that in Heidelberg, as in Rhineland, most firms are mainly platform or diagnostics firms though a majority of these are also active to a lesser extent in genomics, proteomics, etc. This is a sign of risk-spreading practice whereby immaturity in the risky, expensive but potentially rewarding fields is compensated for by capabilities in less expensive, even though more competitive and rapidly changing diagnostics and, for example, bioinformatics fields. With reference to bioinformatics, an interesting case is LION Biosciences that employed 600 in 2001 with sales of €250 million but which in 2006 employed 10 with sales of €25 million as a bioinformatics rather than a multipurpose genomics and bioinformatics company. In line with Table 7.5, Heidelberg until recently had some 27 firms, but a surge of new micro-businesses by no means all in genomics and similar areas raised that number to 46 in 2006. By that date the workforce had climbed to some 1,000 persons. The presence of substantial seed and venture capital resources in Heidelberg partly explains the rapid increase

Table 7.6 Heidelberg diagnostics and genomics biotechnology firms

Expertise	*No. of services*	*A firms*	*B firms*	*A/B firms*
A				
Analytics/Services	10			2
Bioinformatics	6	2		2
Diagnostics	8	3		3
Custom Production	5			1
TOTAL	29			
B				
Genomics	3			2
Proteomics	7		1	
Therapeutics	8		1	1
Tissue engineering	3		2	
TOTAL	21	5	4	11

Source: Baden-Württemberg Biotechnology Guide (2006)

NB: The A/B firms are categorised taking account of their main emphasis. This is usually A with fewer B services being offered (see discussion in text)

in firm formation. A further part is explained by the 'churn' of entrepreneurs returning to business after the post-2000 shakeout at a point when investor reluctance softened. Thus Heidelberg is claimed by its HI innovation centre to have 30 per cent of German venture capital, a sum that rose from €10 million in 1996 to €40 million in 2000 and €34 million in 2005. Moreover, the German High-Tech Foundation Fund – a public–private seed fund of €260 million – has helped improve the business environment for start-ups in Heidelberg as elsewhere. However, in conclusion Heidelberg Innovation perceives few cluster characteristics of vertical and horizontal collaboration among firms because so many are new and extremely small while many established interactions before 2000 were destroyed in the shakeout. So, despite the presence of some key ingredients for successful clustering, like abundant seed and venture capital, a regional innovation support system, knowledge and research institutions, and accommodation in proximity, the local experts do not consider that Heidelberg functions as a cluster. Moreover, the small and immature nature of the sector locally and the loss of some established firms and major downsizing of, for example, LION Biosciences mean it is not very likely to evolve towards cluster status in the foreseeable future.

Munich

The organisation responsible for managing development of biotechnology, BioM, is located at Martinsried, in south-west Munich. The area has become a biomedical research campus with 5,500 researchers working in biology, medicine, chemistry and pharmacy located there. BioM AG is a one-stop shop with seed financing, former administration of BioRegio awards and enterprise support under one roof. Seed financing is a partnership fund from the Bavarian state government, industry and banks up to €600K per company. BioM's investments are tripled by finance from *tbg* (public investment fund) and Bayern Kapital, the special Bavarian financing initiative. The latter supplies equity capital as co-investments. The fund has €200 million for supporting biotechnology activities. Bay BG and BV Bank-Corange-ING Barings Bank have special public/private co-funding pools, and a further eight (from sixteen) Munich venture capitalists in the private-market sector invest in biotechnology. By 2003, numerous start-ups had been funded to the tune of €120 million with a third of this coming from BioRegio sources. DBFs increased from 36 to 120 between 1996 and 2001 (Kaiser, 2003). BioM is a network organisation, reliant on science, finance and industry expertise for its support committees. It also runs young entrepreneur initiatives, including development of business ideas into business plans and financial engineering plans. Business plan competitions are also run in biotechnology.

While at the 2002 peak there were 99 listed firms in Munich, by 2004 there were 93, over a quarter of the total number of core biotech companies in Germany (according to Ernst & Young's report, there were 350 such companies in Germany in 2003). By 2005, of these, 59 were core (therapeutics) biotechnology firms compared to 63 in 2002 (Table 7.7). In 2004, there were 13 insolvencies in Munich including several well-established companies, such as Axxima Pharmaceuticals, a drug discovery company founded in 1997. Axxima has now merged with GPC. BioM's report mentions that venture capital for biotechnology decreased in Germany from €243 million in 2003 to €226 million in 2004 when only some €50 million were invested in Munich. Many young companies had completed the first two rounds of financing but have found it increasingly difficult to obtain further capital. The number of biotechnology jobs in the Munich region decreased to 2,230 (compared to some 2,600 in 2003). The average number of staff per company decreased from 27 to 24.

Seven start-up companies were founded in the region in 2004 (compared to only three in 2003). The bigger companies, Morphosys, Medigene, and GPC were successfully co-operating with pharmaceutical companies such as Novartis, Boehringer Ingelheim, or Eli Lilly, a development that enabled them to continue their research. The big pharmaceutical companies continued to show interest in the results of Munich's biotechnology research. Roche Diagnostics, for example, invested over €140 million in enlarging its R&D facility in Penzberg near Munich. Seventy-three new drugs were in various phases of clinical development, with five of them in phase III and one drug by Scil Technology in the approval phase. Medigene, which acquired the smaller Munich Biotech company, was the first German company to bring a new drug onto the market. The Munich DBFs continue to focus mainly on therapeutics. Their turnover was expected to rise to €250 million in 2005, compared to €170 million in 2004. The consolidation phase in the region continued, but know-how and expertise, especially in cancer research, and a well supplied pipeline were factors justifying the biopharmaceuticals sector's cautious hopes of cluster recovery.

The science base in Munich is broad, but with special expertise in health-related and, less, agro-food biotechnology. There are three Max Planck Institutes of relevance, in biochemistry, psychiatry and the MPI Patent Agency. GSF is the Helmholtz Research Centre for Environment and Health,

Table 7.7 Core pharmaceutical biotechnology companies in the leading German clusters

Location	*2002*	*2003*	*2005*
Munich	63	63	59
Berlin	55	50	54
Rhein–Neckar	31	27	27
Rhineland	29	28	26

Source: Kaiser and Liecke (2006a)

and the German Research Institute for Food Chemistry is at Leibniz Institute. There are three Fraunhofer Institutes, one of Germany's four Gene Centres, two universities and two polytechnics. The main research-oriented big pharma company following Sanofi's absorption of Hoechst AG as a subsidiary is Roche Diagnostics (formerly Boerhinger Mannheim). The work areas of the broader bioscience community include 3D structural analysis, biosensors, genomics, proteomics, combinatorial chemistry, gene transfer technologies, vaccines, bioinformatics, genetic engineering, DNA methods, primary and cell cultures, microrganisms, proteins, enzymes and gene mapping. In 1995, the state of Bavaria, the administrative district of Munich and local authorities of Planegg, together with the IZB Innovation and Start-Up Centre for Biotechnology Martinsried mbH, established one of the first private business enterprise companies in Bavaria. This is mostly a public initiative, with the IZB as a combination of incubator and technology park in proximity to the Gene Centre and two of the Max Planck Institutes conducting biotechnology research.

In common with the other BioRegio winners, the vertical networks from science through (public) funding to start-up are, in principle, strong, though, as elsewhere, given what looked a low-risk funding régime, due to public co-funding, the numbers of start-ups are not overwhelming, perhaps because of the quest for 'quality' start-ups in whom substantial sums may be individually invested. A further explanation for conservatism is that BioM AG, set up as a corporation, makes investments with its shareholders' (state, industry and banks) money. Most of the shares are held by banks seeking to earn high returns. They also use this method to learn about biotechnology, its risks and prospects. Thus, an already rather conservative system is further protected from risk by the influence of banking culture to ensure, as far as possible, risk avoidance. Hence, while BioM is the network face of the biotechnology cluster in Munich, its activities are ultimately orchestrated indirectly and directly by the banks, abetted by a fairly risk-averse, mostly publicly funded, venture capital industry and the local pharmaceutical and chemical companies (see Giesecke, 2000).

As to whether Munich's biotechnology constitutes a cluster, the answer is probably positive although there are conflicting reports as to whether three key firms commercialising biotechnology from Max Planck Institutes are interacting, collaborating companies or not. Dohse (2001) suggested that despite their common origin they were not strongly linked. But Clarke (1998) noted that two of them, MorphoSys and Micromet were collaborating on the development of an antibody-based treatment for micrometastatic cancer. MorphoSys was the first firm to receive a BioRegio grant and had previously collaborated successfully with Boeringer Mannheim on the development of a diagnostic reagent. MorphoSys' business strategy was to focus on the development of horizontal networking. They had no plans then to develop therapeutics themselves, aiming to remain a science discovery firm, but to let partners carry the risk of drug development. Thus MorphoSys worked with a variety of companies, minimising its risk profile, but potentially benefiting from substantial injections of capital from research funding, milestone payments and royalties. Already in 1999 MediGene planned to become a fully integrated biopharmaceuticals company, an ambition that is unlikely to be fulfilled, although as we have seen, MediGene marketed the first new biotechnologically derived drug since Reteplase from Boehringer Mannheim many years ago. MediGene was a spinout from Gene Centre in 1994 and initially raised €8 million venture capital, state and federal funds. Its expertise is in gene therapy for cancer and cardiovascular diseases, with a research partnership at the time with Hoechst. Academic-clinical partnerships included the Munich Gene Centre, the Munich University Hospital, German Cancer Centre at Heidelberg and, in the US, the National Institutes of Health and Princeton University. Its co-founder Horst Domdey gave up a chair at Munich University to become head of BioM. Another firm, Mondogen, was spun out of the virus research department at the Martinsried Max Planck Institute for Biochemistry. Its founder Peter Hofschneider was director of the department and had co-founded Biogen, one of Boston's oldest biotechnology firms, in the late 1970s. IZB and BioRegio, plus a McKinsey Business Plan competition, led to the founding of Mondogen. Martinsreid was said by Hofschneider to be unlike MIT and Cambridge as a cluster, but to have the 'seed crystal' for high-tech firms: the main obstacles were the different cultures between German scientists and venture capital 'speculators'.

The Berlin agglomeration

As the tables have shown, Berlin is a more significant 'critical mass' than anywhere outside Munich. The city of Berlin ranks today as Germany's number two biotechnology cluster. The emphasis, however, is not so clearly on pharmaceuticals as in Munich, but rather on several different areas such as bioinformatics, glycobiotechnology, DNA chips and regenerative medicine. To the region's disadvantage, however, may be its relative geographical distance from the principal actors in the industry. Yet the Berlin region (including Brandenburg) is host to seven founder- and technology-parks and more than 16 different research institutes. One comparative advantage of the city's location might be the existence of the largest research university hospital in Europe, the 'Charité', which raises more than €100 million annually in third-party funds and which finances more than 3,000 scientists working on some 1,000 projects. Development of the life sciences is a strategic priority for Berlin. Key institutions include Berlin's three universities: Free University Berlin, Humboldt University and Technical University Berlin. As noted, Charité is the largest university medical centre in Europe with 15,000 employees, 3,500 beds, and an annual budget of €1 billion. Charité resulted from the merging of the US-founded Benjamin Franklin Medical Centre at the Free University in the former West Berlin, and Charité, part of Humboldt University in the former East Berlin.

There is also a network of internationally recognised life sciences institutes, such as the Max Delbrück Centre for Molecular Medicine, the Max Planck Institute for Molecular Genetics and the Max Planck Institute for Infection Biology. Berlin also is home to the Robert Koch Institute (Germany's version of the US Centres for Disease Control and Prevention), the Federal Research Ministry, Federal Health Ministry, and life science-related trade organisations such as the German Association of Research-Based Pharmaceutical Companies. The number of dedicated biotech firms is comparable to that in Munich, which is to say about 109 firms in 2004. VC investments reached a peak in 2000 with €150 million invested, followed by very meagre years, with investments falling to €25 million in 2002, €45 million in 2003 and €53 million in 2004. In 2005, however, the accrued total of venture capital and other capital investments (e.g. IPO Jerini) was again expected to reach more than €100 million. The number of product candidates remains significantly lower than in Munich. The number of new firms being established in 2004 was also comparable to Munich, with about eight new start-ups.

Biotechnology clusters in the UK

The main UK biotechnology clusters concentrate close to globally significant research universities and at Cambridge, Oxford and more regionally, Scotland. Cambridge has reached the status that numerous international biopharmaceuticals laboratories have located nearby. Examples include Amgen, Beacon, Chiron, Genzyme, Medivir and Millennium, among others. However, while significant big pharma research infrastructure is less than that in, for example, Cambridge, Massachusetts, some are located within a 40-kilometre radius, such as Glaxo, Bristol-Myers Squibb, Dow Pharma, Merck, Sharp and Dohme, Novartis and Organon (Akzo-Nobel). Oxford's assets are mostly home-grown, but US firms Chiron, OSI Pharmaceuticals, Genzyme Therapeutics and Gilead are present in the cluster. Both Cambridge and Oxford, as will be shown, have leading public research institutes that complement the excellence of their university research centres and institutes. Scotland's cluster consists in the main of Centres of Research Excellence in universities and independent research institutes such as the Roslin Institute for transgenics research. Scotland's overseas life sciences companies include Quintiles, which is a US CRO. Key industrial bioinformatics roles include drug discovery and arrays work in companies such as CXR Biosciences, Organon and Arrayjet. It is noteworthy that firms are willing to pay premium technology park land rents in Cambridge, Oxford and Edinburgh of double the rate of equivalent quality space within a 40-kilometre radius. This is due to localised knowledge spillovers of various kinds, that even non-collaborator firms value highly (Cooke *et al.*, 2007).

The Cambridge cluster

In 1998 there were some 37,000 high technology jobs in the Cambridge area and these comprised 11 per cent of the Cambridgeshire labour market. South Cambridgeshire had about 66 per cent of these jobs while the city of Cambridge accounted for most of the remainder. The main high-tech activity remains R&D, supplying 24 per cent of total high-tech employment, electronics has 17 per cent; computer services, 13 per cent; scientific instrumentation, 8 per cent; and next comes biotechnology, fifth in line at 7 per cent. Probably the estimate of some 2,600 employees in biotechnology for the biocluster at the time was reasonable. The Eastern England region has a biotechnology industry association – ERBI – whose Biotechnology Sourcebook at that time showed the number of core biotechnology firms there to be 36 in 1998. By 2000 that number had increased to 54 and in 2004 to approximately 70. Among these, some 30 biotechnology firms specialised in post-genomic biopharmaceuticals, as shown below (Casper and Karamanos, 2003). The most recent (2006) ERBI firm estimate is that Cambridge is home to over 185 biotech companies (of which 109 were core therapeutics firms while 47 were biopharmaceuticals drug discovery firms active in genomic and post-genomic drug discovery). ERBI also announced that Cambridge hosts 20 per cent of the world's Nobel prize winners in medicine and chemistry, 17 of the UK's publicly quoted biotech companies and a quarter of the public biotechnology firms in Europe.

So, we may conclude that the core biopharmaceuticals industry consists of nearly 50 firms and the broader cluster (venture capitalists, patent lawyers, etc.) probably consists of more than 200 firms, including the core biotechnology firms. The Cambridge biocluster specialises in healthcare biotechnology. The two categories of 'biopharmarceuticals including vaccines' and 'pharmaceuticals largely from chemical synthesis' registered 14 and 9 Cambridge firms respectively in 1998, reaching 47 and 20 in 2006, evidence of the rapid rise of biotechnology over fine chemistry in the pharmaceuticals industry more generally. Examples of the former are Acambis, Alizyme, Amgen, Cambridge Antibody Technology (acquired by AstraZeneca for $1.1 billion in 2006), Domantis (acquired by Glaxo for $500 million in 2006) and Xenova, and of the latter, Argenta, UBC, and Mundipharma. In addition to these two key categories are direct biotechnology services like clinical trails, diagnostics and reagent supply (Table 7.8).

Detailed research on Cambridge's genomics sector has revealed the collaborative aspect of biotechnology innovation to be strong (Table 7.9). Regarding co-publication in journals, Casper and Karamanos (2003) showed only 36 per cent of firms were 'sole authors' while the majority partnered with firm founders, current incumbents and/or their laboratory. Academic collaborators are equally shared between Cambridge and the rest of the UK, with international partners a sizeable minority. Hence, the sector is associative in its interactive knowledge realisation, at least in respect of the

Table 7.8 Shares of Cambridge biotechnology and services functions

Biotechnology firm distribution		*Biotechnology services distribution*	
Biopharmaceuticals	41%	Sales & Marketing	29%
Instrumentation	20%	Management Consulting	23%
Agro-food Bio	17%	Corporate Accounting	15%
Diagnostics	11%	Venture Capital	15%
Reagents/Chemicals	7%	Legal & Patents	8%
Energy	4%	Business Incubation	10%

Source: ERBI

Table 7.9 Core pharmaceutical biotechnology companies in the leading UK clusters

Location	*2000*	*2004*	*2006*
Cambridge Core Therap.	54	70	109
Cambridge Genomic	–	30	47
Oxford Core Therap.	46	50	63
Scotland Core Therap.	24	30	38

Sources: ERBI; Oxford Biosciences Network; Scottish Enterprise

all-important publication of results that firms will likely seek to patent. Moreover, they will, in many cases, either have or anticipate milestone payments from 'pharma' companies with whom they expect to have licensing agreements. Yet of all known genomics DBF (dedicated biotechnology firm) foundings 1990–2002, totalling 30, only 9 were spinouts from Cambridge University laboratories, a further 5 were spinouts from the Medical Research Council's Cambridge Molecular Biology Laboratory, while others came from outside universities such as Imperial College, London University (3), the University of Wales, Cardiff (2) others and industry (6). Casper and Karamanos (2003) hold that Cambridge functions as an 'ideas market' with a good scientific image and much scholarly collaboration as well as academic membership of DBF boards and advisory committees. Yet a third of interactions are with academic and entrepreneurial partners elsewhere in the UK, and a further third are abroad, mainly in the US.

The infrastructure support for biotechnology in and around Cambridge is impressive, much of it deriving from the university and hospital research facilities. The Laboratory of Molecular Biology at Addenbrookes Hospital, funded by the Medical Research Council; Cambridge University's Institute of Biotechnology, Department of Genetics and Centre for Protein Engineering; the European Bioinformatics Institute; the Babraham Institute and Sanger Institute with their emphasis on functional genomics research; and the Babraham and St John's incubators for biotechnology start-ups and commercialisation, are all globally recognised facilities, particularly in biopharmaceuticals. However, in the Eastern region are also located important research institutes in the agro-food biotechnology field, such as the Institute for Food Research, John Innes Centre, Institute of Arable Crop Research and National Institute of Arable Botany. Thus in research and commercialisation terms, Cambridge is well placed in biopharmaceuticals and with respect to basic and applied research, but perhaps less so commercialisation, in agro-food biotechnology.

As we have seen, within a 40-kilometre radius of Cambridge are found many of the big pharma or specialist biopharmaceutical firms with which commercialisation development by smaller start-ups and R&D by research institutes must be co-financed. But these are mostly production not research sites, except one Glaxo facility, which is its global headquarters in London. Firms like GlaxoSmithKline, Merck S&D and Bristol-Myers Squibb in the 'big pharma' category are represented, and in the specialist biopharmaceutical sector are Amgen, Chiron, Genzyme and Gilead *inter alia*. Most of this is R&D, labour costs being too high for production in Cambridge or the London region. Thus on another of the criteria for successful cluster development, namely access within reasonable proximity to large customer and funding partner firms, Cambridge is also fortuitously positioned. Finally, with respect to agro-food biotechnology firms, Sanofi-Aventis, Agrevo, Dupont, Unilever and Novartis are situated in reasonably close proximity to Cambridge. Numerous contracts for outsourced knowledge from Centres of Excellence and discovery biotechnology firms flow among these and other large-scale customers accordingly.

Cambridge is relatively well blessed with science and technology parks, though the demand for further space is significant. At least eight of ERBI's 'biopharmaceuticals including vaccines' firms are located in Cambridge Science Park itself. St John's Innovation Centre, Babraham Bioincubator, Granta Park, the Bioscience Innovation Centre and Hinxton Science Park are all locally available. Most of the newer developments occur within short commuting distance of Cambridge itself, on or near main road axes. This is evidence of the importance of *access* for research applications firms to centres of basic research, reinforcing also the point that not everything concerning biotechnology must occur 'on the head of a pin' in the city of Cambridge itself. The final important feature of the biotechnology landscape in Cambridge and the surrounding Eastern bioregion is the presence of both informal and formal networking between firms and research or service organisations and among firms themselves. Cambridge Network Ltd was set up in March 1998 to formalise linkages between business and the research community, connecting both from local to global networks in a systematic way. It is mostly IT-focused, though some of this spills over into biotechnology, given its demand for IT equipment and opportunities for IT-delivered patient and clinician services through, for example, telemedicine. It is modelled on the San Diego CONNECT and BIOCOM networks that knit together that cluster. The Massachusetts Biotechnology Council operates similarly, as does Oxford BioSciences Network in the UK. In Basel–Strasbourg–Freiburg 'Bio Valley' performs this function. Elsewhere, as in Germany, formal cluster networking, financing and promotional organisations like BioM in Munich, Life Science Agency GmbH in Cologne and Biotechnology Centre Heidelberg (BTH) have this function. Of more direct relevance to the biotechnology community are the activities of ERBI.

This biotechnology association is the main regional network with formal responsibilities for newsletter, organising network meetings, running an international conference, website, sourcebook and database on the bioscience industry, providing aftercare services for bio-businesses, making intra- and inter-national links (e.g. Oxford, San Diego), organising common purchasing, business planning seminars, and government and grant-related interactions for firms. It was established before the East of England Development Agency was created in 1999, but ERBI was originally 50 per cent funded by the UK Department of Trade and Industry, the rest coming from membership subscriptions. Today EEDA part-funds ERBI's activities, EEDA's earliest contribution to Cambridge's biotechnology cluster. It is the architect to Eastern England's Healthcare Enterprise Hub, creating a knowledge pool of ideas and innovation for the 67 health trusts across the region with the aim of helping to commercialise intellectual property, working in close association with the region's medical device sector. EEDA supports regional ventures like BioConcept, which helps accelerate the development of early-stage concept businesses and has links with the Papworth bioincubator as part of a regional life sciences enterprise hub. EEDA's contribution is €1,000,000 over three years to boost Babraham Bioscience Technologies Ltd's innovative BioConcept in-house *accelerator* venture. EEDA also contributed €2.1 million to the €3 million cost of the Papworth Hospital bioincubator.

The Oxford cluster

The UK's once most successful indigenous biotechnology firm, British Biotechnology (BB) in 2003 was acquired by OSI of New York and Vernalis of the UK (as shown in Table 7.9 above). It was originally a spinout from the American firm Searle (part of Monsanto, subsequently Pharmacia, now Pfizer) when Searle closed UK operations in 1985. Two research directors established BB and by 1992 it had become the UK's first publicly floated biotechnology company. Its site is close to other Oxford-based biotechnology ventures such as Oxford Glycosciences (in 2003 acquired by Celltech, in turn acquired by UBC in 2005), Oxford Molecular and Xenova. By 1997 BB had become Europe's largest biotechnology company in terms of market capitalisation and R&D costs, and second to Qiagen of Germany in employment, with 454 employees. Thereafter Serono (now Merck, Sharp & Dohme) of Switzerland significantly outpaced both of them to become the world's third-largest biotechnology firm after Amgen and Genentech. In the late 1990s BB suffered a €2 million stock-market decline because of delays in gaining approval for its two leading products. Confidence was badly hit everywhere in Europe by this setback to its leading company. At the time, Celltech also received a big setback with Bayer's announcement of withdrawal from support for its septic shock treatment, leading to a 48 per cent price drop. Investor confidence was further damaged by poor drug trial results from Scotia Holdings and Stanford Rook. The BB clinical trials head, who left after questioning the effectiveness of the firm's cancer treatment in public, moved to Oxford Gene Technology (OGT) Operations, a commercial offshoot of Oxford University biochemist Edwin Southern's pioneering research in DNA biochip technology. In 1999 OGT was in legal dispute with Affymetrix of San Francisco over the invention of the DNA biochip, which the American firm had patented (settled out of court in 2001, Affymetrix agreeing to license OGT's technology). OGT, as noted, is a spinout from Oxford University which retains a 10 per cent stake in the firm, set up in 1995 to manage income from Southern's DNA microarray patents. Other Oxford firms of significance in biotechnology include Oxford University spinout Oxford GlycoSciences (acquired ultimately by UBC), once the world's leading proteomics firm when partnered with Incyte Pharmaceuticals of California, Oxagen (functional genomics) based in nearby Abingdon, and therapeutics company Oxford Molecular. Other firms in the extended Oxford cluster, which is aligned down the A34 highway corridor, include the former Oxford Asymmetry (now Evotec) in bioinformatics and Cozart BioSciences (immunodiagnostics).

Figure 7.3 shows how lucrative to the founders some of the best known Oxford biotechnology firms were in terms of the market prices for which

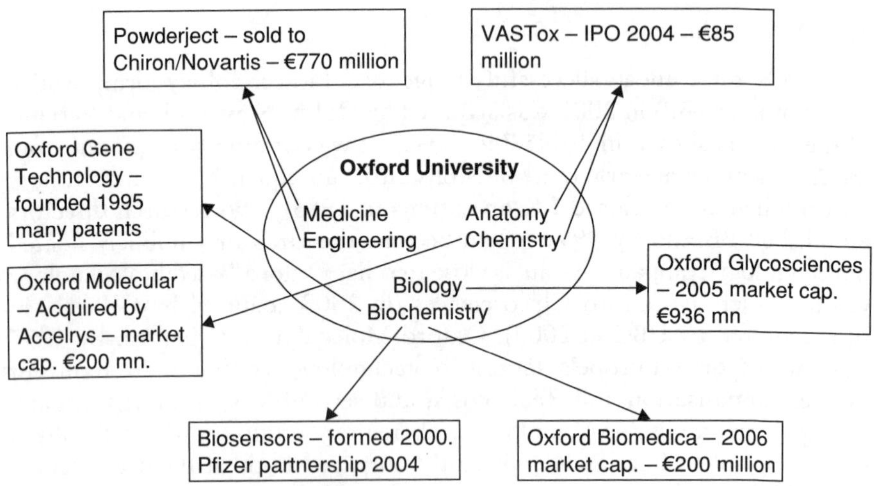

Figure 7.3 Bioscience millionaires from Oxford University

they were sold or otherwise engaged with the market. A number of newer firms are also located at Oxford Science Park, including Progenica (diagnostics), Oxford Therapeutics (drug development), Oxford BioResearch, Kymed (biopharmaceuticals) and Evolutec (drug discovery). Other centres such as the Medawar Centre and Abingdon Science Park also house biotechnology firms. One of the most successful Oxford DBFs is €1.5 million sales business Oxford Ancestors, a gene bank company that has famously traced humankind back to the 'Seven Daughters of Eve'. A genealogical company using DNA to track ancestry and explore surname links, researchers study human evolution and patterns of population origins using mitochondrial DNA and, more recently, Y-chromosome DNA, which are inherited down maternal and paternal lines respectively. They were the first to recover DNA from ancient bones and have been at the forefront of human fossil analysis.

Oxford University specialises in genomics with 50 per cent of university research groups in life sciences working in the field. The Nuffield Department of Clinical Medicine is rated one of Europe's leading centres in functional genomics. Interdisciplinarity is pronounced and biochemistry, pharmacology and engineering, for example, conduct joint biotechnology research. Lawton Smith's (2004) survey found that while most of the bioregion's firms are relatively new and small, the cluster is well established around a core of companies formed in the 1980s. The industry grew rapidly from the early 1990s. Sixty companies were established between 1991 and 2000 (70%) compared to twenty-five between 1981 and 1990. The cumulative annual growth rate is over 14 per cent. Two-thirds employ between one and fifty people, but a quarter employ more than that. Some 17 per cent are still at the start-up and seedcorn stage, but over half (52%) are in their second round of private funding. The firms are R&D-intensive with 42 per cent spending more than $1.8 million on this annually. Almost half anticipated doubling in R&D expenditure in the following year. The primary focus areas are biopharmaceuticals and diagnostics (34% and 17% respectively). The older companies mentioned above, like Oxagen and Xenova each of which employs over 250 people, trebled their employment between 1997 and 2001. The majority of these companies are independent while the rest include other independent companies, some of which are spinouts from other universities including London University, Bath and Surrey. Others are foreign companies such as Genzyme. Some one-third of Oxford's DBFs are foreign-owned, with the majority owned in the US (19 firms). Others have parents in France, Japan, Germany and Denmark.[11]

The science base, mentioned earlier, is a key attraction and Lawton Smith (2004) sees this operating in three ways. First, Oxfordshire's clinical research strengths in medicine and medical research are echoed in the biopharmaceuticals and diagnostics DBF specialisations. Diverse service firms occupy niche roles within pharmaceutical research, development and production, and some drug development DBFs conduct this alongside service functions. Those in the diagnostics sub-sector report little competition between co-located firms. Second, a quarter of the companies originated in Oxford University. Other firms like *Prolifix* were spinouts from elsewhere, in this case the National Institute for Medical Research in London. Through Isis Innovation, the university's technology transfer office, Oxford University has spun-out 17 firms in biopharmaceuticals out of a total of 32 spinouts by 'star scientists' including Professors Edwin Southern and Raymond Dwek (biochemistry), Brian Bellhouse and Mike Brady (engineering) and John Bell, Nuffield Department of Clinical Medicine. Of these, 3 are medical diagnostics and 14 are biotech firms.

The third advantage is that the Oxford cluster enabled DBFs to be close to a top university and other research institutes possessing 'talent'. The availability of clinical trial facilities was not a major sector challenge for Oxford DBFs, which hints that access to hospital units is satisfactory. However, research shows local information sources not being the most important source of information; conferences score highest and other important sources were the Internet, published sources and trade fairs. Oxford DBFs clearly place a high valuation upon codified knowledge, which fits its reputation as a strong clinical biotechnology bioregion, whereas Cambridge is thought of as stronger in medical biopharmaceutical, especially post-genomic research and exploitation. Clinical research is more patient-focused, including patient trials, diagnostics, software, bioinformatics, biochips and technology-oriented, as Oxford's DBF profile tends to show. Universities are thus not the prime source of information. They were ranked ninth along with local sector networks, national trade associations, technology transfer departments and independent research organisations. Least important were education and training councils, general business services, consultancy services, regional sector networks and government research associations. Yet in spite of the evidence of interaction, firms themselves saw proximity to Oxford University and the local research base as an unimportant factor in the development of interactions, other than those of an informal nature. The local labour market was the next most important attraction factor, especially the availability of highly qualified human capital. DBFs in such surveys appear schizophrenic, locating near research excellence but denying it by stressing higher-status global connections.

In addition to some already noted, the university has at least 14 interdisciplinary centres in the biomedical field. Moreover, strong links between the research base and hospitals, such as the John Radcliffe, ensure easy access for the clinical studies that are necessary for putting research into action. Oxford's hospitals have a long record of providing excellent 'testing grounds' by hosting commercial trials as well as UK National Health Service (NHS) trials. Healthcare, science and a culture of 'patient-centred' healthcare and clinical examination mean that Oxford occupies a leading position. The Institute of Molecular Medicine at the John Radcliffe Hospital,

Oxford (part of Oxford University's Clinical School) is a leading research institute which spins-out new firms in oncology and AIDS/hepatitis vaccines, in partnership with Isis, the Oxford University technology licensing and spinout support organisation, private investors and venture capitalists. Oxagen, in nearby Abingdon, was a spinout from the Wellcome Trust Centre for Human Genetics in Oxford, and Prolifix was spun out from the Medical Research Councils' National Institute for Medical Research in London. Yamanuchi Research Institute from Japan was the first privately funded biotechnology research institute to be established in the area (1990) but it closed in 2003.

In 2000, Oxfordshire BioScience, a network association for the industry, was established. Oxford had by 2004 some 50 biotechnology firms and 200 supply, service or intermediary firms and organisations. It has most of the features of a cluster, though it is still relatively small, including rising costs of industrial and domestic property, congestion and shortages of venture or other kinds of investment capital (partly caused by the negative British Biotech effect upon investor confidence). In a study by Mihell *et al.* (1997) it was shown that of 40 core biotechnology firms identified in 1995 (46 in 2000), 9 out of 12 interviewed were spun-out from the university or other public research base, and all firms interviewed had grown swiftly in employment and revenues in the previous five years. Collaboration among local firms and with both the local science base and more distant big pharma are central to firm strategy, though local networking between firms was not as developed as the other links, signifying the comparative immaturity of the cluster. At the time of the survey, in 1996, 2,200 were employed in the 40 firms identified, and new firms were forming at a rate of 3 to 4 per year. More recent research reveals the following information about the Oxford bioregion: 14 diagnostics/biomedicals/platform development firms, and 46 true biotechnology companies in 2000. It is worth noting that such distinctions were more blurred by 2006 since treatments often required new diagnostics and their therapeutic interdependence had evolved substantially. Each of Oxford's DBFs was principally involved in the biopharmaceuticals sector and employment overall was some 3,250 people (Lawton Smith, 2004).

Scotland's biotechnology cluster

The biotechnology industry in Scotland is growing steadily, with more than 100 biotechnology companies employing some 3,600 individuals. The vast majority of activity is highly geographically concentrated in the Dundee/Edinburgh/Glasgow triangle called the Scottish Biotech Cluster. Scottish Enterprise, the Scottish government's economic development agency, established a biotechnology team in 1994, and has pursued a focused policy of developing the bio-cluster on a number of fronts. The Scottish Enterprise model has involved supporting research institutions in their efforts to attract commercial research and to familiarise academic researchers with the requirements and rewards of working with company funding as opposed to government funding. In addition, Scottish Enterprise supports the technology transfer infrastructure by identifying promising technologies at home and abroad, supporting inward investment, within academic institutions and developing exploitation routes for these technologies. By 2002, Scotland had already lured a few Boston, Massachusetts firms to its corridor, including Braintree-based Haemonetics; Bedford-based Millipore; and Inverness Medical, a subsidiary of Inverness Medical Technology (itself a spinoff of Waltham-based Inverness Medical Innovations sold in 2002 to Johnson & Johnson).

Scottish biotechnology has numerous centres of excellence. The biocluster has benefited from leading research universities, which are world leaders in some areas of research, such as Oncology at Dundee University, Neuroscience and the Beatson Institute for Cancer Research at Glasgow University, and the Centre for Genomic Technology and Informatics, and ESRC Innogen (bio-innovation research) at Edinburgh University. Other important research organisations include the Roslin Institute (nuclear transfer technology) and the Moredun Institute (veterinary biotechnology), as well as research hospitals such as Ninewells Hospital in Dundee. Here Professor Sir David Lane discovered the p53 'guardian angel' anti-cancer gene. The Scottish life sciences industry specialises in pharmaceuticals/drug development, medical devices, diagnostics; agro-food biotechnology (animal), waste management; industrial biotechnology; bioinformatics; basic discovery/R&D; field/clinical trials and manufacturing. This 'cluster' attracted financial investment estimated at $35 million in the year 2000, rising to over $200 million in 2005. There were 18 life science companies specifically focused on the drug discovery and development process in Scotland. Key areas of strength included:

- The production of new antibiotics,
- Development of innovative technologies for the diagnosis and treatment of cancer,
- Methods to accelerate drug development and reduce product attrition,
- Development of human antibody-producing cell lines for use against infectious diseases, and
- Products for psychiatric and neurological disorders.

Some examples of where these leading-edge activities are being undertaken:

- CXR Biosciences – accelerate drug development, reduce product attrition, and to attempt 'rescue' where appropriate for molecules with unexplained preclinical observations;
- Auvation Ltd – develop innovative technologies for the diagnosis and treatment of cancer;

- ProStrakan – engaged in the research, development and commercialisation of therapeutic advances mainly in the treatment of bone and skin disease;
- Cyclacel – develop novel therapeutics for treating cancer, by disrupting the cancer cell development lifecycle; and
- Organon – carry out research in the areas of psychiatry, artherothrombosis, anaesthesiology, and analgesics.

Other key companies include: Ardana BioScience, ExpressOn Biosystems, Geron Bio-Med, Stem Cell Sciences and Scottish Biomedical Ltd. This set of companies has a remarkable support platform and network including six drug delivery and formulation companies; over 40 pharmaceutical clinical trials support and contract research organisations; and over 200 organisations involved in the support and supply of the drug discovery and development process.

The main reasons for bio-cluster growth are the following. First, Scotland presents a conducive infrastructure with the establishment of relevant incubators and science parks such as Edinburgh Bioparks and BioAdventures near Glasgow. Second, there exists a combined science and business talent pool that is supported by up to 30 private bio-investors, notably the business angel network which is co-funded through the Scottish Co-investment Fund, accordingly outperforming many other business angel networks (see Chapter 10). In addition to 'angel' investors, Scottish and UK government grants and loans assist capital investment, and with the economic development agency Scottish Enterprise having responsibility for cluster policy, providing equity investment and talent recruitment and formation, Scotland has a well-integrated innovation and enterprise support system. Finally, a collaborative attitude among universities, between them and with R&D centres and among firms prevails in Scotland, considerably assisting the evolution of the bio-cluster. Examples of Scottish Enterprise support are the €40 million 'Proof of Concept' fund – which enables professors and other research leaders to access funding to 'buy out' their academic teaching and administration time to concentrate on the 'proof of concept' for the idea they are seeking to commercialise – and the Intermediary Technology Institutes.

The Intermediary Technology Institutes (ITIs) were announced in 2003 by Scottish Enterprise as a €600 million package over ten years. Their brief is to leverage the excellence of the Scottish university research base by increasing the rate of commercial exploitation of research knowledge. This may be achieved through registering and licensing intellectual property rights (IPR), selling such IPR (trade sale), or establishing a start-up business to take such IPR to market, as appropriate. Three ITIs were established, focusing upon life sciences, information and communication technologies (ICT), and energy. These had been in operation for some thirty months as of March 2006. Each has a budget of €20 million per year for which an annual operational plan is produced within the framework of a rolling three-year strategic plan. Each ITI has a Board of Directors and they have a joint Executive Committee of the three ITIs plus the Chief Executive Officer (Operations). Reporting to Scottish Enterprise is through these structures, with the main interface person being the Board Chairman, although links also exist with key contacts within the Scottish Enterprise managerial structure as appropriate. Thus the ITIs are arm's-length bodies funded by Scottish Enterprise but with a high degree of independence from that agency.

ITIs are a tailored institutional innovation aimed at contributing in three key sectors, identified principally by Scottish Enterprise as growth industries of the future in which Scotland has some global competitive advantage. Promotion of stronger innovation capability in these sectors fits with the key aspiration of the Scottish Executive's vision of a 'Smart, Successful Scotland', the basis for which is seen to lie in the country's knowledge-based economy. It is widely agreed that the three sectors identified for hosting an ITI are among Scotland's strongest in terms of global competitiveness. In Life Sciences, transgenics, stem cells and cancer treatments are clear comparative strengths. The ITIs have the intended institutional role of knitting together important parts of Scotland's innovation system that hitherto had been perceived as rather disconnected.

The manner in which this is improved differs somewhat according to the focus of each ITI. One innovative aspect of the ITI approach is to 'walk the talk' regarding *cluster-building*, a policy approach of considerable provenance since Scottish Enterprise's pioneering adoption of this economic development methodology from the early 1990s. The cluster idea involves taking policy actions that enable businesses to gain from complementarities, collaborations and 'knowledge spillovers', especially where related firms operate in geographical proximity. From some viewpoints this may smack of exclusivity or even construction of quasi-monopoly conditions where a group of firms enjoy 'club' benefits. But building the cluster may, in fact, mean disrupting such equilibrium conditions by, for example, introducing a competitor. This occurred in life sciences and caused anxieties from incumbents, from which new knowledge of the problems of cluster-building arose, but nevertheless continued to be justified in terms of 'building the cluster' rather than protecting established interests. In other respects, the cluster perspective, while being perceived as a worthwhile business development model, is also seen as a policy retro-model, meaning it is based upon 'a compartmentalised 1980s picture' rather than a future-shaping one. Platform technologies like biotechnology, displaying pervasive characteristics, may not fit the cluster idea as practised in regional economic development as well as a network metaphor (Helpman, 1998). This lends greater credibility to the diverse interactions involved in complex, multi-use systems than more conventional sectoral-spatial notions such as clusters.

Thus the ITIs are already performing valuable roles in the spheres in which they operate. We have seen also how the market-facing model of innovation support differs significantly from that traditionally promoted by the development agency 'sector and/or cluster' model of intervention. Intermediary agents perceive that model as due for an overhaul, not least because it is inherently conservative, even backward-looking when innovation is a matter of shaping futures through market transactions. A slight anxiety in the possibly over-heated promotion of market thinking is that issues of timing can be overlooked and those of equity absolutely so. Clearly, Scottish Enterprise adopts the policy stance it does because it has a brief for safeguarding and generating employment. In all likelihood, the ITIs and the entrepreneurs they interact with in academe and the market have that as a lesser priority. Nevertheless, a future-orientation is not necessarily at odds with longer-term economic growth aims. For example, life sciences in Scotland are seen very clearly to be in a comparable position to that of San Diego in 1980 – before that city's rise to global status as a biotechnology cluster. This is because there are good-quality scientific centres (considered globally competitive scientifically), entrepreneurs are 'battle-hardened' because the business environment in Scotland is less benign than that of San Diego two decades ago, and the institutional environment is supportive. But, there is inadequate dedicated life sciences venture funding, the 'cluster' is small and needs to be grown, and by comparison with southern California, Scotland can appear 'a tad parochial'.

In 2000 Scotland had some 50 biotechnology firms, and by 2006 that figure had risen to 100. Of these 100 in 2006, the biggest geographical concentration is in Glasgow, but there are strong science and spinout firms also in Dundee and Edinburgh as well as near Aberdeen. The sector is thus seen as occupying a 'biotechnology triangle' between Dundee, Edinburgh and Glasgow. The role of the public sector had been important in Scottish biotechnology in three ways. First, as elsewhere, the finance for basic scientific research in the universities is mainly provided by UK Research Councils and, to a lesser extent and influenced by the measures of scientific legitimacy conferred by the high public ranking of biosciences schools, university hospitals and the like, this also attracts private sector funding from big pharma or charitable trusts. Second, the sector initially benefited from the adoption of a cluster strategy by Scottish Enterprise, the development agency for Scotland. Scottish Enterprise commissioned Michael Porters' consultancy, Monitor, to conduct a scoping exercise and provide intensive, back-up training for developing four pilot clusters, one of which was biotechnology. It is noteworthy that this methodology, of which the whole cost to Scottish Enterprise was between $1 and $2 million, by means of which cluster-building was established (see Figure 7.4), placed great emphasis on the processes of scoping, picturing and resourcing a 'vision' of the cluster rather than on traditional programming methodologies. Then, armed with the cluster-vision, stakeholders and leaders willing to show strong commitment brought together key actors to engage in interactive learning. Only then, at the third stage, did data-gathering,

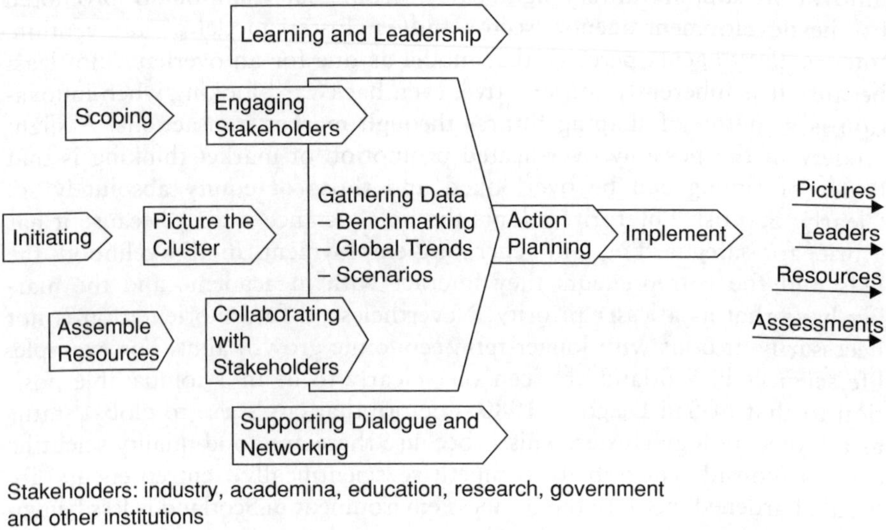

Figure 7.4 The Scottish Enterprise cluster approach in biotechnology

Source: Cooke (2001a)

benchmarking and scenario-building begin. This led to action planning based on consensus and concrete agreements followed by implementation guided by cluster pictures, leaders, expenditures and evaluation. This was the Harvard Business School and Monitor market-led model of public enterprise based on the ideas of 'picture, manage and monitor', rather than 'survey, analysis and plan'. But, as noted, the ITI view was that 'clusters' were already seen as a dated approach from an industry viewpoint, where the 'platform' concept and practice are preferred. In reality, there need be no real conflict between the two since clusters usually manifest 'related variety' and some business complementarities. However, both concepts are inimical to the truly dated notion of 'sectors'. The third way in which public intervention has a generic impact on biotechnology in Scotland flows from the consensus agreement on the cluster strategy at Scotland-wide level, and the collaboration by governance bodies to pool funding to assist in the process of supporting, in this case, the biotechnology sector.

Scotland is globally known as the home of the first transgenic animal, Dolly the sheep, developed at the Roslin Institute near Aberdeen. Other specialities include drug discovery, evaluation and clinical trial management in cancer research, cystic fibrosis, Alzheimer's and Parkinson's diseases. Scottish biotechnology also has a significant presence in agro-food biotechnology, with animal health and breeding, veterinary medicine, crop yields and pest control. Firms deploying environmental biotechnologies are also present. In all, Scottish Enterprise claim a 'cluster' of some 180 core

Table 7.10 Composition of biotechnology sector in Scotland

	Number of firms (Core activity)	
Activity	*2000*	*2006*
Biopharmaceutical therapeutics	24	38
Biopharmaceutical diagnostics	18	25
Biopharmaceutical clinical trials	10	15
Biopharmaceutical contract R&D	14	23
Bioprocessing	17	15
Environmental bioremediation	3	6
Environmental diagnostics	7	6
Environmental waste treatment	5	4
Agro-food therapeutics	1	6
Agro-food plant breeding	2	2
Agro-food diagnostics	4	5
Agro-food contract R&D	2	3
Supplies	23	47
Support services	26	181

Source: Biotechnology Scotland Source Book, 2000, 2006

and supply or service firms engaged to some degree in supporting the 'cluster'. In truth, the core was, by 2006, supported by a support services industry of some 181 firms while the core biopharmaceutical therapeutics firms number 38, as may be seen from the relevant Scottish Enterprise annual data sourcebook summarised in Table 7.10. The industry is quite geographically dispersed throughout Scotland but the core and many diagnostics, CRO and contract R&D firms are located in proximity to their scientific home bases. It is clear that biopharmaceuticals is the strong, core element, with a substantial number of firms in therapeutic product development, fewer in diagnostics, research and clinical trials (many of the contract R&D entries are universities, some with firms attached, others not). There is also a reasonably well endowed supplies (reagents, chemicals, etc.) and support services (legal, consultancy, etc.) infrastructure. Hence, as a whole, Scotland has a robust basis for future growth in biotechnology, but it may lack, at present, the interactive capacities and more sophisticated private support arrangements found more extensively in Cambridge and Oxford. It is clear from much of the preceding material on Scotland's biotechnology cluster that it is both polynucleated and dependent on more public intervention and support than in Cambridge or Oxford. This means that even in a former industrial city like Dundee, there are a number of highly rated bioscience departments in the university, and a Wellcome Trust (the world's largest medical charity) funded biotechnology institute. Cyclacel, described below, and Shield Diagnostics, a manufacturer of immunoassays in cardiovascular diseases, are Dundee University spinouts, the first located in the nearby innovation park of Dundee University, the second having graduated beyond it as it has grown in size. The Ninewells Hospital Medical School has some 320 bioscience researchers to contribute to the total of 1,000 life scientists in Dundee. Of these, 240 are in the Wellcome Trust Institute, which opened in 1997.

Dundee thus has the science base for possible future cluster development, but as yet lacks critical mass of firms. Cyclacel is a contract R&D spinout specialised in drug discovery for cancer genomics; the key connection is with the Ninewells Medical School, formerly headed by Sir David Lane (co-owner of Cyclacel) who, as noted, discovered the p53 anti-cancer gene. Venture capital of €4 million was accessed from London-based Merlin Ventures, headed by biotechnologist Sir Chris Evans, who founded Chiroscience, one of the UK's early biotechnology firms. Chiroscience merged with the UK's first biotechnology firm Celltech, subsequently acquired by Belgian chemicals giant UBC. Cyclacel contracts-in research from American head-quartered Quintiles, one of the world's leading biotechnology contract research firms from Boston, its subsidiary based in Edinburgh. Quintiles entered Scotland by acquiring Innovex, an upstream biosciences spinout from academia. Another firm, located in nearby Perth, Quantase was present in proximity to Dundee because of Shield Diagnostics based at the Technology Park at Dundee. Quantase conducted neonatal screening and was acquired by Shield in 1994. Shield specialises in cardiovascular neonatal diagnostics. In 1997 Quantase was subject to a management buy-out, and employed eight people, of whom three were research scientists and won UK government innovation awards (SMART, SPUR) to support development of its PJU Screening Home Test. Expansion was funded by venture capital from UK firm 3i and bank loans. Quantase, like Cyclacel and Quintiles, found the environment appropriate for its neo-natal screening business activities. So much so that it was, in 2002, acquired by Californian firm Bio-Rad. As with many smaller UK biotechnology companies, the fact that local linkages were strong, even as a supplier, proved less important than even stronger linkages to a global client with whom there had already existed a four-year strategic alliance.

Probably the strongest biotechnology corridor or axis within the polynucleated Scottish biotechnology arrangement is that focused particularly upon an Edinburgh–Dundee group of interactions that are local but also involve direct contracts with UK as well as global pharmaceuticals customers such as AstraZeneca, Glaxo, Merck and Abbott. Connectivity thus involves a substantial variety of non-Scottish partners ranging from other UK universities and hospitals, to US research institutes or laboratories, to biotechnology firms in the other UK clusters, such as Oxford BioMedica, or in, for example, Germany, as with Evotec. One interestingly innovative non-proximate nexus in the Scottish biotechnology cluster was the relationship between Cyclacel and Xcyte (Chapter 10, Table 10.12). It was seen as an exemplary case of a very recent process observable in the UK whereby high-quality basic research was not being followed up sufficiently quickly by commercialisation – part of a wider European innovation problem in biotechnology. Hence Cyclacel reverse-merged with Seattle-based Xcyte primarily to achieve a NASDAQ listing and access to greater quantities of potential shareholder investment.

Swedish healthcare biotechnology

The first commercial exploitation of modern biotechnology in Sweden was based on technology from Genentech, licensed by the Swedish company Kabi in 1978. Kabi merged with Pharmacia in 1990. Pharmacia later merged with two US companies, Upjohn and Monsanto, to form Pharmacia Corporation. In the spring of 2003 Pfizer, the US pharmaceutical company, acquired Pharmacia Corporation. The other major pharmaceuticals company in Sweden, Astra (now a Swedish–UK firm AstraZeneca) has headquarters in Södertälje near Stockholm with Swedish research based in Södertälje, Gothenburg and Lund which started using recombinant DNA technology in the 1980s. From then, and increasingly in the 1990s, new DBFs were founded in Sweden. Most of these new companies were either spin-offs from university research or from existing large pharmaceuticals companies. Swedish biotechnology places fourth in Europe in terms of number of companies and number nine in the world according to the Swedish Trade Council in 2002.

The number of Swedish DBFs increased by 35 per cent from 135 in 1997 to 183 in 2001, and the number of employees increased by 48 per cent to about 4,000 (VINNOVA, 2003). The two pharmaceuticals companies AstraZeneca and Pharmacia Corporation were the dominant large companies engaged in biotechnology activities. Many Swedish DBFs serviced them in the biopharmaceuticals application sector, but also in such industries as food processing and agriculture.[12] These DBFs are highly research-led and knowledge-intensive. Between 10 and 20 per cent of employees in these companies have a doctoral degree. Of company presidents responding to VINNOVA's questionnaire, 93 per cent stated that their companies collaborated with academic research groups. From our earlier results, this appears to differentiate Swedish DBFs somewhat from those in the US and UK, where research interactions are more among firms or distinctively among PROs. This is possibly an indicator of the relative immaturity of many Swedish DBFs, formed as we have seen in the 1990s for, according to the VINNOVA study, a majority of companies were small in 2001, that is, had fewer than 200 employees. Almost 90 per cent of the companies had fewer than 50 employees, and a good half had fewer than 10 employees. However, the category of small and medium-sized biotech companies is growing such that in 2001 Swedish DBFs totalled about 4,000 employees, a 35 per cent increase since 1997.

These are mostly found clustered in Sweden's metropolitan regions and in cities with large universities conducting substantial medical research.

There are 56 DBFs located in the Stockholm region, followed by the Lund/Malmö and Uppsala regions, with 36 and 31 respectively. Uppsala specialises in platform technologies, having GE Healthcare as a key presence. Stockholm, home to the renowned Karolinska Institute, is a key drug discovery centre. Twenty-four DBFs are located in the Gothenburg region. The smallest cluster is in the Umeå region, with fewer than ten biotech DBFs. The Swedish pharmaceutical industry annually spends around 25 per cent of its revenues on R&D, higher than the global standard of 17.5 per cent. This high percentage by international standards mainly reflects AstraZeneca's large expenditures in its Swedish research centres, with around one-third of the group's total R&D investments, $3.1 billion, occurring in Sweden. Of this some 20 per cent or $540 million is spent extramurally in Sweden (Benner and Sandström, 2000). Stockholm–Uppsala, in particular, contributes to Sweden's relative strength in biotechnology research, mainly through the Karolinska Institute, Uppsala University, Stockholm University, SLU[13] and the Royal Institute of Technology. These produce annually some 8,000 publications, co-host some 4,000 PhD students and employ some 2,200 scientists.

However, on publication interactions, Sweden's collaborations revealed 70 per cent of co-authorships 1986–1997 were with other Swedish PROs, while 12 per cent were with US institutions and an equivalent share with UK and German co-authors together. Regarding R&D projects, McKelvey *et al.* (2003) found the opposite, that is, of 215 collaborations made by 67 actors (firms, universities and research institutes), 52 were between Swedish institutions and the rest involved overseas partners, these again being mainly with the US and UK. This undoubtedly reflects the relative thinness of the pharmaceuticals and DBF market, particularly the former in Sweden, which perforce stimulates links for industrial research with larger economy incumbents. This is underlined to a limited degree for patenting, where Swedish biotechnology patents registered in the US involved 62 jointly with US inventors. However, 202 were products of Swedish-only collaborations: a similar pattern prevailed for pharmaceuticals joint-patenting (VINNOVA, 2003).

Hence, Sweden may be said to display a relatively typical European introversion in much of its exploration or basic research activity, including to some extent patenting, but it is more outgoing where applied research with industry is concerned. Stockholm–Uppsala is the stronger with 56 firms and 1,126 employees, at least five strong research universities and around 1,640 employees in smaller biotechnology businesses in four Science Parks. However, Copenhagen–Lund–Malmö's Medicon Valley is a larger concentration at 104 than Stockholm–Uppsala. Medicon Valley is well placed in this respect, being home to 11 universities, 70 biopharmaceuticals companies (60% of the turnover of the Nordic pharmaceuticals market), and 26 hospitals. Its academic institutions include, among others, Copenhagen and Lund universities, the Danish University of Pharmaceutical Sciences, and the Royal Veterinary and Agricultural University, as well as centres for diabetes and stem cell research, and a multidisciplinary centre for stem cell biology and cell therapy. There are also 135,000 students in the area, more than 300 research groups, 4,000 life science employees, and 27 local venture capital firms (Dorey, 2003). Pharmaceuticals firms have been key to growth on the Danish side through the likes of Hansen Laboratories, Carlsberg Laboratory and Novo Nordisk, a global insulin supplier and partner of Biogen in early bioengineering of human insulin. Medicon Valley Academy (MVA) was formed in 1997 to 'create, transfer, and exploit knowledge' with the aim of making it the most attractive bioregion in Europe by 2005. MVA is largely responsible for creating a sense of identity in the region, arranging networking events and also building relationships with other clusters in Scandinavia such as the Stockholm/Uppsala region.

Switzerland's biotechnology clusters

Switzerland is another small country in which multinationals generally, and in 'big pharma' specifically, are a notable feature of the economic landscape. Roche (Hoffmann La Roche) and Novartis (formerly Sandoz and Ciba-Geigy) are indigenous Swiss multinationals, the latter in the global top ten by market capitalisation, the former having slipped down the rankings in later years rather like Bayer and, to some extent, Aventis among former European majors. The Swiss government too in 2002 announced in favour of embryonic stem cells research while banning embryo-creation purely for research purposes. Apart from Novartis and Roche, Switzerland also hosted one of the world's largest biotechnology companies Serono (acquired by Merck of Germany in 2006), which in 2001 had a market capitalisation of $18 billion, ranking it third behind Amgen and Genentech. Serono and Amgen signed a licensing and commercialisation deal for Serono to sell a multiple sclerosis drug in the US it had developed with Immunex, a Cambridge (MA) firm subsequently acquired by Amgen. Other prominent firms are Actelion, Cytos, the Genetics Company, bio-T, CELLnTEC, Debiopharm, GeneBio and Solvias. In addition to Serono, Actelion, Berna Biotech, Debiopharm and Basilea have several pharmaceuticals in clinical testing. Lonza Biologics is also Swiss and one of the largest biosynthesis firms in the world. Debiopharm funds cancer research projects at Tulane University, New Orleans.

Of the 200 Swiss biotechnology companies listed in the Swiss Life Sciences Database (www.swisslefesciences.com) in 2005 around 40 were pure biotechnology firms (DBFs), the others being instrumentation and services firms that nevertheless linked to many of the 40. Some 22 per cent of the 200 were located in the Geneva–Lausanne 'BioAlps' region, approximately 26 per cent were in the Basel 'BioValley' region, and about 35 per cent were in the Greater Zurich region. Zurich has a Functional Genomics

Research Centre. Since 2000, 45 new biotechnology businesses were established, 15 of which were spinouts from the Swiss Federal Institute of Technology in Lausanne and Zurich and 10 were spinouts from other Swiss universities. The remainder came from domestic and foreign subsidiary industry. Of the thirty or so public companies, many like Genedata (Basel), Cytos (Zurich) and GeneProt (Geneva), a proteomics DBF, have long-term collaborative research, opinion and licensing agreements with the likes of Novartis and Roche. International collaborations extend to the partnership between the University of Minnesota and the Swiss Federal Institute of Technology focused upon medical technology. This arises in part from Minneapolis devices firm Medtronic's subsidiary located in Switzerland and its collaboration with the likes of Disetronic, a leading Swiss insulin-pump manufacturer. In conclusion, Switzerland is a small, capable knowledge-intensive biosciences economy. It has leaders in 'big pharma', global DBF capability and numerous smaller DBFs and spinouts concentrating on leading edge proteomics and other post-genomics treatments. It is highly connected globally, but especially to the US megacentres through 'big pharma' and its leading clusters in Zurich, Basel and Geneva–Lausanne.

Biotechnology in France

France's research model differs significantly from most of those discussed in this chapter so far. French universities are not engaged in basic research to the same extent. Moreover, the role of state intervention in delegating and coordinating research is extremely high, as are its interventions in business affairs across all economic sectors, but especially biopharmaceuticals where the state guided merger of Sanofi-Aventis in 2005 produced the world's third-largest pharmaceuticals company, at the expense of effacing German partner Hoechst and resisting a better offer from Novartis of Switzerland. France also lacks entrepreneurs forming new, innovative companies that elsewhere originate from public research institutions or larger corporations. This is partly due to the malfunctions of the public stock market as an exit channel for investors. As a result, not a single Biotech-IPO occurred between 1998 and 2005.

In late 2005 it partially reopened as two smaller French DBFs (ExonHit Therapeutics and BioAlliance Pharma) floated on the Euronext technology market. Despite this, the French network organisation France-Biotech, ranks France number three in Europe in terms of its number of DBFs, comparable to Britain or Germany with some 300 firms, of which only 6 were public companies with only a small drugs pipeline. In truth, French biotechnology seems to be slipping back from a formerly fairly stagnant position. This is undoubtedly because of the nature of its economic governance and relative absence of swiftly moving private sector exploitation institutions and intermediaries with a correspondingly over-interventionist and lumbering state apparatus in support of biotechnology. To try to improve firm formation, several supporting policies have been established. The key ones were the implementation of a 'young innovative company' standard and the establishment of geographical research focal points, known as *Genopoles*. Given France's dearth of entrepreneurs, it is remarkable to observe the multi-level scrutiny through which new entrepreneur candidates are put before a small trickle is allowed into each *Genopole* (Cooke *et al.*, 2006). Moreover, there is a debilitating Parisian élitism associated with inclusion of other regional incubators in representation of French biotechnology at, for instance, international trade exhibitions, reinforcing the imagery of *Genopole* as a somewhat exclusive 'club good'. Thus if a firm has qualified for the young innovative company status (so-called JEI – *Jeune entreprise innovante*) it is completely exempted from contributions to the social security system for R&D employees. The *Genopoles* were established to enhance the quality and output of biotechnology research on the one side, and to facilitate and promote spin-offs on the other side. The oldest, largest and most successful is situated at Evry, south of Paris. It was founded in 1998 by the French research ministry, private foundations, the University of Evry and several regional authorities. In order to promote spin-offs, *Genopoles* act as network-nodes by providing various consulting services to scientific founders, like framing up business-plans, or legal and subsidy consultancy. Other such 'poles' exist, such as those at Strasbourg, Lille and Grenoble, but equally well performing ones such as the one at Nantes *Atlanpole* struggle for recognition in the club-like atmosphere of French biotechnology. Firm founders gain privileged access to a special public seed-fund, and they are further allowed to use laboratory space of the public research body for up to 24 months. Overall there are today seven *Genopoles*, established since 1999 as a key part of the *Génomique* support programme. By 2006 the most successful *Genopole* at Evry employed some 1,800 scientists working in 25 laboratories and 58 biotechnology firms or organisations at the 'pole'.

Table 7.11 shows that French biotechnology firms are comparable to German DBFs regarding R&D productivity, namely not very successful. Both are considerably less successful in later stage projects than Swiss or UK firms, but again like German firms they manage to introduce a significant number of product candidates into early phases of the drug development pipeline. In France these are considerably less innovative molecules under development than in Germany. However, in comparison with Germany, France has a better innovative performance in pharmaceutical product candidates that are currently under development within domestic big-pharma companies (Table 7.12). This is significant not least because it testifies to the relatively mature fine chemicals traditions of both countries, qualities that are elsewhere somewhat in retreat in light of the molecular biology and genomics revolutions that have transformed healthcare and medicine from the 1990s onwards.

From limited data on the performance of the French pharmaceuticals biotechnology sector at regional level, the French Ministry for Youth,

Table 7.11 Key financial indicators for French biotechnology

	2001	*2002*	*2003*	*2004*	*2005*
Number of firms	255	239	246	300*	300*
Raised capital	254	238	179	268	165
Venture Capital	242	238	126	242	93
Follow on or IPO	12	0	53	26	72
Number of IPOs	0	0	0	0	2

Source: France Biotech; Kaiser and Liecke (2006b)

* These numbers appear in the very recent sector report. They are much higher than in earlier sector reports. Presumably the mode of survey has changed, as today also cosmetic and animal health firms are included. Ernst & Young counts only 228 French dedicated biotechnology firms for 2004.

Table 7.12 Product pipeline of public French biotechnology companies 2002–2003

Pipeline	*2002*	*2003**
Pre-clinical	89	111
Phase I	27	18
Phase II	24	22
Phase III	8	13

Source: Rochepeau (2004) p. 29; Kaiser and Liecke (2006b)

* Estimation by Rochepeau

Education and Research (MJNER) published a study in 2001 on the R&D intensity of different bioregions (Table 7.13). According to this study the most developed French biotechnology clusters exist in the three regions of Île-de-France, Rhône-Alpes, and Alsace (MJNER, 2003, p. 4). These three regions concentrate the highest number of biotechnology firms as well as the largest share of R&D employees. In effect, it may be deduced, the French

Table 7.13 Biotechnology in French regions

Regions	*Number of firms*	*R&D employees*
Île-de-France	68	940
Rhône-Alpes	45	374
Alsace	20	248
Midi-Pyrénées	16	159
Auvergne	14	126
PACA	22	123
Bretagne	21	110
Others	90	394

Source: MJNER (2003) p. 4

biotechnology healthcare sector may be characterised in part as a case of arrested development, over-dependence upon public intervention, weakly supported by market transactions and heavily focused in the capital city region. Grenoble, and Rhône-Alpes more generally may be a slight exception to that generalisation, but its existence as a technology pole of consequence is testimony to the past wisdom of French politicians in decentralising advanced technology industries away from Paris to the regions, something that appears less pronounced in policy terms now than hitherto.

Conclusions

It is clear from this chapter and much of the foregoing analysis that the US is well ahead of the rest of the world in its development of biotechnology. This should not be surprising since the technology began in the US, specifically San Francisco with the Boyer-Cohen patent and the engagement of venture capitalist Robert Swanson to establish a firm, Genentech, to exploit it. Arguably at that time Europe led the bioscience from Crick and Watson's discovery of DNA to Köhler and Milstein's discovery of monoclonal antibodies, all at Cambridge. Thus Europe started biotechnology activity, in the shape of the UK, Sweden and Switzerland, some ten years after the US and much of Europe began to take the field seriously only ten years after that. Many accounts show large economies like Germany, the founding country of the modern pharmaceuticals industry led by the likes of Bayer in the 1890s, being completely flummoxed by the onset of the molecular biology age. France too, home to not only Pasteur but Lavoisier and Laplace, seems latterly to have lost its way in the world of biotechnology.

What we see in Europe and North America is the importance of strong, relatively independent and open science, most notably at university centres of research excellence, with scrutiny of knowledge claims by peer review predominating in research and publication, much of which is initiated by public or not-for-profit foundation sources of research funding. Thereafter, market processes, which nevertheless embody their own forms of scrutiny regarding what looks like a sound candidate for commercial drug or diagnostic exploitation, generate the innovative treatments or tests. All the best performing locations have this combination of generous public research funding by arm's-length funding bodies, peer-reviewed knowledge claims, venture capital and its early and late stage financial instruments, and vigorous DBFs seeking markets for innovative and effective treatments. Notice that direct state or big pharma involvement are absent from these elements, something the less well performing European economies have failed until recently to recognise. Biotechnology marks something of a crisis in large-scale modes of governance of innovation for both public and private sectors.

8 Healthcare biotechnology in developing countries

Introduction

Thus far the theoretical framework explaining the exaggerated knowledge asymmetries associated with knowledge economy development processes has worked extremely well. It has shown that not only is biotechnology highly skewed in its distribution between countries, but that this applies to bioregions within countries. Not even the US, where the first modern biotechnology commercialisation was achieved, has an even regional spread of the activity. Moreover, we see specific bioregions predominating in particular knowledge domains – Boston–Cambridge globally strong in genomics, Silicon Valley strong in bioinformatics, for example. Moreover, as we shall see in Chapter 9, agro-food biotechnology has its own socio-spatial logic, St Louis, Missouri being the world's largest agro-food biotechnology cluster, not least because of its agro-food research institutions and the influence exerted there by genetically modified organism (GMO) specialist, Monsanto. Despite this typical combination of high-grade public research facilities and commercial organisations for localised knowledge exploitation, which as we have seen, characterises such knowledge-based clusters, most national and many regional governments nevertheless aspire to develop a biotechnology industry or at least cluster. Simultaneously, Ministers of Science and high functionaries in multilateral agencies such as the EU and OECD downplay the chances of this happening, arguing it is a waste of scarce resources to try to build a successful biotechnology cluster. Why this contradiction?

As we have seen, biotechnology is certainly expensive. We saw that US pharmaceuticals firms spent over $50 billion on R&D in 2004, although with NIH research expenditure of some $40 billion around the same time, by no means all of this was earmarked only for biotechnology research. New biotechnology drugs are expensive for patients or healthcare systems to purchase. Some, in their early years, patent-protected, with no competitor in sight and targeted at a rare disease like Gaucher's, which debilitates the patient by its combined attack on liver, spleen and bones (osteoporosis), could command a price in the hundreds of thousands of dollars for a year's treatment. More modestly, Herceptin, a treatment for certain kinds of breast cancer, costs £25,000 ($50,000) per year administered to sufferers in the UK. In the US, a high price for a Gaucher's treatment is still cheaper than a year's intensive care treatment, which at a market rate of, say $2,000 or more per bednight totals some $730,000. Private health insurance services are thus clearly incentivised to reduce overall costs by approving the use of expensive biotechnology treatments. It can thus be seen how that kind of system is also innovation-enhancing. But in public health insurance systems and, perforce, developing countries where there are only financially weak or in effect no 'systems' of organised healthcare, such costs constitute major barriers in the former case and absolute barriers in the latter.

In the UK, the assessment body responsible for performing cost-benefit and drug affordability analysis, the National Institute of Clinical Excellence (NICE) seems regularly to find itself under fire from patient groups for not recommending access by National Health Service patients to new drugs like Glivec, a leukaemia treatment, and Herceptin (see Chapter 11). In practically all cases these adverse recommendations are overturned by patient pressure being exerted upon the responsible authorities through protests that of course attract headline media coverage. Thus the politics of biotechnology often rise to the top of the political agenda, fading away only until the next negative cost-benefit recommendation by NICE.

In developing countries, governmental pressure has at last begun to prick the consciences of big pharma. Thus in December 2006 Merck (US) ramped up the Accelerating Access Initiative it had already joined with pharmaceuticals firms such as Abbott, Boehringer-Ingelheim, Bristol-Myers Squibb, Gilead, Glaxo, and Roche and relevant United Nations agencies to supply AIDS drugs at cost, in the form of antiretroviral cocktails, to more than 716,000 patients in developing countries, half those then being treated. Merck announced a cut-price anti-HIV drug for Thailand, the significance of which is that Thailand is a middle-income country with a pharmaceuticals industry, and had threatened a 'compulsory licence' on Efavirenz to overturn Merck's patents. Previously Merck had established such philanthropic programmes as the $50 million African Comprehensive HIV/AIDS Partnership in low-income Botswana. Pfizer established a 'fellows' partnership scheme with donations of Diflucan, an HIV drug for developing countries. Gilead works with the Bill and Melinda Gates Foundation, NIH and Bristol-Myers Squibb similarly on the antiretroviral Atripal (Jack, 2006). But South Africa has been key to this 'largesse', threatening court action and suspension of intellectual property rights (IPR) if nothing was done substantially to reduce prices (Kazmin and Jack, 2006). Hence, to the extent that new codes of practice and pricing have been developed; otherwise in countries with appropriate production capabilities, patent-expiry dates were prone to being ignored and generic copies of proprietary treatments produced. But reverse-engineering advanced treatments for HIV-AIDS, for example, is not always easy or possible. So, in such circumstances and where the effort can be afforded, the sound of Western ministers and multilateral bodies advising against evolving biotechnology in developing countries goes, understandably, unheard, ignored or greeted with uproar.

In the following sections, we review some of the most interesting cases of the evolution of biotechnology in developing countries, largely by such countries' own devices. We start with the case of Singapore, not so much because it is a developing country, because it passed that condition on the way to reasonably full economic development status decades ago, but because it is seen as a role-model by other rapidly developing countries, notably oil-rich ones, that have ambitions for economic diversification and the resources to seek to 'build a biotechnology cluster'. Countries such as Saudi Arabia and Abu Dhabi have these aspirations, for example. But it would be remiss in the extreme not to refer also to the remarkable evolution of biotechnology in Cuba, a country hamstrung by US sanctions for nearly a half-century, but which prioritises its limited public spending on education and healthcare, and in the latter case on healthcare R&D of global significance. Finally some large country examples from India, China and Brazil are highlighted for their distinctive routeways in evolving a presence in biotechnology activities.

Singapore's Biopolis

Singapore's government biotechnology initiatives started in 1987 with the establishment of the Institute of Molecular and Cellular Biology at the National University of Singapore, but became industrially serious within the 2000–2004 period. The aim was to build a biotechnology cluster around foreign direct investment (FDI), a policy that worked well in previous developmental stages such as the policies in support of petrochemicals, electronics and ICT. Four new Institutes in bioinformatics, genomics, bioprocessing and nanobiotechnology now exist at a cost of $150 million to 2006. Public venture capital of $200 million has been committed to three bioscience investment funds to fund start-ups and attract FDI. A further $100 million is earmarked for attracting up to five globally leading corporate research centres. The Biopolis is Singapore's intended world-class R&D hub for the georegion. The Biopolis is dedicated to biomedical R&D activities and designed to foster a collaborative culture among the institutions present and with the nearby National University of Singapore, the National University Hospital and Singapore's Science Parks. Internationally celebrated scientists have also been attracted, such as Nobel laureate Sidney Brenner, Alan Colman, leading transgenic animal cloning scientist from Scotland's Roslin Institute, Edison Liu, former head of the US National Cancer Institute, and leading Japanese cancer researcher Yoshaki Ito.

These are 'magnet' appointments meant to attract talent and create cluster conventions and practices among research centres and DBFs. The sector now numbers 38 firms, of which 15 are indigenous start-ups and 23 FDI

R&D, manufacturing, clinical research organisations (CRO) and other services. Johns Hopkins, MIT, Duke University, Columbia University and the Indian Institute of Technology have established facilities in Singapore. Singapore's Bioethics Advisory Committee advised acceptance of embryonic stem cell but not human cloning research, which is also a globally attractive locational factor, shared as we shall see with Israel, among others. Pharmaceuticals firms from overseas manufacturing in Singapore include Glaxo since 1989, Schering-Plough (1997), Genset (now Serono – bought in 2006 by Merck of Germany) (1997), Aventis (2000), Merck (2001), Wyeth (2002), and Pfizer (2004). R&D centres of the following firms are also present: Genelabs (1985), Becton Dickenson (1986), Oculex (1995), Perkin Elmer (1998), Sangui (1988), Cell Transplants (2000), Schering-Plough (2000), Eli Lilly (2001), Surromed (2001), Affymetrix (2001), Novartis (2002), ViaCell (2002), PharmaLogicals (2002). Finally, CRO services are available from Quintiles (1995), Novo Nordisk (1999), Covance (2000) and Pharmacia-Upjohn (now Pfizer) (2000). Joining ViaCell in stem cells are indigenous DBFs ES Cell and CordLife, a few genomics firms like APGenomics and Qugen, and a variety of drug discovery, bioinformatics and diagnostics firms mostly established since 2000. In brief, Singapore is host to a large number of mainly US and, to a lesser extent, European R&D laboratories of 'big pharma' businesses. It has strength in public research activity and small signs of growth in stem cells exploration and exploitation activity. The benign regulatory environment allowing embryonic stem cells research is undoubtedly an attraction which, contextuated by Singapore's celebrated 'developmentalist state' capabilities, will stimulate cluster growth as an 'offshore' research and production platform targeting the burgeoning Asian market (Finegold *et al.*, 2004).

Biotechnology in Cuba

Cuba is one of the few remaining Communist countries, relatively small in population, poor, agricultural, and exposed since 1963 to nearly fifty years of the US trade embargo. The collapse of its main trade partner, the Soviet Union, in 1990 further seriously damaged Cuba's export markets. Despite these handicaps, and due to the prioritisation of healthcare as an expression of the ruling anti-capitalist ideology, Cuba has developed a sophisticated healthcare biotechnology sector. Among the large range of products developed, perhaps the most significant has been the world's first meningitis B vaccine, which was based on Cuban research at the Finlay Institute and appeared in the late 1980s following a mid-1980s epidemic. More recently, at the Synthetic Antigen Laboratory of the University of Havana, the world's first vaccine with a synthetic antigen was produced to protect against Hib type b infection, which causes pneumonia and meningitis in very young children. This was a collaboration with the University of Ottawa Hospital which led to a joint patent, royalties from which are waived on sales of the vaccine in developing countries. Other research targets are vaccines for Dengue fever, AIDs, cholera and a cancer treatment vaccine, as well as drugs, diagnostic tests and natural product drugs such as anti-cholesterol product Policosanol (PPG), which is derived from sugar cane.

Although Cuban biotechnology began by copying Western drugs like Interferon-α 2b, in recent times the industry has developed a rather strong drug pipeline, with some 500 patents filed on the European Patent Office's databases. Others have been filed elsewhere, including the US, Brazil, India, China and South Korea, and products are exported to some 50 countries worldwide. Exploitation has predominated over publication or co-publication of academic research papers, hence Cuba has been under-represented in such comparative metrics. A list of leading examples of Cuban biotechnology products is shown in Table 8.1 below. The associated research institutes for the producers are as follows: Vacunas Finlay (Finlay Institute); Heber Biotec (Centre of Genetic Engineering and Biotechnology); CIMAB (Centre for Molecular Immunology); Laboratorios Dalmer (National Centre for Scientific Research); and Tecnosuma Internacional (Centre for Immunoassays). All of these are concentrated in Havana, with many in Western Havana at the West Havana Scientific Pole, and were mostly founded in the 1980s and 1990s. Thus research institutes often have commercial manufacturing arms within them. Clearly links between exploratory, examination and exploitation knowledge are extremely close, effective and relatively efficient. Notably, clustering occurs in biotechnology despite the completely opposite socio-political régime under which Cuban science and exploitation operate compared to the West. This was deliberate government policy as the terminology 'pole' suggests – a cluster of research

Table 8.1 Selected Cuban healthcare biotechnology products at 2004

Sector	*Type*	*Application*	*Producer*
Vaccines	Purified meningococci	Meningitis B & C	Vacunas Finlay
	Recombnt. surface antigen	Hepatitis B	Heber Biotec
	Synthetic HiB	Pneum./Meningitis	Heber Biotec
Therapeutics	Recombinant streptokinase	Cardiovascular	Heber Biotec
	Recombinant IFN-α	Viral/Oncological	Heber Biotec
	Recombinant epidermal	Burns/Ulcers	Heber Biotec
	Recombinant granulocyte	Leukopenia	Heber & CIMAB
	Monoclonal Ab to CD3	Organ transp.reject.	CIMAB
	Recomb. Erythropoetin-α	Anaemia	CIMAB
	Mab anti-epidermal	Head & neck tumours	CIMAB
	Ateromixol (PPG)	Anti-cholesterol	Labs. Dalmer
Diagnostics	Immunosorbent assays	AIDS test	Tecnosuma Int.
	Mabs cancer markers	Cancer imaging	CIMAB
	Immunosorbent assays	Syphilis	Heber Biotec

Source: Thorsteinsdóttir *et al.* (2004)

institutions, higher education institutes and hospitals were linked in the early 1990s to stimulate from proximity enhanced integration of science, health and education.

Cuba's emphasis on education and health means that its educational level by 2000 was close to the OECD norm and ahead of middle-income countries such as Chile and Taiwan. Although Cuban scientists used to learn abroad, the Cuban Science and Technology training system means that there is a surplus of incomers seeking higher education today. In recent years with some cuts in research institute budgets, stronger collaboration with university research has been encouraged and departments like University of Havana Chemistry Faculty have contributed significantly to discoveries, notably the synthetic antigen discussed earlier and noted in Table 8.1. However, innovation in biotechnology is not a simple matter in Cuba's command economy, as the difficulties of partnering even one of the US's leading biotechnology vaccine research firms Chiron (now fully acquired by Novartis) testifies. In 2004, the US Treasury fined Chiron $168,500 for illegally exporting to Cuba. Between 1999 and 2002, Chiron exported five vaccines for infants and children to Cuba from its plants in Germany and Italy. Chiron held a licence to export one vaccine through UNICEF, the United Nations Children's Fund, but had no clearance for the other four. A routine audit in 2002 uncovered the oversight, and Chiron voluntarily disclosed the error to the Treasury Department and negotiated a settlement, recognising it as a violation. Thereafter Chiron received permission legally to export all five vaccines, including polio, haemophilus influenza, flu, rabies and a vaccine for measles, mumps and rubella. The fine was the second-largest civil penalty issued in 2004 by the Treasury Department's Office of Foreign Assets Control, or OFAC, which also fined Alpha Pharmaceutical $198,711 for similarly illegally exporting goods from 1998 to 2003. Under federal regulations, sales of US products, technology or services to Cuba are not allowed. Exports of medicine and medical equipment are legal, but only when companies or individuals had received appropriate clearance from OFAC. In recent years, under the Bush administration, experts observed that the US Treasury Department has clamped down on such political misdemeanours.

Nevertheless, the Cuban experience demonstrates that advanced biotechnological innovation occurs under the kind of political régime many thought to be inimical to innovation, as suggested by commentaries on the collapse of the Soviet Union due in large part to its failure to innovate outside the highly expensive space-programme that had absorbed up to 70 per cent of Soviet R&D in the 1980s. The absence of free exchange of ideas and knowledge with Western scientists must seem like an insurmountable barrier to progress along with the non-returning emigration of scientists. Nevertheless, by focusing on niche markets of relevance to typical developing country diseases, which are largely ignored by Western big pharma, Cuban innovation prospers and thrives. Cuba has devoted substantial resources to healthcare and its biotechnology, resulting not only in Cuban self-sufficiency in some therapeutic and diagnostic areas, but in world leadership in a few. Despite its isolation Thorsteinsdóttir *et al.* (2004), on whose research much of this account draws, note strong international and commercial partnerships between the Marie Curie and Pasteur Institutes in Paris, Heidelberg University and Harvard University. On the commercial side, Finlay Institute licenses Glaxo to distribute its meningitis vaccine in Europe and north America, CIGB has a partnership with Panacea Biotec of New Delhi to manufacture hepatitis B vaccine, and CIMAB (immunology) has a joint venture with YM Biosciences of Mississauga, Toronto to market and develop cancer therapeutics. The joint venture CIMYM has US Treasury approval to conduct joint development and licensing by Californian DBF CancerVax of Cuban cancer vaccines.

Biotechnology in India

India also started its post-1947 existence as a socialist economy, with science and technology visualised as the key vehicle to bring a better quality of life to its people than had been experienced hitherto. Building upon the base established then, including some research and higher education institutions like the Indian Institute of Science (Bangalore, 1909) dating from the Raj, modern India has developed expertise in agro-food biotechnology, healthcare biotechnology, and possibly most impressive of all, a thriving pharmaceuticals industry led by the likes of Ranbaxy in New Delhi, Dr Reddy's Laboratories in Hyderabad, and Wockhardt in Mumbai. In terms of biotechnology firm numbers, India ranks twelfth in the world (Kumar *et al.*, 2004). India's largest healthcare biotechnology success was its first domestically produced hepatitis B vaccine, not original in itself but utilising a novel expression technology, a recombinant surface antigen to produce a treatment labelled Shanvac-B. This drug was produced by a remarkable consortium led by 1993 Hyderabad start-up Shantha Biotechnics, with research support from Osmania University (Hyderabad) and the Centre for Cellular and Molecular Biology (CCMB), also at Hyderabad, and New Delhi's Council for Scientific and Industrial Research, and funding from the Sultan of Oman. The drug sells for $0.50 compared to the Western import price of $16 per dose.

India has not yet progressed to the status of Cuba as a producer of indigenous drug innovations. Its biopharmaceuticals strategy is based on producing low-cost, relatively simple imitations that address India's health needs as well as engaging with global markets. International linkages noted by Kumar *et al.* (2004) include Bharat Biotech (Hyderabad) accessing Gates Foundation and European Malaria Vaccine Initiative funding for research and product development. Shantha received certification from the World Health Organisation and a contract with UNICEF for 8.5 million doses of its hepatitis treatment. Panacea Biotec and the Serum Institute of India (both New Delhi) also supply UNICEF. India's publication and patenting both more than doubled during 1991–2002, reflecting a rise in federal and regional government funding programmes and academic research organisations. India's first dedicated private venture capital fund for biotechnology was established in 2004. Since 1980, biotechnology features strongly in the government's Five Year Plans, and a government Department of Biotechnology (DBT) was established in 1986 to support training, infrastructure and development programmes. DBT also funds research in priority areas like vaccines for cholera, rabies, encephalitis, tuberculosis, malaria and HIV. Elsewhere, the Indian Council of Medical Research has a network of 21 research institutes and 6 regional research centres focusing on cholera, AIDS, tuberculosis and leprosy. Other research into plant and mammalian bioscience, bioreactors, molecular drug targeting and predictive markers is funded by India's Department for Scientific and Industrial Research. This department is India's main biotechnology patent assignee, licensing of intellectual property being conducted by Patestate, its IP management division. The Indian Institute of Science at Bangalore, which became a university in 1956, also has bioscientific research and teaching. It ranks highest of the fifteen internationally research-active Indian health biotechnology universities, accounting for some 20 per cent of India's total international journal articles in the field. It specialises in immunology, reproductive biology and plant development related to disease targets like malaria, rabies and tuberculosis. New research institutes were also established during the 1990s such as the National Institute of Immunology, CCMB (Hyderabad – a UNESCO starred 'centre of excellence'), and the Institute of Microbial Technology (Chandigarh).

India has 96 DBFs (Ernst & Young, 2006) and lies second in Asia after China (136). Many of these firms are located in Hyderabad and, to a lesser extent, Bangalore clusters. Other key locations are New Delhi and Mumbai. Biocon, India's 2004 publicly floated biotechnology firm, is in Bangalore, while others like Shantha and Bharat are in Hyderabad. Larger pharmaceuticals companies like Dr Reddy's and public enterprise Indian Immunologicals are also located in Hyderabad, while Ranbaxy is in New Delhi. Many foreign pharmaceuticals majors have a presence in India, notably AstraZeneca with a $10 million R&D laboratory in Bangalore. Eli Lilly is in Haryana near New Delhi, Boehringer Ingelheim is at Sevagram (east of Mumbai), Glaxo, Novartis and Pfizer are all located in Mumbai. Regarding R&D and production outsourcing, Divi's Laboratories (Hyderabad) performs custom chemical synthesis for Merck, Abbott Laboratories and Glaxo and manufactures generic anti-inflammatory and anti-arthritic products for other firms. Bulk drug manufacturer Matrix Laboratories saw its outsourcing business grow fivefold, to $10 million, in one year. In Bangalore, Biocon has 300 scientists doing contract research, up from just 25 in 2000, after contracting with Pfizer, AstraZeneca and Bristol-Myers Squibb. Novartis contracted Dr Reddy's (Hyderabad) to research a diabetes molecule despite an ongoing lawsuit over a generic version of Novartis' antifungal cream Lamisil. Glaxo did the same even after Ranbaxy Laboratories (New Delhi) had successfully challenged Glaxo's patent on the antibiotic Ceftin, later contracting Ranbaxy to conduct research. Table 8.2 shows selected Indian healthcare products and their originators. It can be seen that not only has there been a good deal of imitation of foreign products, but also domestic drug innovations. Nevertheless, the products of these and the larger pharmaceuticals firms have given India a strong competitive base upon which it is rapidly building. Other healthcare segments in which there is Indian presence are bioinformatics such as Strand Genomics and Infosys in Bangalore, Tata in Mumbai and Ocimum in Hyderabad. There are a number of public venture capital companies for biotechnology, the first from 2003 being the Biotechnology Venture Fund, the second a joint venture between Dynam Ventureast Group and Andhra Pradesh Industrial Development Corporation (Hyderabad). In some ways, it can be observed, Hyderabad has its 'knowledge domain' covering the fullest range of assets for advanced biotechnology commercialisation, and true to its ICT and software traditions, Bangalore specialises more than elsewhere in bioinformatics.

Biotechnology in China

Once more, in China we observe a Communist régime with a strong scientific ethic, and latterly since Deng Xiao Ping's decentralising economic reforms after 1992, a liberal market growth experiment of truly global impact. Within this, healthcare biotechnology is of relatively small but not

Table 8.2 Selected Indian healthcare biotechnology products at 2004

Sector	*Type*	*Application*	*Producer*
Vaccines	Recombinant surface antigen	Hepatitis B	Shantha Biotechnics
	Recombinant surface antigen	Hepatitis B	Bharat Biotech
	Recombinant surface antigen	Hepatitis B	Serum Institute
	Polysaccharide Vi	Typhoid	Bharat Biotech
Therapeutics	Recombinant insulin	Diabetes	Wockhardt
	Recombinant erythropoietin	Anaemia	Wockhardt
	Recombinant streptokinase	Cardiovascular	Shantha Biotechnics
	Recombinant interferon α-2b	Cancer	Shantha Biotechnics
	Recombinant streptokinase	Cardiovascular	Bharat Biotech
	Liposomal amphotericin	Leishmaniasis	Lifecare Innovation
	Recombinant granulocyte	Neutropenia	Dr Reddy's
Diagnostics	Immunoblot assays	HIV-1 & HIV-2	J. Mitra
	Immunoblot assays	HIV & Hepatitis C	Bharat Biotech
	Immunosorbent assay	Hepatitis C	xCyton Diagnostics
	Immunosorbent assay	Neurocysticercosis	xCyton Diagnostics

Source: Kumar *et al.* (2004)

insignificant impact as China evolves regional innovation systems capable of translating exploration knowledge into exploitation knowledge and commercial outcomes. For example, China was the only developing country involved in the Human Genome Project, an indication of its expertise in genomics, including the Beijing Genomics Institute and the Chinese National Human Genome Centre, located in Beijing and Shanghai. With state-of-the-art sequencing equipment, these have gone on subsequently to decode the rice genome pf, China's predominant rice species. As elsewhere, China adopted a 'me-too' approach to vaccine, diagnostics and therapeutics development, but on the basis of such a platform has also begun to develop discovery lines related to domestic health pathologies. These involve antibody research, notably related to severe acquired respiratory syndrome (SARS), gene therapy, functional genomics and stem cells. Regarding stem cells, Chinese research institutions are working on adult as well as embryonic stem cells, and discoveries regarding cell reprogramming have been announced (Zhenzhen *et al.*, 2004).

Further research continues on genome sequencing, utilising the resource of China's large and varied population groupings. This enables research into disease genes and functional genomics, which has led to some 150 health biotechnology products being put through clinical trials and an associated 30 per cent increase in healthcare biotechnology patents, albeit from a low base. Similarly, scientific publications in international health biotechnology journals have risen since the late 1990s. An indication of the kind of health biotechnology products emerging from Chinese institutes and firms is given in Table 8.3. It is clear from this table that Chinese therapeutics capabilities are more pronounced than in Cuba and India, while vaccines and diagnostics

Table 8.3 Selected Chinese healthcare biotechnology products at 2004

Sector	*Type*	*Application*	*Producer*
Vaccines	Recombinant surface antigen	Hepatitis B	Shenzen Kangtai Bioproducts
	Recombinant oral vaccine	Shigella dysentery	Lanzhou Institute
Therapeutics	Recombinant interferon α-1b	Keratitis/Hep. B&C	Changchun/Shanghai Insts.
	Recombinant epidermal	Skin injuries	Shanghai Dajiang Group
	Recombinant interleukin-2	Renal cancer	Shenzen Neptunus
	Recombinant granulocyte	Neutropenia	Amytop Biotech
	Recombinant erythropoietin-α	Anaemia	Shenyang Sunshine
	Recombinant somatotropin-α	Dwarfism	Changchun Jinsai
	Recombinant streptokinase	Cardiovascular	China Tonghua
	Recombinant Ad-*p53* gene	Cell carcinoma	Shenzen SiBono
Diagnostics	Immunosorbent assays	Hepatitis C	Shanghai Huaguan Biochip

Source: Zhenzhen *et al.* (2004)

emphases are less evident, signifying relative strength in research compared to the other two developing countries examined thus far. An example of Chinese advancement in gene therapy is the Shenzen SiBono GenTech treatment for head and neck squamous cell carcinoma licensed for sale by the Chinese Food and Drug Administration in 2003. SiBono was the first firm in the world to receive such certification for a recombinant gene therapy, named Gendicine, and utilising the p53 'Guardian Angel' cancer protection gene discovered by Sir David Lane at Ninewells Hospital, Dundee before he moved to Singapore's Biopolis.

Of key importance in funding research of the kind that led to Gendicin are a variety of central and regional administrative divisions. Centrally, the State Council and Ministries of Defence, Health, and Education and, for basic research, the Chinese Academy of Sciences are important. But decentralisation means that some regional funding sources are also important for applied research; thus Guangdong has a Provincial Biotechnology Research Institute specialising, for example, in SARS research. Nevertheless, the Chinese Academy of Sciences is the platform for most of the more prestigious bioscientific and biotechnological research in China. It indirectly supports the Beijing Genomics Institute, founded in 1999 by overseas Chinese scientists, while the Shanghai Institutes for Biological Sciences are a direct responsibility of the Academy. These in turn consist of the Institutes of Biochemistry and Cell Biology; Nutritional Science; Neuroscience; Materia Medica – one of the oldest in China for drug research; and the Health Science Centre. The research areas covered by these institutes include functional genomics, proteomics, bioinformatics, brain development and functions, and biotechnology *per se*.

Since the mid-1990s and with some assistance from the Canadian Industrial research Centre, China has developed a national innovation system approach to bioscience in which industrial firms have been moved more closely to the centre of biotechnology activity, research institutes were encouraged to become more like pharmaceuticals firms than pure research institutes in some cases, and healthcare biotechnology commerce has expanded accordingly. By 2004, some 500 biotechnology firms, half founded in the preceding five years, had around 50,000 employees, of which 60 per cent were engaged in healthcare biotechnology. Firms are divided between state-owned enterprises evolved from the hitherto public research institutes, which are relatively well funded with advanced equipment; and, small DBFs, many spun-out as businesses from the public research institutes or established by returning scientific migrants. Also present are foreign 'big pharma', mostly in Beijing (Pfizer, Glaxo, Novartis) and Shanghai (Merck). Collaborations and joint ventures exist between foreign and indigenous businesses and research institutes.

Chinese biotechnology faces a number of challenges in terms of rapidly transforming its 'command economy' hierarchical, silo-designed, linear organisational mode, more reminiscent of Western big pharma in the 1970s.

Thus, for example, collaboration of a lateral kind between divisions or institutes in complementary research areas is yet habitually to be practised. So much so that special top-down initiatives have been promoted by the State Council, such as the National Engineering and Technical Research Centre in Beijing, and the Research Centre for Engineering and Technology for Medical Bioengineering in Guanzhou. Furthermore, several DBF incubators for commercialisation of research have also been established. Together these have doubled the level of collaborative biotechnology activity in scientific publication practices from about 14 per cent in 1991 to 30 per cent in 2002. The evolution of biotechnology has been, perhaps surprisingly, somewhat hampered by financial bottlenecks. Although government is the main research funder, it has been criticised for short-termism and budgetary constraints. More understandably, China had to develop a venture capital sector, with the state setting up the National Science and Technology Venture Capital development Centre in Beijing in the early 1990s. Finally, there is a sizeable Chinese bioscientific brain-drain to contend with. Some 300,000 Chinese students were studying abroad in 2004, of which one-third were bioscientists. Some have returned, often to set up biotechnology businesses, but while they remain in demand in the West and can command salaries far greater than those paid in China, most were expected to remain abroad. Accordingly, the Chinese government took measures to attract expatriates home, with incentives for scientists to establish laboratories and start-up programmes for entrepreneurs. The number of return migrants doubled between 2000 and 2002 as a result of such inducements, many of these having established significant scientific or entrepreneurial reputations abroad.

Biotechnology in Brazil

Brazil has remarkable biodiversity and strengths in agro-food, and in environmental and energy-related biotechnology. Some aspects of this biodiversity act as spillovers into healthcare biotechnology. Moreover, Brazil has certain health imbalances that make a focus on certain diseases somewhat imperative, as elsewhere in developing countries. According to Ferrer *et al.* (2004) Brazil has the eighth highest world ranking for diabetes and seeks to reduce, this and other health deficits, the cost of drug imports, which ballooned from $170 million in 1981 to $2 billion in 2002. Hence a highly considered Brazilian health case concerns its recombinant human insulin product. Collaboration between the University of Minas Gerais at Belo Horizonte and biopharmaceutical company Biobrás of São Paulo resulted in this diabetes treatment being developed and US patented in the 2000. Subsequently, Biobrás was acquired by Danish human insulin pioneer firm Novo Nordisk, present at the dawn of modern biotechnology through its alliance with Cambridge, Massachusetts start-up of the 1980s Biogen, in its struggle with Genentech, the world's first modern

Table 8.4 Selected Brazilian healthcare biotechnology products at 2004

Sector	*Type*	*Application*	*Producer*
Vaccines	Recombinant surface antigen	Hepatitis B	Butantan Institute
	Attenuated virus	Yellow fever	Bio-Manguinhos
	Autologous whole cell	Prostate cancer	FK Biotecnologia
Therapeutics	Recombinant insulin	Diabetes	Biobrás
	Recombinant erythropoietin-α	Anaemia	Butantan Institute
	Monoclonal antibodies	Immunotherapy	FK Biotecnologia
Diagnostics	Recombinant antigens	Chagas disease	Bio-Manguinhos
	Phosphatase immunoassays	Antibody detection	FK Biotecnologia

Source: Ferrer *et al.* (2004)

biotechnology firm, to produce human insulin rather than modify that which had prevailed hitherto, which was derived from pigs and cattle. Hence Biobrás became only the fourth supplier of human insulin to the global market at that time.

Like other developing countries, Brazil has focused on diseases relevant to its own population. Hence, in addition to its successful focus on improved diabetes treatment derived from mimicking human rather than mammalian insulin, there have been successes with drug developments in a variant of hepatitis, anaemia and antibody detection. Vaccines have successfully been produced, notably that derived from gene sequencing the plant pathogen *Xylella fastidiosa*, which is a platform for tackling targets such as Chagas disease (*American trypanosomiasis*) and cancer variants. Some diagnostic devices have also been developed as shown in Table 8.4. However, international comparative change in co-publication and patenting activity has not been significant since the mid-1990s when evidence of movements were first observable. This is despite the fact that government support for biotechnology development was rather early in Brazil. Thus, even in the 1970s the national Research Council pioneered two research programmes, the Integrated Genetics Programme and the Integrated Tropical Diseases Programme. In 1981 a new initiative, the National Biotechnology Programme, aimed to network the various research institutes concentrating on agricultural, energy and health biotechnology, then followed up with a general Scientific and Technological Development Programme for capacity building through scientific training, infrastructure and more networking initiatives, of which 17 per cent was devoted to biotechnology. The provincial states also have programmes (States Research Support Foundations), a good example of which is that in São Paulo where financing of biotechnological projects has been pronounced. Nevertheless, the impression given by observation of the outcomes of the Brazilian biotechnological effort is that despite an early start it has been less impressive than either the relatively well sustained public effort on the part of the

Cuban and Indian authorities or the recent 'catch-up' incentive system put in place by China's State Council.

Why should this be so, especially since Ferrer *et al.* (2004) make the point that despite frequent governmental instability in Brazil, biotechnology received uninterrupted public support from the beginning and, it might be added, Brazil has a relatively liberal stem cells régime? One reason is clearly the difficulty experienced by several government initiatives to get collaboration among the key actors. Another is certain restrictions upon commercialisation of research findings, something that has affected the liberal intentions of such stem cells legislation in the past. Added to which, university professors may not be employed by industry, a major restriction on start-up and spinout DBF activity. Moreover, while some bioincubators exist, the best is in Belo Horizonte, a different location from centres of research or commercialisation. Possibly this and the relatively narrow research base in health biotechnology yielding up not much clustering help us understand the situation. Thus, while naturally the Federal Ministry of Health is located in Brasilia, the capital, a leading laboratory, the Oswaldo Cruz Foundation (FIOCRUZ), is in Rio de Janeiro with its production arm Bio-Manguinhos. The Butantan Institute is in São Paulo, as is the Ludwig Institute, an important private laboratory. But FK Biotecnologia, which in 1999 became the first Brazilian biotechnology firm to receive venture capital for the development of its innovative technologies, is located in Porto Alegre. Having said that, networking and collaboration among the institutes seem to have improved, as well as their interactions with universities of consequence, according to Ferrer *et al.* (2004). Of Brazil's 354 biotechnology firms at 2001, about one-third were in healthcare biotechnology. Foreign firms with presence in Brazil include Glaxo, based in Rio de Janeiro, Sanofi-Aventis and Roche, but they have low R&D, innovation and patenting rates in Brazil.

Probably the most important challenge concerns continuously improving network linkages in what is clearly a vast country with a rather distributed knowledge base. For example, Ferrer *et al.* (2004) conclude that private sector healthcare biotechnology firms lack linkages and knowledge transfer is perceived to have suffered in consequence. Until recently, as we have seen, co-operation among other institutes and university centres has been limited in the extreme, requiring numerous government initiatives to try to kick-start collaborative actions. That this should be so, given what is well addressed in the literature about healthcare biotechnology being one of the most interactive, to some extent classically Mertonian 'open science' and 'open innovation' contexts, it is a little surprising that Brazil with evident resource and knowledge assets should remain somewhat institutionally individualistic about its biotechnology conventions. Patenting of pharmaceutical, let alone biopharmaceutical products, also came late to Brazil, only in the 1990s, further disincentivising innovation by those with potentially appropriate knowledge to commercialise. Brazil remains interesting because it has the greatest natural bio-resources in the world, it has good, if somewhat isolated research institutes and evidence of some locally relevant if somewhat 'me-too' innovations. The challenge to build stronger networks, to build even stronger scientific capacity and to adjust its regulatory régime to be somewhat more risk and innovation friendly, possibly by pursuing a cluster strategy in order to exploit and evolve its existing and future assets, seem to be imperatives for Brazilian healthcare biotechnology if it is not to suffer ever greater import costs from what Brazilians know to be highly expensive Western, mainly American, biotechnology based therapeutic treatments for many of their clinical ailments.

Conclusions

This has clearly been only a relatively brief outline of the status of healthcare biotechnology in some of the more prominent developing countries. Other accounts dealing with countries such as South Korea, South Africa and Egypt are to be found in the useful *Nature Biotechnology* (2004) supplement on which much of the account presented in this chapter drew.

What can be said about these cases in relation to our general theoretical framework of asymmetric knowledge development, knowledge domains and knowledge capabilities as driving current global healthcare biotechnology configurations? The most striking feature, with the partial exception of Brazil, is that the success of this complex scientific and economic activity rests upon relatively high degrees of embeddedness, proximity and platform building in geographically fairly concentrated spaces in developing countries just as much as it does in accomplished settings in the developed world (Mytelka, 2006). Cuba's healthcare biotechnology is concentrated in one main area, the West Havana Scientific Pole. In India, the already high technology ICT and software cities of Bangalore and Hyderabad house biotechnology activity, especially that involving bioengineering, bioinformatics and so on. In China, Beijing, Shanghai and Shenzen are major concentrations of different combinations of relevant knowledge ranging from basic research to commercialisation. In all cases it is not surprising that much depends on the quality of publicly funded research at specialist research institutes since this, too, is the pattern in the West.

Distinctively, it might be thought, research and commercialisation focus on the quest for remedies for indigenous diseases, until it is remembered that this characterises research in the West too, where cancer, cardiovascular, auto-immune and neuro-degenerative disease is a product of those environments as much as hepatitis, anaemia and auto-immune illnesses are severe problems in developing countries. Of interest is the manner in which drug-line solutions and vaccines or at least diagnostics have been pursued and sometimes fulfilled in areas which have been of little or no interest to Western big pharma. Or if they had been of interest, extortionate prices that could not really be afforded were having to be paid. The facts of endogenous healthcare biotechnology innovation in developed countries has undoubtedly been a factor in forcing a relaxation in the grip of big pharma over drug accessibility in developing countries. To that extent it has been money well spent by those governments involved.

However, it is striking that much of this effort has been, expensively, duplicated. Notice how in many cases slightly different or retro-engineered techniques have been utilised to evolve treatments for diseases that affect many developing countries. The most obvious cases are hepatitis-related. There is scope for better international co-operation in strategic public healthcare biotechnology disease targeting. The multilateral organisations, notably the World Health Organization but other appropriate ones under the UN aegis come to mind, have an efficiency-enhancing and effectiveness-promoting role to play in optimising the use of often extremely scarce resources in developing countries. Cuba, in its semi-isolation ironically offers the best model whereby drug treatments are provided to other developing countries at favourable cost – not surprisingly at complete odds with the tight-fisted traditions of Western big pharma who, in a sense, force developing countries to go it alone to exert market power over them until they can threaten pharma IPO. The extent to which developing country pharmaceuticals production has begun modifying Western power asymmetries was testified to in the New Year outburst from Jean-François Dehecq, chairman of Sanofi-Aventis, the world's third largest pharma (Jack, 2007). Dehecq claimed to be 'scandalised' by developing country pharma companies selling cheap generic (off-patent) drugs to developed country health systems, holding that they should desist, otherwise high pricing strategies by Western pharma would be undermined. Such staggering patronisation provides an insight into the gold-plated mentalities that often reside in the higher reaches of big pharma, as elaborated at length in Law (2006). The lessons from this brief journey into the asymmetries in health-care biotechnology in the developing world is that greater co-operation among them, with or without the intervention of the multilaterals, who may in their own ways be in thrall to pharma, will produce win–win outcomes comprised of more efficient and effective spending on the one hand, and on the other, greater leverage upon pharma to reduce prices on a far larger scale for drug treatments they produce for diseases inordinately suffered in developing countries. The savings developing countries might thus make might then be available for focus on the niche or 'orphan' diseases developing countries continue to suffer, in which Western pharma is never likely to be interested.

9 Environmental, energy and agro-food bioregions

Introduction

Although bioregions are generally dominated by biopharmaceuticals DBFs and specialised research institutions, they often contain firms in other segments of biotechnology too, as we saw in Cambridge, UK. However it is less usual for such complexes to be dominated by non-biopharmaceuticals activity. Four of these are profiled in the following, one predominately containing environmental DBFs and related institutions, and three comprising substantial agro-food specialisation, with one also concentrating upon agro-food, biosynthesis, vitamins, and chemicals. The environmental one is Rhineland in North Rhine Westphalia, Germany; then three biofuels regions are profiled in Brazil, Sweden and the UK, followed by five agro-food bioregions in Canada (Saskatoon), Finland (Helsinki and Turku), Sweden (Skåne) and the Netherlands (Wageningen-Nijmegen). In all cases, as with biopharmaceuticals, there is geographical proximity and institutional embedding of the relational variety in the structure of these bio-platforms.

Rhineland environmental bioregion

We saw in Chapter 7 how the Rhineland was selected as one of Germany's top bioregions in the BioRegio contest. One reason was that Rhineland has a mixed biotechnology platform. It will be recalled that various networks were established, including a venture capital network of some fifteen local but also internationally operational firms and groups, a competence and training network, and a management and coaching network. BioGenTec had established Biocentres at various locations and organised an annual international meeting called the Bio-Gen-Tec Forum. In 2000 an international forum on nanobiotechnology was organised. The agro-food research strengths of the *land* include the Max Planck Institute for Plant Breeding research at Cologne, which has become the centre of agro-food biotechnology, around which larger firms like GMO specialists Monsanto, biologics specialist DSV and Agrevo and smaller firms are clustered. In 1998 a letter of intent was signed between the governments of NRW and Saskatchewan, Canada to improve collaboration in the field of agro-biotechnology. We will see that Saskatoon is one of the leading agro-food biotechnology centres, something recognised early at BioGenTec. Also in Cologne is the Max Delbrück Laboratory, an affiliate of the Max Planck Society specialising in plant genetics. There is a Fraunhofer Institute for Environmental Chemistry and Ecotoxicology at Schmallenberg. Altogether the *land* has some 167 research institutes, many employing relatively small numbers of researchers, but with representation across the environmental, agro-food and pharmaceutical biotechnology spectrum, as well as many other fields not related to biotechnology.

It is worth noting that, in terms of the sectoral distribution of *companies* working in the field of biotechnology, in 2000, 22 per cent were in diagnostics, 12 per cent in pharmarceuticals, 7 per cent in agro-food biotechnology, 18 per cent in environmental protection, 9 per cent in filtration engineering and 10 per cent in bioanalysis. The last three are primarily engaged in *environmental* biotechnology activities, making that, at 37 per cent, easily the largest category. Thus, although it is seen as a 'Cinderella' part of biotechnology and one in which it is hard to get university-derived start-ups underway, it is a strength of this region's biotechnology profile. The origins of this commercial expertise are interesting and reflect well on the restructuring efforts of the NRW government following the decline of coal and steel. Because of German rules on co-determination, involving management and unions in strategic decisions concerning, for example, company restructuring, management could not simply decide to close down redundant plants but were required to explore alternative trajectories that firms might seek to move along. Because of expertise in the mining and steel industries including, for example, filtration and ventilation technologies, it was recognised that adapting these for environmental clean-up, initially in the Ruhr itself, but later in former east Germany and into Central and Eastern Europe, would meet a huge potential market need. Moreover, it was known that the EU was set on introducing tougher environmental legislation in 1990, and on advice from NRW politicians the federal government introduced an equivalent German version in anticipation, thus augmenting market demand for environmental clean-up technologies. Between 1984 and 1994 some 600 firms turned partly or wholly in this direction and a number of new start-ups or spin-offs also emerged. Over 100,000 jobs were found in this new industry, which itself has been shown to have a cluster-like character (Cooke and Davies, 1993; Rehfeld, 1995).

Finally, to what extent can clustering be said to be a feature of the Rhineland BioRegio or areas adjacent to it? From responses elicited to the question it seems that the key conditions are in place: a strong science base, expanding numbers of firms, qualified staff, available physical infrastructure (e.g. the Rechtsrheinisches Technologie Zentrum, RTZ in Cologne, a 5,000 metre square biotechnology incubator with plans for a C4 quality central laboratory), availability of finance, business support services, a skilled workforce, effective networks and a supportive policy environment. However, the number of biotechnology start-ups was at a peak of 26 in 1990–1991, and in decline to a figure of 12 in 1994–1995, recovering over 1996–1997 after the announcement and including the first operational year of BioRegio (BioGenTec, 1998). However, by 2002 the Rhineland BioRegion had outperformed the Munich and Heidelberg ones, as published data show.[1] This is connected to the effect of the *Neuer Markt* collapse that discouraged new firm formation in biopharmaceuticals, in which Munich and Heidelberg specialise, but which left bioenvironmental activity relatively unscathed. The ITS Expert Mission report (DTI, 1998) reported eleven start-ups and eight company expansions since 1996 in biotechnology in the Rhineland BioRegio area. However, from 1997 to 2002 the total number of biotechnology firms in Rhineland rose from 45 to 140, while in Munich the equivalent statistics were 55 to 120, and Heidelberg 45 (1999) to 55. Thus regional shares of biotechnology firms by 2002 were Rhineland (37%), Munich (33%) and Heidelberg (16%). Thus despite BioGenTec's caution (the 'quality' criterion) and disappointment in the environmental biotechnology centre of Aachen and Jülich where start-ups had been hard to stimulate, a positive outcome resulted. It is also worth remembering that QIAGEN, based in this BioRegio, remains Germany's best firm, leading networks such as the Gene Alliance. In 2000 DSM Research commissioned the Gene Alliance to determine the DNA sequence of the fungus *Aspergillus niger*, one of the most important organisms used in enzyme production at DSM. Thus far the Rhineland BioRegio has confirmed it had the appropriate conditions for stimulating the development of reasonably large numbers of new biotechnology firms with characteristics of a cluster of growing biotechnology firms on the evidence presently available.

Brazil's bioethanol industry

Brazil's effort to develop alternative fuels occurred with the 1970s oil crisis. At the time, Brazil was ill-prepared to absorb the negative economic impacts of rising fuel costs with associated inflation. Ironically, Brazil is now known to possess the second largest oil reserves in South America after Venezuela. Proalcohol was created in 1975, bringing blends of ethanol and petrol in high volume, aimed also at stimulating production of ethanol-fuelled vehicles. Subsequently some vehicle fleets utilised 100 per cent ethanol fuel evolving latterly into hybrid fuel fleets using petrol-ethanol blends. In 1997, the government partly privatised the state-owned oil company, Petrobras. Like several other oil companies they are active in both traditional and biofuels markets domestically and abroad. The Proalcohol programme impacted on the Brazilian energy market since it required all petrol stations to supply bioethanol. It also stimulated growth in availability of bioethanol vehicles through a 5 per cent tax reduction for such vehicles.

Bioethanol sales and prices have fluctuated with changes in fossil fuel prices, notably oil. But in the 2000s, global climate change concerns, the Kyoto Protocol and rising oil prices led to a resurgence of the ethanol market in Brazil. In 2003 hybrid vehicles were available for sale in Brazil and by 2006 a variety of models from Citroen, Fiat, Ford, GM, Peugeot, Renault, VW were available to run on ethanol, petrol or a blend. Much of this biofuel vehicle and equipment production is centred on São Paulo, although there are some 347 sugar mills and distilleries throughout Brazil in locations that farm sugar cane. For example, a large Cosan refining plant at Costa Pinto is approximately one hour from São Paulo. Cosan is Brazil's largest sugar refiner and with 16 other plants produces a billion litres of ethanol worth $200 million per year. In recent years demand has grown from Europe, the US, and India to utilise bioethanol fuels. In the US corn and soybean are seen as potential but expensive biofuels. But greater effciency is thought likely to arise from the use of biowaste both in biomass power stations and for producing biofuels. Nine General Motors bioethanol vehicle models are produced in São Paulo and 95 per cent of GM output in Brazil is hybrid petrol ethanol fuelled.

Pump prices for Brazilian ethanol are approximately half those for petrol in Brazil. Exports are 11 per cent of the ethanol Brazil produces and a strategy is being implemented to triple ethanol production by 2010. Brazil also consumes about 40 billion litres of diesel annually, of which 10 per cent is imported. This is a stimulus for its emerging biodiesel market, which was planned to expand biodiesel annual production from 156 million litres to some 520 million litres during 2006. Biodiesel is produced with animal and plant fats and is subsidised with tax credits and incentives for consumers, producers and farmers, and supported in federal and some state codes with renewable energy policies. Policies at national and regional levels are important to the evolution of Brazil's biofuel platform, including the supply chain and intermediaries. The Proalcohol programme requires bioethanol and biodiesel producers to access feedstock from production networks of small farmers, which is meant to prevent agribusiness monopolies from dominating supply. Next-generation developments include policies for total biodiesel production from crop farming, to fuel production and transport to filling stations to utilise biofuel and bioelectricity.

Sweden's bioethanol cluster and market

Sweden has become a world leader in the development and processing of renewable fuels, primarily low blended ethanol in petrol with a leading edge tax subsidy and globally leading research programmes. Due to policy evolution over a number of years, Sweden is one of the largest markets for ethanol fuels used in vehicles. Only Brazil is more advanced in terms of development, and Brazil and the USA in terms of volume. All ethanol used for fuel is processed from wood or other renewable sources, a defining feature of biofuel. Thus Sweden has had buses running on pure ethanol for more than 20 years, and approximately 34,000 cars run on low ethanol blends (e85). Low ethanol blends reached 6.2 per cent of the petrol used by vehicles in 2006. Elsewhere in Europe there are some test cars but few buses running on pure ethanol.

The northern Swedish town of Örnsköldsvik has become a cluster of ethanol processing, research and development. Sekab, Etanolteknik (Etek) and the BioAlcohol Fuel Foundation (BAFF) are all based there. In May 2004, a pilot plant for industrial-scale ethanol production was inaugurated in Örnsköldsvik. An ethanol cluster arose in Örnsköldsvik due to a pre-existing sulphite pulp mill processing sulphite for sugar as a by-product from which ethanol is produced. Etek, for example, is locally involved in a research project on New Improvement of Lignocellulosic Ethanol (NILE) funded by the EU and budgeted at €12.8 million. Lund University of Technology was one of the initiators of this project, encompassing 21 universities, institutes and companies, coordinated by Institut Français du Petrol. The Etek role in the project is primarily to conduct verification of laboratory process development tests on a large scale. Etek's pilot plant produces 13,000m^3 of cellulose-based ethanol. Sekab and Agroetanol are leading producers in Sweden with customers being global oil companies. Despite these exports, ethanol fuel is also imported from Brazil and the US. Thus Sweden is also a substantial market for ethanol fuel. In 2006, 377 petrol stations supplied e85 and the number is increasing by some 200 per year on existing trends.

Although the ethanol fuel market is strong in Sweden, biodiesel is less evolved, accounting for only 1 per cent of total Swedish diesel consumption. Sweden's few biodiesel pumps are concentrated in the three biggest cities and very few vehicles are driven by this product alone. In 2005 some 9,000m^3 of low blended biodiesel and 1,600m^3 of pure biodiesel were sold in Sweden. It approximated to two thousand ordinary cars according to Miljöfordon in Gothenburg, an organisation working towards a more environmentally friendly vehicle stock in the city. Similar organisations exist in Stockholm and Malmö financed by each municipality respectively. It is anticipated that the supply of biodiesel will increase in Sweden, since two new processing plants were opened by 2007. Tax incentives mean that expectations for further companies to utilise biodiesel in future are positive. In 2004, the Swedish government allocated €133 million in their tax strategy for alternative fuels. In 2006, the Swedish government announced tax reliefs for 20 Swedish companies retailing fuels with ethanol or biodiesel. These companies gain relief from CO_2 and energy taxes when producing or importing biofuels. The Swedish government further announced that biofuels would continue to receive supportive biofuel tax relief until 2013.

Northeast England biofuels cluster

Building on work begun in 1997, Northeast Biofuels (NEB) is dedicated to identifying and nurturing new biofuels business opportunities. Biofuels include all crops grown for fuel either as biomass, which is mostly waste vegetation, or farmed biofuels like sugar cane in Brazil, corn and soya in the BioBelt of the USA and the birch forests of Sweden for methane or ethanol that fuels power plants and cars, or even waste oils collected in community schemes from homes and restaurants. As a case in point, NEB's work has focused mainly on the liquid transport fuel sector, making NEB distinctive among other publicly supported initiatives operating in renewable energy and fuels, as well as non-food uses of farm crops. 'One NorthEast', the regional development agency, supports this initiative financially and in networking activities. NEB is also distinctive in its cluster approach by building a 'vertical' cluster group. Thus NEB brings together companies and public sector organisations who represent all levels of the new supply chains necessary to develop sustainable biofuels businesses.

A vertical cluster group differs from many that exist between cooperating companies and organisations that work together in a specific business area to improve the efficiencies of the industry and to promote innovation. Thus when a new industry is evolving, as with biofuels, there is a need to establish new supply chains. These span the source of production of raw materials, for example the crops in farmers' fields, through to the distribution of the finished biofuel, to intermediate and final consumers. Many of the companies involved in such new supply chains can be unfamiliar with the practices and problems of other firms in the chain. NEB members collaborate in exchanging knowledge and expertise among the group. In turn they reap advantage from networking and learning opportunities for 'open innovation'.

Members of NEB include companies already involved in the production and distribution of biofuels, including biodiesel production companies, Biofuels Corporation PLC, and D1 Oils PLC, as well as the major diesel refining and marketing company Petroplus. The emphasis is upon production of feedstock for fuel production on regional farms; various farmer-owned and agri-businesses are involved as members. These include Farmway, GrainCo, Agrovista, Monsanto and Terra. Related variety companies come from engineering, utilities, bulk storage and logistics, including K Home International, SembCorp, Simon Storage and Vopak. One NorthEast, northeast England's regional development agency, supplies financial support and expertise to the group alongside various independent bodies, such as Renew Tees Valley, NEPIC and North East Community Forests. In multi-level terms, these are further linked through direct membership to a network of local, UK and European policies and business associations involved in biofuels.

Saskatoon, Saskatchewan, Canada

Agro-food bioregions are less widely present, less developed and less researched than those in biopharmaceuticals. However, as Table 9.1 makes clear, while many claims are made, only a few can be considered significant agro-food bioregions, especially when judged by the important criterion of the percentage of agro-food biotechnology businesses. Having discriminated mainly on those grounds, it is worth noting that large chemical or food corporations are often more important as innovation leaders in agro-food biosciences. In the former group, corporates like Monsanto (now Pfizer), Aventis and Bayer stand out, with, as we have seen, Novartis also active. In the latter category Unilever and Nestlé, for example, are responsible for leading much agro-food biotechnology research.[2]

Saskatoon is an interesting case of a remote cluster, the origins of which lie in the activities of large-scale organisations that have given rise to specialist spinout innovators and related supplier industries. These act as 'cluster anchors'. Saskatoon's primacy grew based on public research into rape-seed oil (in 1978 copyrighted in Canada as canola) over the 1940–1970 period conducted by Agriculture and Agri-Food Canada (AAFC). By the 1980s the federal National Research Council–Plant Biotechnology Institute (NRC–PBI) engaged in private partnerships with agro-chemicals firms like Monsanto to exploit the genetically modified variants that were beginning to have apparent commercial potential. The University of Saskatchewan, AgWest biotech, SABIC and the Saskatchewan Canola Development Commission are other key institutional actors assisting in sustaining the bioregion's agro-food commercial activities, about a third of which involves core agro-food biotechnology exploration, examination and exploitation. Much of the last-named commercial activity occurs elsewhere in seed and plant science firms in North America and further afield. Hence Saskatoon's is a scientific research-driven cluster in the main, where proprietary technologies are imported, assembled into new crop varieties, then exported as germplasm or as intermediate product to global markets. Collaborative scientific publication equally have 30 per cent of NRC collaborators located outside Saskatchewan. The research emphasis is reflected in new projects such as that initiated by Genome Prairie, 'The Abiotic Stress Project', on ethical, social and environmental aspects of the genetically modified organism (GMO) debate. Most importantly, the cluster network has recently been joined by Canadian Light Source Synchrotron Inc. (CLSI) that enables synchrotron capabilities to be available for needed bioimaging. CLSI anticipates attracting a further 2,000 scientists.[3] Among the most important members of the Saskatoon cluster, measured by IPR portfolios, are the following private actors: Aventis, Biostar, Dow Agrosciences, Enviro Test Laboratories, Fytokem Products, Monsanto, Performance Plants Inc., Philom Bios, Pioneer Hi-bred International, and VIDO. Key public organisations are: AAFC, NRC–PBI, Saskatchewan

Table 9.1 Selected agro-food bioregions

Countries	*Bioregion*	*Brand*	*Actors**	*% Ag-Bio*	*Market focus*
Canada	Saskatoon (Sk.)	'Innovation Place'	115	29	Canola, flax
	Guelph (Ont.)	'Agrifood Quality'	41	49	Corn
USA	Connecticut	'Bioscience Cluster'	110	1	Corn, fruit
	Raleigh-Durham	'Rsch.Triangle Pk.'	145	3	Corn, soybean
	St. Louis	'BioBelt'	1,183	24	Corn, soybean
	San Diego	'Biotech Beach'	700	3	Forestry, fruit, veg.
Europe	Scotland	'Innov. Triangle'	428	2	Transgenics, potato
	Sweden	'Skåne Food Cluster'	60	25	Functional foods
	Fr-Ger-Switz.	'BioValley'	459	6	Cereals, cotton, livestock
	Seinajöki, Finland	'Foodwest'	25	5	Diagnostics
	Netherlands	'Food Valley'	48	60	Food genomics
Australia	Brisbane (QL.)	'QBio'	43	5	Forest, aqua, hort.
	Sydney (NSW)	'BioHub'	28	18	Livestock, cereal
	Melbourne (V)	'Bio21'	24	4	Plant/an.genomics
	Adelaide (SA)	NA	25	44	Wine, plant/an.gen.
	Perth (WA)	NA	27	20	Wheat, lupins

Source: Ryan and Philips (2004); Svensson-Henning (2003); Invest Skåne (2004); Tulkki *et al.* (2001); www.plant.wageningen

*NB: Food producers; R&D institutes; raw materials and ingredients suppliers; packaging firms; industry institutes; government agencies; food organisations

Research Council, and the university of Saskatchewan Crop Development Centre. Between them, these actors hold some 375 patented innovations, of which 232 are protected only in Canada (Ryan and Phillips, 2004).

Finland's agro-food bioregions

We saw in Chapter 6 how Finland displayed a distinctive mix of regional cases worthy of exploration. Through Tekes and the other national support organisations for research and innovation, Finland developed numerous Centres of Expertise, and later Centres of Excellence in its regions. As we noted earlier, these were meant initially to be secured by relatively few, leading research centres in the range of biotechnology applications. But political demands resulted in many more centres being established, diluting somewhat the effect of the policy. Two things are striking regarding the Finnish biotechnology policy experiment. First, despite a relatively large share of the Finnish R&D budget being committed to biotechnology, the desired result of the rapid development of a biotechnology 'national champion' like Nokia in ICT simply did not happen. This has subsequently caused some significant soul-searching within the policy apparatus, to the extent that Tekes and the wider system of innovation support have been further 'regionalised', not least because the 'one size fits all' policy tendency in over-centralised innovation systems is ineffective and inefficient, as indicated arguably by the 'pork-barrelling' around the Centres of Excellence initiative. The second noteworthy feature of the Finnish approach to supporting biotechnology is that Finland's relative excellence in agro-food biotechnology was profiled and strengthened while gains in healthcare biotechnology were relatively modest.

To recap, Finland's key policy aim had been to build on the research potential of four university-affiliated regional centres of biotechnological expertise by 1992. Helsinki, Turku, Kuopio and Oulu were the chosen centres. The programme was evaluated and continued in 1996, then extended to 2000. Financing came from the Ministries of Trade and Industry, Agriculture and Forestry, and Social Affairs and Health as well as Education. In 1993, the Ministry of Education had set a new Centres of Excellence standard, seeking to identify ten 'top units' (Academy of Finland, 2002). By 2000, 26 had actually been established, of which 9 were in biosciences and biotechnology. It is rather complex to estimate with accuracy biotechnology's share of this funding, and harder still to assess the share of that going to agro-food and other non-healthcare biotechnology applications.

Tekes invested 27 per cent of its annual budget of approximately $90 million in biotechnology. The Centres of Expertise programme received $4.1 million annually from Tekes and the Academy of Finland, equivalent to 40 per cent of their combined budgets in support of biotechnology. A further annual amount of $13.5 million in biotechnology came from a national biotechnology programme dating from 1988. Finland's Genome and Cell Biology research programmes spent $6.3 million annually. Then from 2001 an average of a further $50 million annual co-funding came from Tekes, the Academy of Finland, Sitra, the Finnish national R&D fund, Finnish Bioindustries, and a substantial group of pharmaceuticals companies to 'Medicine 2000', addressing biomedicine, medicine development and pharmaceutical development research and technology. At the peak, therefore, it is likely that some $164 million of mainly public research and innovation investment supported Finnish biotechnology annually. Other centres were added such as Tampere and Seinajöki.

Of the 26 centres, it was noted in Chapter 6 that 9 covered aspects of biotechnology and of these Helsinki and Seinajöki specialised in agro-food biotechnology, with some agro-food R&D also performed in Kuopio, Oulu and Turku. Hence there was belated recognition of the relative importance to Finland's future economy of its relatively advanced agro-food biotechnology expertise. Finland's agro-food firms produce nutraceuticals innovations such as the claimed anti-cholesterol product Benecol. The responsible firm, Raisio, is based in south-west Finland near Turku. Raisio is a specialist in plant-based nutrition and in ensuring food safety with, as noted elsewhere, claims to be a forerunner in functional foods. Raisio has four divisions: food, animal feed and malt, ingredients and diagnostics. Regarding, for example, diagnostics at Raisio, this involves pathogen testing, utilising modern rapid testing methods, the demand for which from the food industry continues to rise. Products include those that respond to demand for swift analysis of the main food pathogens, notably salmonella and listeria, automated solutions for the screening of large samples, and enzymatic analysis methods that allow for diagnostic automation in the dairy, wine and soft drinks industries. Raisio employs some 1,400 people at 12 locations in 6 countries, about 70 per cent of employment being in Finland.

Valio is headquartered in Helsinki and specialises in functional foods in the form of lactobacter drinks and UHT infant food. It has been a pioneer in probiotics research and exploitation. The claimed health effects of lactic acid bacteria attracted the interest of Valio's R&D department in the early 1980s, leading to production of the early range of lactic acid products called Asidophilus. Further product development followed discovery of the bacterial strain *Lactobacillus GG (LGG)*, isolated in the US by Gorbach and Goldin of the Friedman School of Nutrition Science and Policy at Tufts University, Boston. Valio made contact in 1985 and in 1987 secured an exclusive global rights licence under which Valio obtained the right to sell and re-license *LGG*. It is the most widely studied lactic acid bacterium in the world. It is claimed to prevent and cure intestinal disorders, to protect the intestines from harmful microbes and toxins and to reduce allergic reactions.

Xylitol is the first Finnish functional food ingredient whose health impact has been scientifically proven. It is mainly used as a sweetener in chewing gums and pastilles. Finnish Professors Scheinin and Mäkinen proved the inhibitory effect of xylitol on dental caries in the early 1970s. Later, two Oulu-based researchers, Uhari and Kontiokari, also proved that short-term use of xylitol over some two to three months significantly reduced the risk of inflammation of the middle ear among day-care children during the test period. The Finnish Xyrofin company is the market leader among xylitol sweeteners. Xyrofin's predecessor Suomen Sokeri patented the industrial manufacturing method for xylitol in 1972 and began production of xylitol from birch chips at its Kotka, south-east Finland plants. Today, Xyrofin xylitol is used round the world, mainly as a sweetener in chewing gums and pastilles. Other applications include oral hygiene products such as toothpaste, fluoride tablets and mouthwashes. The pharmaceutical industry uses xylitol as a sweetener in its products, Danisco Finland being a major producer. It is worth recalling that municipalities have played an important role in supporting innovation in Finland. They have substantial fiscal autonomy and invest directly in capital and revenue costs of innovation. Turku's bioscientific strategy arose from the municipality's efforts to transform a declining port into a BioCity. Also in Finland's south west, Seinajöki, a small market town in an agro-food production area dominated by SMEs, led a group of small cities in the Vaasa–Suomi region to club together and fund 'filial' chairs in food biosciences, earning that location one of the Centres of Expertise although it was without a university (Kosonen, 2006). It also hosts an innovative global–local 25-member partnership 'Foodwest' supplying diagnostics for the food and timber industries.

Sweden's agro-food biotechnology cluster at Skåne

Some 50 per cent of Sweden's agricultural output derives from its southernmost region of Skåne, centred upon the port city of Malmö, the university city of Lund, and numerous smaller settlements in the region. Rather in the manner that in 1995 Germany launched its BioRegio contest, Sweden in the early 2000s launched a cluster support contest. One of the first three winners of VINNOVA's (the Swedish Innovation Systems Agency) VinnVäxt programme support when it began in 2003, alongside Robotics Valley near Västerås, west of Stockholm, and Uppsala Bio, northwest of Stockholm, was the Food Innovation at Interfaces proposal from the Skåne Food Innovation Network. This association has been in existence since 1998 and represents about 33 per cent of all food companies in the region. The two main categories of agro-food represented in the cluster are conventional foods, notably meat and fruit packaging and other food preparation, chilling, packaging and transportation companies, and of greater interest to this book, the *functional food* industry. The latter is represented by a whole panoply of research interests from Lund University and Lund Technical University, the IDEON science park, and social and natural scientific researchers ranging from management to nutrition and food technologies.

The early history of the product that exists on supermarket shelves in Sweden, ProViva, has been outlined earlier but bears a little repetition and greater elaboration here. Thereafter, some attention is devoted to two other research lines that have yet adequately to penetrate consumer markets. One of these is an oats-based functional food substitute for milk, known currently as Oatly. The other involves the Mona Carota carrot-juice drink, a programme that has received numerous setbacks that are indicative not only of the 'health claim approved' standard of 'healthiness' but also the sheer complexity of bringing new functional foods to the market as innovative products. We begin with the history of ProViva. This product began in the laboratory in the mid-1980s as an attempt to improve the post-operative health of surgery patients in hospitals. Lack of food during the pre- and post-operative period led to healthy bacteria leaking away from the intestines. A solution was proposed by Kåre Larsson, Professor of Food Technology at the Lund Institute of Technology, based on oatmeal. This has high protein nutrient, a good amino acid profile, and is high in fatty acids and betaglucans. It was thought capable of re-balancing levels of intestinal flora, thus aiding recuperation as long as appropriate lactobacilli bacteria to ferment the oats could be identified. The best candidate was *L. plantarum* 299v, and patent applications on the manufacturing process and three lactobacilli were lodged. Thereafter, energy turned to the formation of a firm to act as the commercial exploiter of the research, resulting in a firm called Probi being selected, with a dairy firm Skånemejerier being invited to take a 23 per cent share in Probi. Their interest was attracted by a sharp rise in consumer interest in health food in the late 1980s, a field Skånemejerier had yet to find a product to place on offer to. One firm had opened with a probiotic sour milk called Verum while another introduced one called Dofilus. Hence a non-milk alternative was of interest, especially one with a history of assisting nutrition improvement in post-operative surgery patients. Eventually, in 1994 a new product called ProViva, consisting of a cocktail of bacteria and aimed at a supermarket audience, was produced and marketed. Because of its alleged health gain properties – for example, relieving then-fashionable irritable bowel syndrome – it was sold at a premium price. In the UK the marketing slogan by 2000 was 'a healthy start to a busy day'. In Finland and Sweden it was 'a good lifestyle starts in your stomach'. Thus a stereotype of workaholic Brits needing to overcome their bad food habits contrasted neatly with a more laid-back, open-air-minded Nordic consumer market message. For reasons to do with knowledge transfer from functional food producers to supermarket managers, this campaign was directed to more receptive Finland, a small market clearly. The specialist nature of the product and the small marketing budgets available to the producers made it hard to sell outside specialist markets, though ProViva is sold in Swedish supermarkets reasonably successfully.

The Oatly case, as the name suggests, also owes its origins to the eponymous cereal, one that thrives in Skåne and comparable northerly climes where others do not. Human lactose intolerance had been identified as a medical syndrome, and then later medical research connecting oats with anti-cholesterol health gain sparked further consumer interest in health foods and producer interest in a potentially fast-growing market. The key problem was similar to that faced by ProViva – how to make an oats drink palatable to the consumer. Blueberry and rose hip flavouring popular in Sweden had assisted that process for ProViva while limiting the market somewhat. Once again Lund Institute of Technology was the pacemaker, particularly scientists at its Department of Food Chemistry and Nutrition. But it was aided substantially by the interest of the regional agricultural research fund, Skånska Lantmännen Research Foundation (Lagnevik *et al.*, 2003). This agricultural research fund, established in 1985, received numerous research proposals on oats following the publication of positive bioscientific research results on the possibly benign effects of oats upon human health. Then IDEON Agro Food, a network linking industry and research, presented at a seminar at the science park a vision of healthy consuming based on their concept of oats milk. This would aim to yield lactose-free milk products, contribute to the lowering of blood cholesterol, and be palatable to the consumer. Patenting was applied for and a commercial vehicle for realising the vision – Lundagrion – was founded. Two industrial partners, local packaging innovator TetraPak and the aforementioned Skånemejerier showed interest, but ultimately withdrew from the project.

Other large firms from producers to supermarkets were approached, but found the oat drink unpalatable until one decided it might make a successful dietary supplement. A joint venture firm Ceba was formed between the agricultural foundation and the researchers and development work on a new slimming drink (Complätt or Complete) began. A supplier to Friggs the main contractor, with an aseptic production line, was identified and production was set in train. However, when Friggs sold Complätt the new owner did not support the product, removing it from the market. So Ceba returned to the oat milk concept and the new Oatly product was test-marketed in the UK in 1996, once again looking for maximum publicity at the least feasible cost. But neither in the UK nor later in the US could the product gain acceptance by the supermarket giants until French major Danone took the product on, but shortly afterwards they withdrew it due to poor sales. By 2003 sales of Oatly were 3 million items of which 1 million were sold in Sweden. Oatly is also on sale in organic supermarkets such as the UK's Fresh & Wild (in 2006 acquired by US organics chain Whole Foods Market, see Chapter 11). Clearly, this functional food like ProViva appeals mainly to a niche Nordic market.

Finally, what about Skäne's next and third effort to produce a blockbuster functional food? This involved a long shelf-life carrot drink called Mona Carota. Such a product does not yet exist because carrots do not have a long shelf life due to its propensity to rot in storage in the short-term.

Once again IDEON Agro-Food was the network handmaiden, leading to interaction between scientists at Lund University's Department of Food Engineering and the Skånska Lantmännen Research Foundation. Chilling and freezing technology firms were involved and the Department of Plant Biochemistry became active, as did Skånemejerier. However, the evolution of this product has been hampered by, among other problems, once again, problems of taste. Testers have complained of carrots tasting of earth, the nematode worms that can infest them, and even the steel on the mill utilised for milling the carrots. Two scientists drove to Switzerland to find an appropriate, tasteless milling machine, but had to make do with one that was not purpose-built. Some progress was made on a visit to a German research centre on the return journey, which was also a competitor. Amidst fears of industrial espionage, the researchers discussed the palatability issue with their counterparts. On hearing of the taste problem, which the German scientists had solved some time ago, the response was to the effect that the Lund scientists might try a traditional boiling of the carrots before turning them into juice, since boiling is an age-old method of removing impurities.

These qualitative sketches are drawn from a remarkable study of the Skåne functional food industry by Lagnevik *et al.* (2003). This acts as a neat supplement to the more quantitative data necessarily employed in accounting for the biotechnology platform worldwide today. Lagnevik *et al.* indicate the difficulties and setbacks experienced by researchers and businesses in trying to generate market activity, even in the relatively benign conditions of an increasingly health-conscious consumer market. That the functional foods industry remains relatively small is testimony to the obstacles that surround the evolution of innovative activity. It is also a relatively understudied field at this time, though Skåne and agro-food students are doubly lucky in having a second book published that addresses more quantitative aspects of the industry, by Coenen (2006). In late 2006, the first evaluation of performance of this and the other two VinnVäxt clusters was undertaken and some interesting responses to lessons learned were absorbed and changes in the direction of the strategy of each were reported. For the Skåne Food Innovation at Interfaces initiative, as for Uppsala Bio and Robot Valley, there was clear recognition that the cluster perspective had been too narrow and that for the future the leadership teams wished to broaden their perspective to recognise the reality that each was more engaged with a platform of related industries than a specialised cluster.

All three realisations of the weakness of the cluster fashion when put into practice as a well funded policy mechanism are interesting and worthy of summary at this point. Recall, the Swedish Innovation Systems Development Agency (VINNOVA) expressly required a cluster approach from contestants in its cluster-building strategy. This encouraged proposals focusing on highly confined technical fields, such as the functional food products discussed in the Skåne case. But for the reasons discussed, these take lengthy time periods to reach fruition, there is insufficient variety to enable risk-spreading if one line fails and, in the case of functional foods, markets are fragile because healthy food is a selling point but without scientific support. Thus in the Skåne plan for the future, a number of lateral moves were envisaged. First, more attention was to be to devoted to two aspects of the consumer market that had been understated hitherto: foremost was a recognition that 80 per cent of the Swedish food market is accounted for by institutional customers like hospitals, elderly-care homes, schools and other large public bodies in Sweden's extensive welfare state system, while only the remaining 20 per cent is accounted for by the private market. One startling fact underlining this shift in perception was that within the Swedish school meals system, which is in any case among the best in the world, that of Malmö is the best in Sweden and widely perceived by experts to be in the top three globally.[4] Although the Malmö school meals system was included in a sub-project of the Food Innovation at Interfaces initiative, the project had only tangential aims in relation to the diffusion of innovation within the region. The project aim was to find out how pupils taking school meals evaluated the experience, and why those who took their lunch in fast-food restaurants continued to do so. For the former, the food was well evaluated but the design of the canteen was criticised for its noise, scale and dowdiness. So young designers were hired to create more segmented spaces, cubicles and booths of the kind, ironically, found in fast-food outlets. Remarkably, those using fast-food outlets freely expressed the view that they knew they were daily eating 'bad food' but that, nevertheless, they and their peers preferred a fast-food, 'oppositional' ambience typically found in their favoured fast-food outlet. Thus 'ambience' was discovered to be a much more important variable in the school-age food experience than hitherto suspected.

The second turn in policy perception in the Food Innovation at Interfaces initiative was to come to terms with the politics of Malmö's school meals excellence and pay more attention to what lay at its foundations in future thinking about the Skåne food industry. Hence, what was the reason for Malmö's school meals system being ranked among the top three in the world? It arises from the Food and Environment ('Mat och Klimat') project, the aim of which is to serve 100 per cent organic food to all public school pupils from primary to secondary school and high schools in Malmö by 2012.[5] The project is a part of the environmental profiling programme of Malmö city council. Thus it is a political project and one that is asynchronous to the current predispositions of the conventional food industry that dominates the regional economy. The school meal programme was initiated with the purpose of being a model for other institutions by supporting healthy and sustainable food production. The school restaurants prepare the meals, organise transport and serve the food in local schools with the slogan: school lunch '*more than a meal*'. Apart from good and healthy meals, emphasis is also on sharing, good food experiences, and developing healthy habits. Some 350 persons are employed in the project, which has 12 production units, delivering to 71 receiving units at the schools. Some 35,000 meals are served per day, a total of 6 million meals per year. The cost of raw materials is some €0.75 (£0.50) per meal while the total cost per served meal is some €1.60 (£1.25), and in 2005, 22 per cent of the food budget was expended on organic food products.

Recognition of the influence of such programmes and the purchasing power they represent has meant that the Skåne food 'cluster' has to evolve into a food platform embracing not only functional food and conventional food but also organic food. Even the limitations of the cluster focus on a narrow range of functional products will change with a new, more horizontal approach to making functional food applications in the conventional food production system to seek to improve the nutritional value, durability and hygiene of what is rapidly in danger of being perceived by discerning customers as cheap and not necessarily good-quality food. Thus functional food is more and more being presented somewhat like *haute couture* fashion design – something that would surely be challenged by supporters of organic food – the best of which is copied by retail outlets like H & M. The politics of food are vividly represented in this regional food industry, while the desperation of the food equivalents of big pharma to remain close to innovations in science and technology rather than 'the enemy' in what are still seen by orthodox food scientists and corporations as 'hippy farm food' is striking. Supermarket chains are less biased against the rising demand for organic food, gradually expanding shelf space for such produce. Thus Wal-Mart (Asda in UK), Sainsbury and Tesco (UK), Carrefour's 'Bio' range in France, Edeka and Tengelmann in Germany, and Ahold in the Netherlands all carry substantial organic food ranges. Meanwhile, in some countries like Germany, with Rewe's Vierlinden Naturmärkte (including its dedicated organic food label Fällhorn), Denmark, where market-leader Aarstiderne is owned by a partnership of four organic farms (2003 turnover $22 million) and, as we shall see further in Chapter 11, the US and UK through Whole Foods Market, dedicated organics-only supermarket chains are growing apace. Easily the largest two global markets for this growing, natural food variant are the EU and US with some $11 billion each and growth rates of 15–20 per cent in the early 2000s (Gibbon, 2006; Humphrey, 2006).

Finally, and briefly, the kind of platform perspective evolved in Skåne through recognition that a politically inconvenient part of the regional customer base and the organics part of the regional food chain needed to be spotlighted to a greater extent in assisting the further construction of regional advantage was shared in different dimensions by the two other pioneer cluster-building initiatives in Sweden. Robot Dale or Robotics Valley evolved its awareness that although a globally competitive lead manufacturer such as ABB headquartered in Västerås was an important anchor for the proposed cluster, the future lay in involving high-level engineering expertise in robotics alongside large firms in related industries like heavy earth-moving equipment from Örebro university and companies with SME market-demand for new kinds of smaller, more flexible robots, for which they could act as a test-bed in and around Eskilstuna. This had become a more strongly entrenched perspective on behalf of the firms and innovation intermediaries involved in this initiative as experience of the difficulties of cluster-building evolved in the first three years of the programme's existence. Similarly, Uppsala Bio had learnt that it was necessary to engage with Stockholm's large firms like AstraZeneca and Pfizer, its therapeutic treatment and diagnostics DBFs, and world-famous medical research complex, the Karolinska Institute, in what had also come to be perceived as a broader regional platform of complementary expertise. This was also recognised as important by the Stockholm partners in what was evolving first into a possible joint-marketing venture combining the two geographically proximate but hitherto industrially and culturally distinctive, asymetrically scaled cities. The clustering approach thus performed a consciousness-raising service rather than a solution for intermediaries and key actors that enabled them to perceive the need for a broader, more platform-based regional growth policy model in all three cases.

The Netherlands agro-food and biologics bioregion

Representative of the shift in pharmaceuticals from synthetic chemistry to 'biologics' and molecular biology more generally is the strategy of DSM. This leading producer started life as Dutch State Mining, the concern responsible for mining the Netherlands' coal until it ran out in the 1960s. Thereafter it is a textbook example of 'reinvention' as it shifted first into fertilizers, then petrochemicals, performance materials and now life science products. The firm's capabilities cover mechanical and chemical technology, polymer technology, materials science, organic chemistry and biotechnology. In 2002 it greatly expanded its life sciences capabilities with the purchase of Hoffmann La Roche's Vitamins and Fine Chemicals division for €10 billion, a purchase assisted by the €5.7 billion sale of DSM's petrochemicals division. The company is strongly active in Nijmegen, among other Dutch and foreign bioregions. Of importance to the emergent cluster, anchored by DSM, are the joint industry–university professorships practised by leading DSM executives in university departments and centres, and the Dutch and EU governmental policies supporting partnership in biosynthesis research in both instances and, in the former case, between specialist agro-university Wageningen, on the one hand, and entrepreneurial Twente University, on the other, to create critical bioscientific research mass in the Eastern Netherlands. This comes after disappointment at the failure to collaborate on joint biomedical research between élite Amsterdam and Utrecht universities in West Netherlands.

DSM's shift in focus has been from physics and chemistry to biology because of the rise in importance of food and healthcare markets. Life science products are 33 per cent of the business, performance materials 33 per cent, industrial chemicals 20 per cent, and agro-foodstuffs 14 per cent. DSM works closely with academic research, and many of its scientific staff in Maastricht, Delft and Nijmegen in the Netherlands and sites in Austria are industry professors in those universities. The industry–academe division of labour consists of DSM initiating *what* is to be researched and academe initiating *how* it is to be researched, a typical industry view of its status as financier of contract research. A recent project is called 'Cluster Project Antibodies' and involves six Dutch university bioscience research institutes aiming at making penicillins and cephalosporins at large scale and low cost. India and China host DSM's main competitors.

This project involves innovating new 'green' biocatalysts that are low in waste production, by integrating process design and intensification. This constitutes so-called third generation processing, i.e. fermentation of non-natural products leading to 'rational enzyme development'. Here, clearly, is a distinctive form of rational bioproduct design, which had a budget of €10 million in 1996–2001 and was supported during 2002–2008 with a Dutch national research programme valued at €20–30 million. The Integrated Biosynthesis (IBOS) programme aimed at integration of biosynthesis and organic synthesis of enzymes and molecular compounds for therapeutic treatments. In this regard Dutch research and DSM specifically are global leaders. Key to these new ambitions is the tenfold increase in the number of experiments that can be conducted due to the advent of high throughput screening for biotechnological targets. But although the industry shift to 'biologics' is structural, it is expected that there will remain a substantial role for synthetic chemistry. And DSM's comment that despite the rise of biotechnology, the world has seen only eight new chemical entities on the market (although in 2002 there were 90 biotechnology drugs, many composed of variants of such entities, on the US market and at least 70 in Europe; by 2004 the US figure was 130 and the European nearer 100) remains food for thought.

Concentrations of research excellence are thus important for industry that increasingly conducts R&D close to bioregions like Boston, to which Pfizer, Amgen, Abbott, Wyeth, AstraZeneca, Aventis and Novartis were all attracted from 1999 through 2003, and even Nijmegen wished to access knowledge spillovers. DSM leverages both internal and external knowledge networks. Internally, this is achieved through a global programme of project and competence management, and internal knowledge management and recording within the organisation. DSM encourages staff to build up personal academic networks and to pursue important part-time positions outside DSM, such as editorships of scientific journals, professorships, and scientific committee memberships. DSM also works closely with innovative R&D partners and exchanges scientific and technological knowledge with some

2,000 university departments. Participating in strong research organisations and networks, such as the Gene Alliance (Germany), the Biocatalysis and Bioprocessing of Macromolecules Consortium (US), the Wageningen Centre for Food Sciences[6] and the Dutch Polymer Institute (The Netherlands). DSM also regularly organises and speaks at relevant symposia and conferences. DSM's students receive insight into corporate activities and R&D culture by organised company visits. Star post-doctoral students have in-house research projects with which to engage. DSM also organises the international Vaalsbroek Scientific Meetings several times a year and invites PhD students and professors to present their work to a wider DSM audience.

Around the Wageningen and Nijmegen campuses are numerous bioscientific firms with market focus towards agro-food biotechnology. One leader is Noviant, a supplier to industries using carboxymethyl cellulose (CMC). Noviant's R&D involves close collaboration with customers. R&D is carried out at the pilot plant in Äänekoski, Finland. This plant facility is a technology centre for product and process development, where industrial applications of CMC are conducted. A further R&D group in Skoghall, Sweden, is focused on personal care applications and at Nijmegen is the corporate centre for development of products for the food and the pharmaceuticals industries. Table 9.2 lists other related firms and expertise.

Finally on the Netherlands bioregion in Wageningen-Nijmegen (evolving from 'Wageningen Food Valley') it is important to note the Centre for Biosystems Genomics. This is a Centre of Excellence under the auspices of the Netherlands Genomics Initiative. The network of scientists spans four universities, two research institutes and 15 industry partners in the Netherlands. The €50 million research programme focuses on the potato and tomato as the prime research crops, with arabidopsis as the genetic model plant. The university institutes involved are the University of Wageningen Centre for Genetic Resources, the University of Nijmegen Botanical Garden, the University of Amsterdam Institute of Life Sciences, and the University of Utrecht Arabidopsis Microarray Centre. Firms specialising in potato genetics are Agrico Research, Averis Seeds, HPZC Holland, C. Meijer, Van Rijn, and industry associations AVEBE and VAVI. For tomato genetics the firms are ENZA, De Ruiter, Nickerson-Zwaan, SVS and Syngenta. Local genetics and diagnostics DBFs noted in Table 9.2 involved in the consortium are: Plant Research Institute, Greenomics and Keygene. Hence this network both assists in knitting together the exploration–exploitation dimensions of the bioregion and pulls in academic and business partners from the Netherlands as a whole.

We may conclude this section by noting two key features of agro-food biotechnology bioregions that differentiate them from biopharmaceuticals bioregions and two that compare with them. First, agro-food bioregions, even the largest, which is probably in St Louis, US are smaller in number of biotechnology-focused incumbents than even rather small biopharmaceuticals bioregions. Second they are highly dependent, often on a single or few large

Table 9.2 Agro-food biotechnology DBFs in eastern Netherlands bioregion

DBF	*Location*	*Speciality*	*Employment*	*R&D*
ATO	Wageningen	Food research	320	300
Aurion	Wageningen	Immunology	8	5
BiFactory	Wageningen	Honey wound healing	4	3
Brainquiry	Nijmegen	Cognitive neuroscience	3	1
CatchMabs	Wageningen	Platform technologies	11	10
Chiralix	Nijmegen	Functional genomics	4	3
CBGenomics*	Wageningen	Agro-food genomics	50	50
Elti Support	Nijmegen	Environmental immunoassay	3	3
GeneTwister	Wageningen	Plant biomarkers	20	18
Greenomics	Wageningen	Agro biotechnology	10	8
KeyGene	Wageningen	Gene expression	100	70
MercaChem	Nijmegen	Biosynthesis	40	37
Mercatorial	Nijmegen	Contract synthesis	3	3
Micromit	Enschede	DNA analysis	12	3
Noviant	Nijmegen	CMC	200	35
Plant Dyn'ics	Wageningen	Horticultural bioimaging	1	1
Plant Rsch. I.	Wageningen	Plant biotechnology	2000	700
SPAD-it	Nijmegen	Bio data-mining	1	1
Syntarga	Nijmegen	Anticancer conjugates	3	2
Vitro Chemie	Nijmegen	Diagnostics	1	1
West'g Geno	Wageningen	Gene cloning	7	6

Source: BioPartner (2003), *The Netherlands Life Sciences Report*, www.biopartner.nl

* NB: Centre for Biosystems Genomics is a network of 4 universities, 2 research institutes and 15 firms in agro-food biotechnology in the Netherlands.

agro-chemical firms like Monsanto (Pfizer) or Aventis. We also noted that food retailer-producers like Unilever and Nestlé conduct R&D in-house that would be outsourced to DBFs in biopharmaceuticals. This is because they are scientifically based in biology, not chemistry. By contrast, there are similarities across agro-food and biopharmaceuticals bioregions in that universities are key generators of exploration knowledge for examination/exploitation down the value chain. Public funding is substantial and there is possibly less large-firm investment in research overall, although a few individual large firms are deeply integrated in research and IPR activity with universities, public research institutes and a few DBFs. Finally, agro-bioscience does spawn DBFs but in far fewer numbers and even smaller mean scale if Dutch data[7] are a guide, than biopharmaceuticals DBFs. The key to success lies in combining public and large-firm research and exploitation agendas, possibly in consortia, and spinning out specialist service providing DBFs.

Conclusions

These are distinctive parts of the general biotechnology platform, showing just how broad that platform dimension has become as the technologies have evolved. It is worth dwelling for a moment on the idea of 'platforms' since it marks an important element in the evolution of biotechnology from 'quirky' niche scientific activity to 'general purpose technology' in the thirty years 1976–2006. Hitherto industrial economists have categorised their object of interest into sectors, defined increasingly inaccurately as time passes by statisticians. Towards the end of the last century a discourse of 'clusters' entered the scene, in partial recognition that sectoral thinking was increasingly unrepresentative and invalid since clusters contained supporting industries from other sectors. Nearing the end of the first decade of the following century and after a decade of economic literature and policy advice on the support instruments for defining then designing clusters of firms co-located in proximity and competing globally in a productive manner, it seems that those that were first observed as role models are extremely difficult and inordinately time-consuming to reproduce.

More recent thinking, which was discussed in Chapters 2 and 3 but is germane here, solves many problems arising from the vertical and narrow thinking embodied in sectors and clusters. Why? Because it sees the world in lateral as well as vertical perspective, not least because of the spatial consciousness brought to the task by evolutionary economic geographers who discovered the importance of 'related variety'. In so doing, they were somewhat influenced also by sociological insights regarding relational and structural embeddedness, interactive learning and networks that had been elaborated by economic sociologists like Granovetter (1992). Hence this book attempts to address both these rather heterodox communities since it seeks to synthesise some of the more important theoretical contributions from both wings.

Platform thinking, utilising the concept of 'related variety', seems to approximate to the way firms in the contemporary era think of themselves and their points of engagement with their economic world. One surprising discovery, arising from research reported in this book and elsewhere, is that, by and large, policy makers tasked with intervening in support of industry and entrepreneurship display a very different mindset and associated methodology regarding economic processes. They are wedded to vertical narrowness in thinking of the sectoral and cluster kind, while firms see themselves as involved in pervasive technological advancement on the platform-base they consider themselves to be occupying. Biotechnology displays this almost perfectly with its skills in, for example, genetic engineering applicable across a range of economic activities from food to health, environment and energy, to which can be added security via DNA fingerprinting and biometrics that measure uniquenesses like the iris of the human eye for face recognition, not to mention the sadly long established use of bioscience in military technologies. Latterly the most interesting advances in ICT have embodied bioscientific knowledge and ingredients and, since ICT is the most pervasive contemporary platform for both industry and technology, the trend is clear. Biotechnology is no longer a quirky niche but already a platform technology for itself and related industry activity, as was vividly displayed in the Skåne and related cases in Sweden.

Of course some of these platform characteristics are more evolved than others. The near-perfect sustainability implied by Brazil's next-generation biodiesel production process fuelled at all stages by its own product is as yet an aspiration, but a realisable one. The functional foods produced in southern Sweden are, in some cases alongside their derivatives, on supermarket shelves albeit non-globally and with a rather Nordic niche, while others struggle to overcome technical issues of palatability, durability and feasibility. That is in a consumption context where the positive imagery of healthy food for populations assailed daily by reports of food hazards is a bonus, but the evidence that those that exist have any health effect is uncertain. So difficulties prevail, but once again barriers do not seem wholly imsurmountable even if health gain as such is not proven. The bioremediation work undertaken by firms in the Ruhr, first to clean up toxic and otherwise polluted soil, water and air in the home-base, then to export such skilled services to other polluted areas, notably in Central and Eastern Europe, has been a success. Bioenvironmental innovation has moved up the agenda for those responsible to ensure environmental improvement and, importantly, for those who have to live with the residue of the Industrial Age but may now experience improved conditions arising from the fruits of the knowledge economy. Probably, unlike healthcare, larger companies are more central to knowledge advancement utilising biotechnology in agro-food, fuel and energy production, although DBFs are particularly prominent in functional food technologies, and where that is the case they tend to mimic the small-firm-in-proximity-to-laboratory locational model found in Lund and Wageningen. Many of these seek ways to improve breeding of plants and animals, or even as in BioINN in Hamar, Hedmark County, Norway to use biotechnology to put right the breeding nightmares discussed in Chapter 1. These arose, it will be recalled, from the scale-obsession of the large supermarkets with milk yield that led to physically distorted Holstein dairy herds. Cross-breeding with Norwegian Red Cattle drawn from BioINN's biobank is a solution to this particular problem that is beginning to find markets worldwide.

10 The financing of biopharmaceuticals firms

Introduction

Following a bleak period after the technology stock market downturn in 2001, the financial position for healthcare biotechnology improved in 2005. Compared with 2004, net losses were down by 30 per cent and revenue was up 18 per cent ($60 billion; see Table 10.1). R&D investment overall increased by 4 per cent, from $19.54 in 2004 to $20.42 in 2005, but R&D spending in Europe declined by more than one-fifth (Lawrence, 2006a). Public equity markets were more open to initial public offerings (IPO) than at any time since the downturn. Perhaps surprisingly, the IPO market was strong for European companies and the number of European biotechnology IPOs almost doubled, from 12 in 2004 to 23 in 2005. In contrast only about one-third of global IPOs were by companies based in the United States in 2005 (Lähteenmäki and Lawrence, 2006). Europe's healthcare biotechnology sector appeared to have reached a stage of maturity, where companies need to raise substantial amounts of money to support product development.

Regarding the key industry indicator of number of pipeline products, a long-established picture of US–UK hegemony prevails, with only Switzerland challenging the latter (Table 10.2). Notable in the Swiss case is that only five public companies employ on average 1,600 persons each. This is a

Table 10.1 Global overview of biotechnology companies in 2004 and 2005

Company data	*2004*	*2005*	*Percentage change*
Revenues ($ billions)	53.37	63.16	18%
R&D expense ($ billions)	19.54	20.42	4%
Net loss ($ billions)	6.27	4.39	–30%
Public companies	645	671	4%
Private companies	3,522	3,532	0.3%
All companies	4,167	4,203	1%

Source: Ernst & Young (2006)

Table 10.2 Main international bioscience competitors, 2003

Country	*Companies*	*Public cos.*	*Market cap.* (€ bn)*	*Revenues* (€ bn)*	*Employees**	*Pipeline**
USA	1,457	307	205	27	191,000	872
UK	331	46	9.4	3	22,000	194
Switzerland	129	5	7.3	2	8,000	79
France	239	6	0.5	0.3	9,655	31
Germany	369	13	0.5	0.5	13,386	15

Source: Bioscience Innovation and Growth Team (2004)

NB: * Public Company Data Only

very different profile from the UK whose firm employment profile averages some 500 persons, closer to the US with some 620. Germany's public companies have a more Swiss profile, while France is also similar to Switzerland in employment per firm, but like Germany displaying much lower productivity measured in terms of revenue and pipeline products.

With respect specifically to pipeline product abundance, Table 10.3 supplies detail on this indicator and shows how the US easily outperforms the rest of the world, with the UK leading a badly trailing pack. It is partly a matter of early entry, the first US biotechnology firm being Genentech in 1976. It is partly a matter of very large R&D budgets in the US compared to elsewhere on a per capita or per firm basis. But more importantly for public companies it reveals the importance of private investment by biotechnology venture capitalists. These have the greater impact upon the capabilities of firms actually to realise the fruits of significant injections of

Table 10.3 Product pipeline of public healthcare biotechnology companies

Country	*Pre-clinical*	*Phase 1*	*Phase 2*	*Phase 3*	*Pipeline products*
USA	584	96	148	44	872
UK	65	50	56	23	194
Switzerland	45	12	11	11	79
Sweden	14	8	10	0	32
France	16	8	6	1	31
Denmark	14	5	5	4	28
Italy	9	0	4	3	16
Israel	2	3	6	4	15
Germany	7	4	3	1	15
Norway	8	2	2	3	15
Netherlands	9	1	1	0	11
Finland	9	1	0	0	10
Ireland	2	0	2	3	7
Belgium	2	0	1	3	3
Total Europe	202	94	107	53	456

Source: BIGT (2004)

R&D expenditure. However, this is high-risk investment and approved new drug treatments follow lengthy and expensive trialling of up to ten years and in many cases more. Many of these products fail in trials. Table 10.3 is interesting since it shows that while individual European countries are behind on most key indicators of pipeline performance, when added together (plus Israel) they comprise approximately half the innovation impact of the US. Moreover, proportionally speaking, Europe has more drugs in late stage trialling than the US. Some explanation for this imbalance is given in Table 10.4 which reveals a significant surge of new chemical entities (NCEs) towards the end of the 1990s but a rather stagnant position previous to the late 1990s technology boom.

It was easier to find risk capital to take firms out beyond Phase 3 trials at that point than hitherto or since. Hence the US has many more approved drugs available for sale than Europe. However, Europe may take a larger share of world markets if the requisite funding is made available in the near future. This will be a great relief to many in industry and government who have seen many false dawns in the European biotechnology sector's market performance. A key requisite, in a period when venture capitalists are risk-averse and in other respects not seen as the preferred partner by biotechnology firms, is for small firms to seek scale to finance the late stages of pipeline development. Here, the Swiss model of a few large, successful firms, notably the world's number three in market capitalisation Serono, was something of an exemplar. In 2006 it was acquired by Merck, Sharp and Bohme, the German pharmaceuticals company. That France and Germany also follow this route, presently less successfully, is revealed in Table 10.3. As noted, the UK seems to have a firm size profile more

Table 10.4 Declining pharmaceuticals productivity

Date	*US NCE approvals*	*R&D expenditure $ billions (2001)*
1963	19	2
1967	20	2
1971	18	3
1975	17	3
1979	16	3
1983	16	4
1987	19	5
1991	25	8
1995	25	15
1996	45	17
1997	37	18
1998	23	22
1999	32	25
2001	23	30

Source: BIGT (2004)

comparable to that of the US, showing that the 'varieties of capitalism' thesis retains some purchase (Hall and Soskice, 2001). Nevertheless, UK and US firms were, in the first half of the present decade, also seen to be pursuing scale through numerous acquisitions and mergers. Thus in 2003–4, the UK's Celltech was acquired by Belgian chemicals giant UCB, Powderject by Chiron, and British Biotech by Cold Spring Harbor spinout OSI. Meanwhile US leaders Biogen and IDEC merged. Simultaneously, risk-averse venture capitalists were also challenged by direct funding of promising pipeline spinouts by large pharmaceuticals firms, small firms thus avoiding the venture capital route entirely.

It now remains to incorporate the pharmaceuticals sector of the industry to evolve a number representing the scale of the healthcare plus bioscientific and pharmaceutcals sectors in global terms. Industry estimates for 2003 shown in Table 10.5 put this figure at some $400 billion, excluding biotechnology (Jack, 2005). Hence the resulting value of the three sectors in the global economy measured in 2001–2003 average market revenue is $400 billion + $170 billion + $33 billion = $603 billion. As succeeding sections of this chapter show, while the world constitutes 'the market', that market is well captured as North America, Europe and Japan. However, the originating production locations of that market are overwhelmingly American and European, and within both, they are highly localised in a few 'megacentres' that account for huge portions of the global value-chain.

Key factors in biotechnology firm development

This section will examine a few bioregions among the countries under different economic market systems. As has been emphasised many times already, the US has some of the most advanced bioregions, generally larger

Table 10.5 World trade in pharmaceuticals, 2003 (£ million)

Country	*Exports*	*Imports*	*Balance*
Ireland	8,756	1,258	7,498
Switzerland	9,754	5,044	4,711
France	10,552	6,551	4,001
UK	11,926	8,293	3,634
Sweden	3,928	1,315	2,613
Italy	13,213	11,453	1,760
Netherlands	5,291	4,536	755
Australia	5,464	5,569	–105
Germany	873	1,999	–1,126
Spain	1,482	3,245	–1,763
Japan	2,346	4,176	–1,830
Canada	1,417	3,508	–2,091
USA	9,753	16,998	–7,245

Source: www.abpl.org.uk

and more mature than those in Europe, while the UK leads Europe in biotechnology and, as we have seen, the developing countries such as India and China are just starting their mission (Chapter 8). Successful bioregions have a range of different market and interventionist factors in place which can be seen as critical to their performance. These factors exist to varying degrees in the cases we examine here. Most policies have adopted a cluster approach modelled on the more market-driven form taken by leading bioregions in the US and, as discussed, found to be a particularly pronounced feature of healthcare biotechnology because of the need for firm proximity to basic research, which since the molecular biology revolution is found in the highest-class university research laboratories rather than inside pharmaceuticals firms.

Proof of concept and university challenge

'Proof of concept' funds assist in the earliest phase of possible DBF formation. The first was established in Scotland, where the importance was recognised of buying time for academics to concentrate on possible commercialisation of their research. Normal teaching and administration are known as factors restricting the creation of spinout firms from the higher education sector. The government response to this need was to establish a joint fund involving also a contribution from the higher education funding council and Scottish Enterprise, the regional development agency called the 'Proof of Concept' fund. This has assisted company spinout from universities by freeing interested academics from normal duties to focus on commercialisation activities. 'Proof of Concept' is the stage at which the key principle of the discovery or research line has been demonstrated, thus giving reasonable assurance that the idea can feasibly be translated into a business model. Thus the fund supports the pre-commercialisation of leading-edge technologies emerging from universities, research institutes and general hospital laboratories. The Scottish programme in 2006 was supporting 172 innovative projects worth over £23.7 million and had created some 400 new jobs since inception. Among these were some 40 biotechnology firms that followed on from 'Proof of Concept' funding. 'Proof of Concept' has important impacts on the creation of high-tech spinouts. Northern Ireland's INI investment agency and London's technology development body have imitated this policy to improve their level and quality of commercialisation.

The University Challenge funds represent a unique initiative between the UK government and major research organisations. These provide early stage finance to help scientists exploit the commercial potential of their discoveries. This early funding is the riskiest stage of the venture process, and many institutions experience a funding gap in the provision of finance for bringing research discoveries to a point where their commercial usefulness can be demonstrated and first steps taken to secure their utility. A few

European countries have had comparable schemes to promote the research commercialisation.

Technology transfer

The European Investment Fund's Technology Transfer Accelerator (TTA) report of 2005 analysed the state of commercialisation of research in Europe and designed a programme to improve the commercialisation landscape (European Investment Fund, 2004). *Technology transfer* in their parlance defines the process of transformation of R&D results into marketable products or services. The transformation process can take place in three main ways: through collaboration between research institute and industry, licensing or creation of a new company. Technology transfer in Europe is recognised as changing faster in some countries than in others, with some research institutions having successfully established in-house technology transfer operations. This is especially true in the UK, Ireland and the Nordic Countries, but is by no means universally the case. However, the report also observes the large gap in the rate of technology transfer comparing Europe and US (Table 10.6).

There are distinct technology transfer models. In a traditional transfer model, the university supports technology transfer within the research

Table 10.6 Comparison of licensing revenues at academic institutions in the US and Europe, 2003

	University/Institution	*Revenues (€M)*
US	Columbia University	115.4
	University of California, Stern	65.3
	University of New York	49.9
	Sloan Kettering Institute for Cancer Research	43.3
	Stanford University	40.0
	University of Rochester	33.5
	City of Hope National	31.4
	MIT	30.2
	University of Wisconsin-Madison	25.6
	Florida University	25.2
Europe	Pasteur Institute	32.6
	University of Edinburgh	4.5
	University of Utrecht*	4.0
	INRIA*	3.0
	VIB*	2.7
	Cambridge University	2.3
	University of Bern and Zurich	0.6
	Ludwig Maximilian University Munich	0.2

Note: * Revenue from industry

Source: EIF/EIB Technology Transfer Accelerator (TTA) report, 2005

organisation. In this model, technology transfer vehicles provide funding and management support, in return for share of the technology transfer proceeds, either 'in kind' (percentage of licensing equity) or in revenue. This model is exemplified by the idea of the 'campus company' where the university owns the intellectual property. Another variant known as 'IP2IPO', which as the name suggests takes patented knowledge right through to public flotation of a firm. IP2IPO provides both funding and management support to a number of university or laboratories in, for example, the UK. In return, it receives an equity stake in the spinout created as well as a share of licensing revenues.

A different model is to perform the technology transfer outside the research organisation. In this model, the technology transfer vehicle is an entity that provides a technology transfer service to one or several research organisations. These services cover most of the management of the technology process. In return for its service, the TTA vehicle obtains a claim on the future revenue streams to be generated from technology transfer. Contrary to the previous model, the revenue streams in this model go first to the TTA vehicle, a share then being passed on to the research organisation. BiolineRx, an Israeli company, uses an innovative approach to cut down the length of the product licensing process and is a private sector example of this model. It is referred to as an *accelerated licensing model* of technology transfer. BiolineRx evaluates, acquires and advances projects from lead selection to advanced clinical trials, regulatory approval and marketing. It takes projects from Israeli academia, industry and government research institutes to develop them to an advanced stage to enable a strategic partnership with a global pharmaceuticals or biotechnology company. Bioline is claimed as an efficient model, which reduces the time to develop product licensing through its expertise as an accelerator agent.

The TTA programme also proposed a new generation technology transfer model, adapting venture capital into a portfolio approach (Figure 10.1). In this model, the TTA facility funds portfolios of transfer projects. The technology transfer projects are contributed to by one or several research organisations or individual scientists. In return, contributors receive a stake in the TTA vehicle, which owns the intellectual property. The TTA financing facility provides funding to the TTA vehicle in return for a share in the commercialisation proceeds, which will usually be received in the form of dividends or capital gains. Technology transfer takes place within the various projects and is managed, for the portfolio as a whole, by the TTA vehicle manager. This model amounts to a rethink of the classical 'VC seed' model in that it neutralises the uncertainty surrounding deal flow sourcing. Investors have access to an existing portfolio instead of an uncertain future one and timing is improved whereby the fund starts operating with the existing portfolio instead of having to develop one.

The Scottish Life Sciences Intermediary Technology Institute (ITI) has a similar business mission, namely to translate basic research into commercial

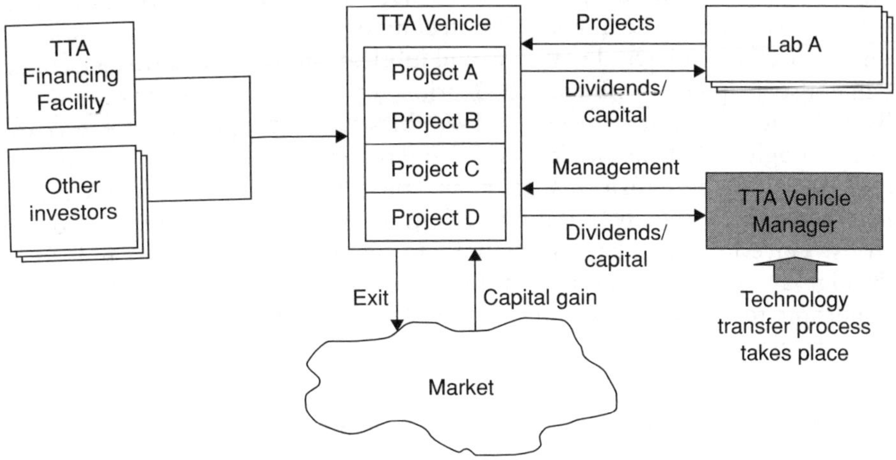

Figure 10.1 New generation model of technology transfer, venture capital funded

Source: EIF/EIB Technology Transfer Accelerator, 2005

outcomes in the form of IP, licensing, trade sale or spinout company. Scottish Enterprise is also responsible for this ten-year €200 million venture. Three ITIs were established in 2003, focusing on life sciences, information and communication technologies (ICT), and energy. They are publicly funded, but commercially driven. Promotion of stronger innovation capability in these sectors fits with the key aspiration of the Scottish Executive's vision of a 'Smart, Successful Scotland', the basis for which is seen to lie in the country's knowledge-based economy. Thus the ITIs are already performing valuable roles in the spheres in which they operate. We have seen also how the market-facing model of innovation support differs significantly from that traditionally promoted by the development agency 'sector and/or cluster' model of intervention. In 2000 Scotland had some 50 biotechnology firms, by 2006 that figure had risen to 100. Of these 100 in 2006, the biggest geographical concentration is in Glasgow but there is strong science and spinout firms also in Dundee and Edinburgh as well as near Aberdeen. The sector is thus seen as occupying a 'biotechnology triangle' between Dundee, Edinburgh and Glasgow.

The UK's ability to develop stronger and more market-led supply chains would be significantly improved through supporting consolidated/new industry-orientated knowledge transfer networks to spearhead informed and effective communication, technology development and knowledge/technology transfer to industrial suppliers, industry users, consumers and other key stakeholders.

It is clearly important that such networks be properly funded to enable proactive 'championing' and representation of UK R&D and business interests in domestic European and international policy. Improved funding for the basic science of industrial biotechnology and for transformation of this science into industrial reality would lubricate and improve the networks while also helping to inform R&D priorities across all key stakeholders. Biotechnology is science-driven and exhibits classic properties of externalities through knowledge spillovers. Private returns to technology developers may be below the return to society, and the return to the initial risk-taker may be competed away if the technology is easily recreated or copied by rivals. Government intervention, both through funding for research and through an enforceable system of intellectual property rights, is, therefore, important to protect ideas and returns for those that incur the costs of their development.

The Small Business Innovation Research model

The Small Business Innovation Research (SBIR) programme was established in the US in 1982 as a tool for small business to compete for R&D grant money. In 2001, the Small Business Administration (SBA) defined a small business as having fewer than 500 employees with a minimum of 51 per cent ownership held by individuals. Under this programme, 2.5 per cent of the external research budget of 11 US federal agencies is set aside for funding R&D in small firms. The programme is widely regarded as effective in encouraging university faculty to set up small companies. A study by Lerner (1999) gave strong evidence to support this point. He examined 1,435 SBIR awardees and a matched sample of firms that did not receive awards, over a ten-year post-award period. Because most of the firms were privately held, he was unable to analyse the resulting valuation or profitability of the firms, but found that firms receiving SBIR grants grew significantly faster than the others after receipt of the grant. He attributed some of this effect to 'quality certification' by the government that enabled firms to convince the investor community of their provenance.

The scheme is not, however, uniformly popular with all government agencies because it takes funding away from specific research areas and because it is felt that the quality of research proposals from industry tend not to be as good as those from universities. Another concern was that companies spent substantial amounts of time and effort on grant applications. Firms also employed experienced grant writers to prepare the grant proposal, so part of the award was used to fund the next proposal instead of research commercialisation. However, many early-stage biotechnology companies have received SBIR funding to assist in the start-up process. This programme is widely seen to have had significant impact on creating new companies in the risk-prone early stage and SBIR has contributed significantly to the growth of biotechnology in the US. In the UK the Biotechnology and Biological Sciences Research Council (BBSRC) has established the Small Business Research Initiative (SBRI) designed to encourage small, high-tech firms to start up or to develop new research capacities. Companies that have a record of working with researchers in universities and Research Council institutes, and who have the capability to turn research outputs into new products or processes, are targeted under the scheme.

Incubation of biotechnology firms – the Bio-Link experience

In 2004 a European Union Framework 6 scientific project reported on its investigations and recommendations for improving incubation services for biotechnology firms. The project aimed to explore whether biotechnology firms might benefit from co-incubation, where services from a platform in one country might be accessed in another through designing a partnership network. It involved BioM from Munich, Oxford BioTechNet, Polaris Bio-Park in Sardinia, Italy and Hadasit incubator in Jerusalem, Israel (Cooke *et al.*, 2006). Based on the Bio-Link project, a Best Incubation Practice (BIP-tool kit) was recommended as shown in Table 10.7. The partners saw the adoption of this tool kit as an opportunity for policymakers in the EU, and at national and regional level, to improve the incubation practice for young biotechnology companies. A key proposal was that the optimal bioincubator should be based on the 'all in one' model. Co-incubation is dependent on the services and the surrounding business environment provided by the host incubator. The analysis of the five participating incubators in Bio-Link demonstrated that portfolio companies belonging to bioincubators which provide vertical services have better chances for co-incubation than those belonging to incubators that provide only part of these services.

1 *Technological infrastructure*. The bioincubator should be able to provide space and infrastructure for a variety of technologies. This includes upstream and downstream technologies related to products derived form various hosts such as bacteria, yeast and mammalian cells. The incubator should be able to support synthetic chemistry activities. The incubator should own part of the expensive equipment and enable the clients to use it.
2 *Access and collaboration with industry*. The bioincubator should establish relations and collaborations with the healthcare industry. The projects in the bioincubator should be presented on a periodic basis to companies in the sector. In addition, the technical support of these companies can be very beneficial (use of equipment, materials, animal facilities, etc.). It is the responsibility of the bioincubator manager to make sure that his projects will get a high level of exposure to the healthcare industry. Bioincubators with strong relations with the industry have a clear advantage in promoting co-incubation and business development as compared to those which do not give enough emphasis to relations with industry.

Table 10.7 Best incubation and co-incubation practice tool kit

Optimal incubation model	*'All in one' model:* – In house biotech technological infrastructure – Access to and collaboration with industry – Access to business development experts – Networking – Access to VCs and financial community – Access to service providers (legal, IP, accounting) – Collaboration with local academic institutes – In-house science and technology guidance – Synergy with next door neighbour – Public relations
Selection criteria for co-incubation	*a) Technology developer* Wide scope Ready for implementation and/or technology transfer *b) Service provider* Certification Technological background material plus website Fast response Accessibility *c) Maturity* Post 'creation hassle' – 2–3 years old *d) Critical mass – not less than 8 employees*
Key success factors for co-incubation	1. *Technology* Beyond proof-of-concept 'Transferable' – at 'operating procedure' level 2. *Incubator* Co-incubation oriented and supportive Business development guidance – from day zero Administrative assistance (TT and MTA template documents in place) 3. *Company's Management* Willingness and openness to collaborate and share (language and culture gap) Managerial awareness and attention Co-incubation project manager 4. *Budget* Financial resources allocated for co-incubation – labour, material, travel 5. *Time* A successful co-incubation requires at least 2–3 years from conception.
Likelihood of co-incubation success	High for service-providing companies Medium for platform technology companies Low for developers of molecules High for companies in commercial stage Medium for companies in transition (pre-commercial) Low for early stage companies

3 *Access to business development experts.* Each bioincubator should have a director of business development whose responsibilities will include partnering, preparation of presentations, deal structuring and analysis assistance in the preparation of business plans. The lack of this function in most of the partnering bioincubators in Bio-Link was the main reason for the lack of co-incubation deals in the first year of the project. The function of business development can also be split between several experts in their own particular fields. It should be remembered that most if not all of the entrepreneurs are scientists who lack the experience and the culture of the healthcare industry and can tend to make many mistakes during the early years of their incubation.
4 *Networking.* This is a joint task of all the management members of the bioincubator. Networking can be split into personal networks which are the result of the relations created during the years between the members of the bioincubator to external organisations, industry, academia, financial community and so on. Another form of networking is the virtual one, which is based on databases provided on the Internet. Access to literature and reports issued by expert companies are part of this virtual networking. We found out that some of the incubators are unfamiliar with the current publications related to bio-business such as *Bio-World Today*. Networking was found to be the major parameter appreciated by client companies in all the five partner incubators of the Bio-Link project.
5 *Access to VCs and financial community.* This is clearly a key factor for the success of each entrepreneurial organisation. Connections to the financial community refer to business angels and private investors, local banks, global venture capitalists (VCs), local VCs and also to the healthcare industry. A basic business model is an essential tool for each company in a bioincubator to enable it to present its business concept to these potential investors. The incubator should arrange periodic meetings with potential investors and be able to sell the three basics of a bio-start-up: 'Technology, Networking, and Dream (Profit)'.
6 *Access to service providers.* The bioincubator should be able to take care of all the administrative tasks which a young client company has to undertake. This includes accounting and book-keeping, legal services, intellectual property services and personnel recruitment agencies. Provision of these services to the client companies should enable the chief executives of the young entrepreneurial companies to focus on their technology objectives.
7 *Collaboration with local academic institutes.* Good relations with academia are an essential service that should be provided by the bioincubator. Such relations contribute to the flow of projects that are forwarded from the bioincubator. They also are essential as a source for scientific experts who can serve as consultants and as members of the scientific advisory boards of the client companies. A relationship with a prestigious academic institute is also a key success factor in fundraising

and for the public relations of the bioincubator. The relations between the bioincubator and the academic institute should be almost symbiotic so that the bioincubator is the default address to which young entrepreneurs from the university are referred by their technology transfer offices. On the other hand, the university should be able to present the bioincubator as its window for the 'applicative world'.

8 *In-house science and technology guidance.* This guidance refers mainly to technological aspects which are part of the development process in the healthcare industry. Pharmaceuticals development, clinical development and regulatory affairs are just part of the guidance required by young biotechnology and pharmaceuticals companies. Guidance in these fields which are usually part of the more advanced stages of a project are essential from the first day since they have a cardinal impact on the company's development plans, business model, likelihood of registration and hence on fundraising.

9 *Synergy with next door neighbour.* Bioincubators should try to form an internal technological cluster. This recommendation refers mainly to basic technologies such as genetic engineering, monoclonal antibodies and so on. Internal clustering can contribute to real sharing of expertise and resources. However, its main advantage is in business development and accumulation of internal expertise in a certain field. Companies with similar technologies can use the same basic development equipment, which is usually quite expensive and cannot be justified for one company, especially if it is not routinely used.

10 *Public relations.* Young biotechnology companies are 'selling dreams'. It is within this scope that the bioincubator must provide them with an optimal exposure to the printed and electronic press. Public relations have a major influence on the financial community, especially on fundraising which is so essential for any young company. Therefore the bioincubator should hire the services of a professional public relations firm as part of its service package to its client companies. Many of these recommendations are applicable to practice in single bioincubators; the co-incubation option is a value-adding aspect of the well networked bioincubator.

Business angel networks

Business angels (BAs) are wealthy individuals who invest in high-growth business in return for equity. Some BAs invest on their own, whereas others do so as part of a network, syndicate or investment club of BAs. In addition to money, BAs often make their own skills, experience and contacts available to the company. BAs typically invest in businesses with an investment need of between £10,000 and £250,000, high potential return and good early stage development or expansion. In the UK, 'business angels' are often a network investing after the seed fund stage and they are very active. BAs tend to invest in the early, pre-venture capital stage as a rule, as they have less to invest than the latter and expect to have a high return. Business angel networks spread their risk in a number of companies while gaining the benefits of a better valuation based on the negotiation power of the whole group, rather than that of individual private investors. The company not only achieves its aim of gaining investment but also saves time and money by speaking to only one individual while attracting investment from many. The relative scarcity of early stage private risk investment means there will be an abiding demand for single business angels and business angel network models for the foreseeable future.

Supplementary support systems include the Swedish Small Business Investment Company (SBIC) programme. Dating from the early 1980s, the Swedish government established the first series of investment companies (along with instituting a series of measures such as reduced capital gains taxes to encourage private investments in start-ups), partly on the US model. By the late 1980s, the government share of venture capital funding was 43 per cent. Recently, the UK has instituted a series of government programmes under the Enterprise Capital Fund (UCF) umbrella, which allocate funds to SME in high-tech and certain regions, as well as guaranteeing some loans to small business. There are also programmes at the European level. At sub-national level, Scotland has once more proven to be innovative in policy design in support of commercialisation of scientific research.

To introduce more investment and reduce the risk for business angel activities, the public sector may act as a co-founder of specific business angel investment. The Scottish Co-investment Fund (SCF) was set up by Scottish Enterprise to increase the risk capital available for investment opportunities in small but ambitious Scottish companies. The equity fund of £45 million can be invested between £50,000 and £500,000 in company finance deals of up to £2 million, and it is open to SMEs based or located in Scotland. This model has proved to be good practice to support SME start-up and follow-on growth. The East of England imitated this scheme to support follow-on growth of SMEs.

The sources of finance options for start-up biotech are listed below:

- *Borrowing*
  - An overdraft, savings or friends and family are among the most common sources.
  - A 'term-loan'. The bank loan option is more secure because of the agreed term limit. However, banks view biotechnology as a risky sector, and a long-term investment (ten years to develop a product). As a result, banks may charge startlingly high interest rates and/or protect themselves with terms that are unfavourable to the company. Many companies seek bank funding but banks are less adept at providing, for example, management advice than, for example, business angels or venture capital firms.

- *Contractual joint venture*
  The company will license out its ideas, technology and possibly know-how, to another company that has the commercial resources to develop the knowledge into a marketable end-product. One typical investor in biotechnology firms is a pharmaceuticals company, which has the resources and expertise to invest in R&D, and is driven to market products from the expanding DBF technology spectrum. Licensing agreements are the set of rules of the arrangement in such a joint venture, which can be carefully crafted to suit the purposes of the parties involved. A properly drafted licensing arrangement will allow flexibility, which is why many biotechnology collaborations come within this category.
- *Equity joint venture*
  The joint venture company will be owned and controlled by the investor company and the biotechnology company, generally as equal partners. They will be bound by an agreement to pursue a common goal, which is to market the product successfully. The joint venture company will be empowered to use its parents' knowledge and finance through licensing and funding agreements. The distinctive feature of the model is the joint venture company as a separate entity. The existence of a separate company has several implications but, above all, it means that the business becomes more independent – both conceptually and financially. The parent companies can lose only the amount they originally put in, because of the joint venture company's limited liability. However, it may be difficult to remove money from the joint venture company.
- *Direct investment for shares*
  A more invasive type of investment is where the investor, typically a venture capitalist, purchases shares in the biotechnology firm. This can be the most psychologically difficult for the biotechnologists, as it involves dilution of his/her ownership of the original company. The greater the percentage of total shareholding the investor company acquires, the more control it has. This control will usually include appointing a director onto the board of the company, and other similar controls, as it seeks to protect its investment. Traditionally, venture capitalists had the purpose of investing in small companies, but many have grown and, like banks, become risk-averse private equity companies. Hence, more investors are coming from within the industry. The wealthier pharmaceuticals companies are acquiring interest in DBFs to access knowledge in exchange for capital to continue their work, and the larger pharmaceuticals company can market the fruits of that work. As biotechnology companies have grown, some have themselves acquired shareholdings in other biotechnology companies. By sharing information, the biotechnology sector shows capability to become more efficient and effective.

Venture capital

Historically, several financial incentives gave venture capital (VC) a boost in the US. In 1979, the Department of Labor reinterpreted provisions that allowed pension funds to invest in VC funds. Today, pension funds contribute to over 50 per cent of US venture funds (Lee and Dibner, 2005). The Economic Recovery Act of 1981 provided incentives to individuals to invest in VC funds by dropping individual capital gains taxes from 42 per cent to 20 per cent. Consequently, total VC funds for biotechnology surged, roughly doubling each year, from $43 million in 1980 to $140 million in 1981. US biotechnology received $3.8 billion venture dollars in 2004 (Lee and Dibner, 2005). The US biotech sector has received $114 billion (VC, IPO, follow-on and debt financing) since 1994. Over the same period, about $27.2 billion has been invested in the European biotechnology sector. This clearly suggests that European biotechnology is still in its early stage, relative to the US. There are many private companies to be pushed into public market (Fazeli, 2005).

The European VC industry is less well developed than the US. It picked up momentum only in the late 1990s (Howell *et al.*, 2003). The European VC industry is not immune to the economic and cultural factors affecting the biotechnology industry in the US. In Europe, most scientists did not pursue commercialising their technology, and institutions lacked support organisations, such as technology transfer offices, to facilitate the development of commercially viable research discoveries. The fragmentation of the tax and legal environments in the EU and the hurdles for entrepreneurial growth in many EU countries are problems. Fragmentation of stock markets among European countries creates a lack of broad exposure for biotechnology companies to sophisticated investors. Lack of liquidity and exit in public markets forces venture capitalists to invest additional funds in existing companies to keep them alive. Governments in Europe are

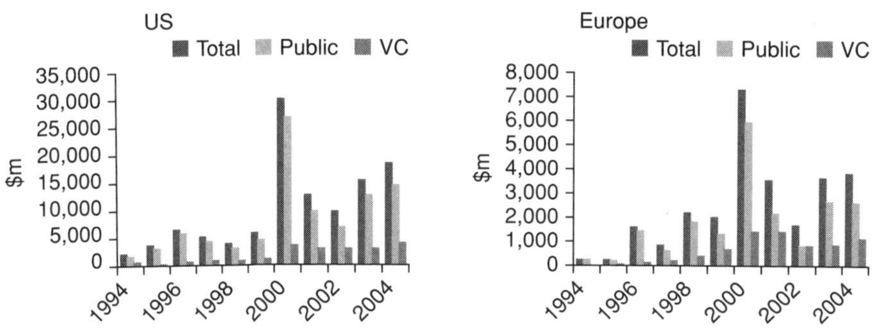

Figure 10.2 Annual investment in private and public DBFs: US and EU

Source: Fazeli (2005)

slowly seeking to respond on tax issues and seeking to foster general entrepreneurial development. For example, Germany's BioRegio Competition supported the growth of biotechnology clusters by providing seed capital for start-ups in selected regions (Chapter 7). In France, companies were entitled to a tax credit equal to 5 per cent of all R&D expense up to $6.5 million per year from the beginning in 2004. In Germany, banks are by far the most important source of finance for the VC industry, and pension funds are conspicuously absent. By contrast, in the UK, pension funds, other institutional investors, and individual investors provide funds to as many companies as banks. Contrary to popular views, government funding plays a more important role in the UK than it does in Germany, and this is negligible in the other countries (Mayer *et al.*, 2005).

In the US, the VC industry consists of a specialised pool of funds (usually from private investors) that are managed and invested in companies by individuals knowledgeable about the industry in which they are investing. Lerner (1995) highlights the important role of investing and monitoring experience from VC in biotechnology's evolution. VCs increase the value of the firm they fund, especially when they are experienced investors. Firms backed by seasoned VC financiers are more likely to have success in the market. This phenomenon, that VC-backed firms grew faster than the non-VC-backed firms, was also observed in Germany during 1991–1998 (Hall, 2005). Lumme *et al.* (1993) found that small UK firms, embedded in a financial market-based economy, rely on equity and less on loan finance and grow faster than small Finnish firms in its bank-centred economy. Black and Gilson (1998) emphasised the contrast between market-based financial systems (e.g. the US and the UK) and bank-centred capital market systems (e.g. much of continental Europe). VC was considered as combining the strengths of the two systems and provides both the strong incentive for the manager-entrepreneur characteristic of the stock market system and the monitoring by an informed investor characteristic of the bank-centred system. They stressed the importance of an active stock market to provide an exit strategy for VC investors and to allow them to move on to financing new start-ups. Thus having a VC industry that contributes to biotechnology development requires the existence of an active IPO market to permit successful entrepreneurs to regain control and also to provide powerful incentive for undertaking the start-ups, which then ensure that VCs themselves are able to use their expertise to help to establish new companies (Hall, 2005).

In a recent survey BioAbility (2005) reported there were 524 VC firms investing in biotechnology in the US, while there were only 260 VC firms investing similarly in Europe. In Europe, Germany leads the rankings with 28 per cent of VC firms (83 firms) that invest in biotechnology, while the UK follows with 16 per cent (47 firms), and Switzerland has a significant number of VC firms (29) (Figure 10.3). Although Europe now has more biotechnology companies than the US (1,976 vs. 1,830 in 2003), they are

Figure 10.3 European biotechnology venture capital sites

Source: BioAbility (2005); Lee and Dibner (2005)

younger and smaller, but as the biotechnology industry in Europe matures, European VC firms invested in biotechnology are predicted to experience parallel growth (Lee and Dibner, 2005). The average venture round in Europe was some $16 million in 2005. The gap in the average venture round between Europe and the US is getting smaller. It is notable that in recent years, the venture capital community invested in fewer European companies, but funded projects received a larger amount than previously (Ward, 2006a). VC investments are increasingly flowing in both directions across the Atlantic, moving between the US and Europe. Most venture capitalists hesitate to invest money in a company whose progress they are not able to watch closely. For example, Oxford Technology Venture Capital Trust reported to the Bio-Link project that like most VCs it invests only in companies within one-hour driving distance so that they can easily get a hand to monitor the invested company. This surprised some public sector agency personnel present at the workshop, and in questioning, the company admitted it sometimes broke that rule. Hence it is an idealisation than a full representation of reality as a visit to the website of one of the biggest VCs of all, Kleiner Perkins Caufield & Byers of Palo Alto, California reveals. While a majority of its holdings are close to Palo Alto, many others are elsewhere in the US.

There are marked differences in patterns of investment between US and European private companies, depending on the maturity of the invested companies (Figure 10.4). In the US, 15 per cent of companies under five years old received equity backing in 2004, while only 9 per cent of European companies in this age group received funding. Companies between 6 to 10 years old were more modestly supported by investors in both the US and the Europe. There was a gap in funding for older companies (over 10 years) between the US and Europe. Yet 1 per cent of companies over 10 years old received funding in Europe, while 6 per cent did in the US. The funding gap between the US and Europe was also reflected at the top of the funding range (Table 10.8). The leader in the US list for venture capital rounds in 2004 was Jazz Pharmaceuticals (Palo Alto, CA), receiving €180 million venture capital investment. Of the European top ten, only anti-infective specialists, Novexel (Paris) and the anti-inflammatory company, Arakis (Chesterford, UK) could match the US top ten.

Hence venture investors have begun concentrating their investments into a small number of later stage companies and broadly refusing start-ups seeking capital (Chaya, 2005). In 2004, 48 companies raised funds in

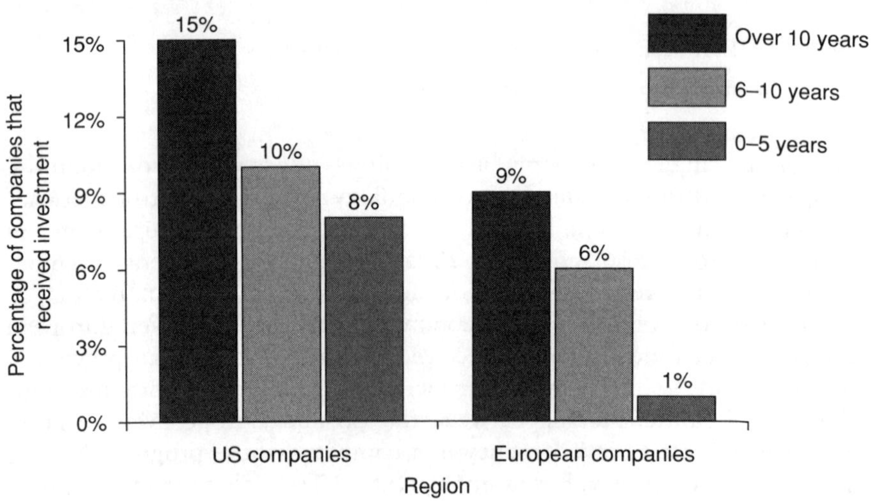

Figure 10.4 Disparity in investment: US and EU DBFs

Source: Hodgson (2006)

Table 10.8 Top ten venture capital rounds, 2004

Company (location)	*Money raised in millions*
US ventures	
Jazz Pharmaceuticals	€184.70 ($250)
Synta Pharmaceuticals	€96.04 ($130)
AlgoRx Pharmaceuticals (Secaucus, NJ, USA)	€48.02 ($65)
Eximias Pharmaceutical	€46.91 ($63.5)
JDS Pharmaceuticals	€45.81 ($62)
ARYx Therapeutics	€40.63 ($55)
MannKind	€36.94 ($50)
PTC Therapeutics	€36.94 ($50)
Dynogen Pharmaceuticals	€36.94 ($50)
PharmAthene	€36.94 ($50)
European ventures	
Novexel	€71.19 ($96)
Arakis	€41.05 ($56)
Arpida	€33.19 ($45)
Addex Pharmaceuticals	€32.67 ($44.2)
ProStrakan Group	€31.14 ($42.1)
Jerini	€31.00 ($41.9)
PowderMed Limited (Oxford, UK)	€28.31 ($38)
Santhera Pharmaceuticals	€27.95 ($38)
CareX (Strasbourg, France)	€25.00 ($33.8)
Ablynx	€25.00 ($33.8)

Source: Hodgson (2006)

Europe, of which 58 per cent of the funds were concentrated in only 11 companies. It is getting harder for smaller companies to survive and hence consolidation becomes increasingly inevitable. Surprisingly, against this, Dimov and Murray (2006), who recently studied 2,950 venture capital firms worldwide in the period 1962–2002, found the larger the VC fund, the more likely it was to invest seed capital and the more capable it was of picking winners. Governments across European countries have given increasing financial support to young start-ups. However, it seems the big financial gap in biotechnology's business development between the US and the UK is particularly at the later stage, as indicated in Figure 10.4.

VC investment differs across the countries in terms of their stage, source and geographical focus (Mayer *et al.*, 2005). Banks are particularly important in Germany and private companies in Israel, and to some extent pension funds in the UK. The Israeli VC industry was established far more rapidly than it was in the US. VC emergence in Israel was 'policy-led'. Israel took advantage of the learning that has already occurred in the US and their historic link with VC in the US. German banks engage in more risk taking than banks in the UK and the US. There is some evidence that UK banks invest in later stages than Germany.

Initial public offering

The global biotechnology industry showed signs of recovery in late 2003 and early 2004 with 52 biotech firms going public and raising $2.1 billion. The recent decrease in valuations and numbers of initial public offerings (IPO) revealed the harsher conditions in the sector. Europe experienced an increase in the number of IPOs, indicating an improved financial environment. The so-called IPO window thus re-opened in the mid-years of the decade. Then, however, it slowed in 2005, with fewer offerings and less money raised (Figure 10.5). One bright spot in the IPO market was the strong sign for international, particularly European-based companies (Lähteenmäki and Lawrence, 2006). The number of European biotech IPOs increased from 12 in 2004 to 23 in 2005. The total amount of money raised by European biotech IPOs reached $916 million, greater than the US ($811 million). In 2005, the UK launched about 12 biotech companies onto the London Stock Exchange and alternative investment market (AIM). AIM has become the primary market in Europe for growth companies. Overall, biotechnology stocks performed solidly in 2005. The Nasdaq Biotechnology Index was up some 5 per cent. Two large DBFs, Amgen and Genentech, each grew their revenues by about $2 billion in 2005 (Table 10.9) (Lähteenmäki and Lawrence, 2006). Genentech grew its revenue with a 44 per cent increase since 2004, leading on the top list of gainers. Meanwhile, some biotech companies decreased their revenue at a fast rate. For example, Regeneron Pharmaceutical reduced 62 per cent of its revenue since 2004.

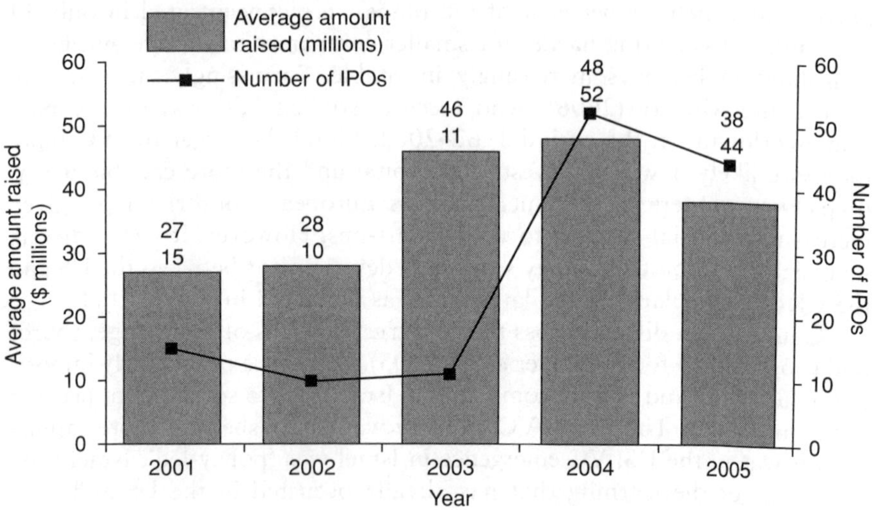

Figure 10.5 Global healthcare biotechnology initial public offerings (IPO)

Source: BioCentury; Lähteenmäki and Lawrence (2006)

Table 10.9 Top ten public revenue gainers and losers, 2005

Company	*Absolute change in revenue ($ millions)*	*2005 revenue ($ millions)*	*Percent change from 2004*
Gainers			
Genentech	2,012	6,633	44%
Amgen	1,880	12,430	18%
Monsanto	871	6,294	16%
Gilead Sciences	704	2,028	53%
Genzyme	534	2,735	24%
KOS Pharmaceuticals	255	752	51%
Biogen Idec	211	2,423	10%
Endo Pharmaceuticals	205	820	33%
Cephalon	196	1,212	19%
Chiron	196	1,920	11%
Losers			
Regeneron Pharmaceuticals	–108	66	–62%
Enzon Pharmaceuticals	–93	74	–56%
Nabl Biopharmaceuticals	–72	108	–40%
Flamel Technologies	–32	24	–57%
Idenix Pharmaceuticals	–31	65	–32%
Cardiovascular Therapeutics	–29	31	–48%
CollaGenex Pharmaceuticals	–26	26	–49%
Third Wave Technologies	–23	24	–49%
InterMune	–18	110	–14%
Aradigm	–18	11	–63%

Source: Lähteenmäki and Lawrence (2006)

Business model shift

In the 1990s the biotechnology business model moved away from a service model to focus on a product-based strategy. Furthermore, companies shifted away from platform technology to clinical development businesses. However, drug discovery and development is a long and expensive process, and it is also high risk. Many management teams see it as the only option to create a more sustainable long-term business (Chaya, 2005). But recently, with the tight financial constraints after the downturn and active ongoing merger and acquisition, some firms began to shift the drug development technology back to a service model or platform technology, predominantly so that investors could obtain their return in the short term. The case of Cambridge Antibody Technology (CAT) is illustrative of such a business model shift.

Cambridge Antibody Technology (CAT) was founded in 1990 with a service business model. The peak market capitalisation was valued of $2 billion in September 2000. Its shares had soared like those of many other biotechnology companies during the stock market rise. In 2002, it was transformed into a fully integrated product developer that aimed to be profitable by 2008 (Ward, 2006b). However, while other antibody companies were taking advantage of strong merger and acquisition deals or raising new funds, CAT spent its time in an expensive and destructive patent fight with Morphosys AG (Germany). This dispute started in the mid-1990s and was not settled until the end of 2003, which scared the potential partners away. Although CAT's pipeline looked full, with ten human monoclonal antibodies in clinical trials, most of these are licensed to others, from which CAT received only modest royalties. In most cases, the targets were identified by the partner; CAT produced the relevant antibody and then licensed it to the partner, who was then responsible for all further development and marketing. Only one candidate drug was indigenous, namely anti-human interleukin-13 (CAT-354) to treat asthma, which had been developed in-house and was owned exclusively by CAT. The antibody companies that focused on developing their own proprietary products rather than running service models have been more generously valued by investors. Developing proprietary therapeutics was perceived as the best way to build a company. When Abgenix was acquired by Amgen necently, it stimulated AstraZeneca to perceive a need to have secure access to antibody technology. This prompted AstraZeneca to acquire CAT in order to build a major international research and development presence in biologicals. CAT decided that there was no near-term event that would drive the company's valuation of £702 million, offered by AstraZeneca, hence CAT became yet another UK biotechnology company to decide that its future was best secured as part of a larger organisation.

Partnering

Biotechnology business partnering activity continues to demonstrate the importance of the financial role it plays in the global industry. As biotechnology companies mature and move into drug development, there is an ever-increasing strategic need to build clinical expertise, business development and commercial capabilities to drive greater returns on their products and technology. As a result, increasing numbers of partnering deals occur, in particular for late stage DBFs with clinical products (Chaya, 2005). A review of biotech product alliances during the first quarter of 2004 illustrates a growing number of alliances for clinical and marketed products, from 38 in 2002 to 54 in 2003 and 68 in 2004 (Figure 10.6). Pharmaceuticals companies are increasingly viewing biotech start-ups as a source of clinical compounds to sustain their thinning pipelines. The reduction in the number of discovery deals from 2002 through 2004 suggested that pharmaceuticals companies are more interested in later stages, although we can still see Roche's $178 million deal with Syrrx. This also indicated that strong and distinctive platform technologies still have the ability to secure large deals. Overall biotech start-ups are benefiting from growing interest from pharmaceuticals companies and biotech firms at discovery stage can benefit from partnering.

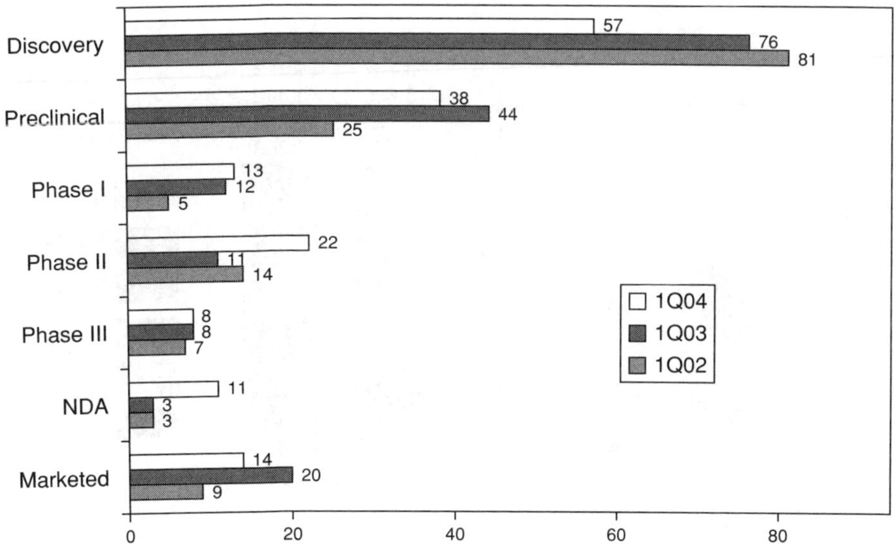

Figure 10.6 Number of biotechnology product alliances by stage of development

Note: 1Q04 etc. is the first quarter of 2004

Source: Burrill and Company (2004)

The total value of $17 billion in 2005 made partnership deals both the largest and fastest growing source of financing for DBFs. It accounted for almost half of funds raised by them globally (Figure 10.7). The value of partnership deals is getting bigger and is dominated by a few big pharmas and larger-sized DBFs (Table 10.10). In 2005, there were eight biotechnology licensing deals that exceeded half a billion dollars. Public biotechnology companies were the main players, as nine of the top ten partnership deals involved public DBFs. In recent years, the number of biotech–biotech deals has been accelerating, while the number of big pharma–biotech deals has been relatively steady (Figure 10.8). Bigger DBFs such as Amgen and Genentech have begun to compete with big pharma on more or less equal terms.

Consolidation through mergers and acquisitions

Merger and acquisition (M&A) is another important exit for biotechnology companies. This mechanism is attractive because a trade sale can produce more capital instantly than an IPO or licensing. Consolidation can create value with cost savings, diversification of pipelines and optimisation of management. It can also help start-up companies achieve the critical mass that is important in justifying a sustainable business. Through M&A, big

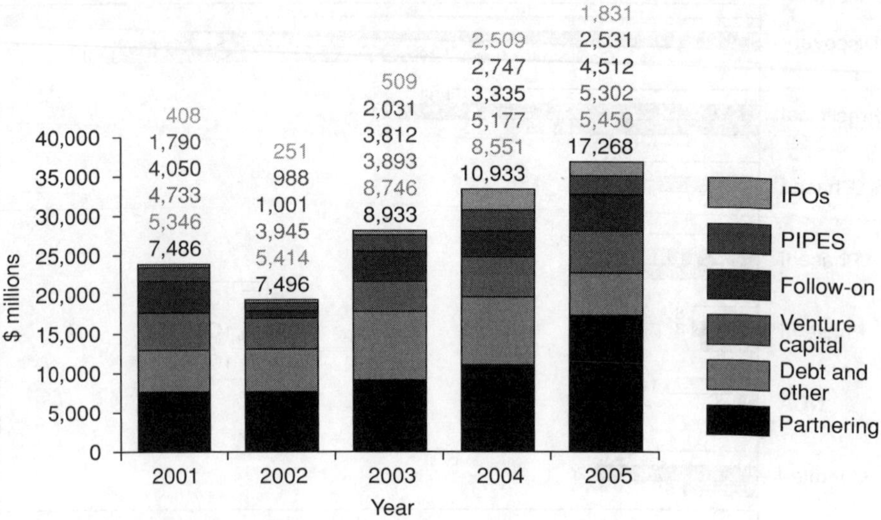

Figure 10.7 Global biotechnology industry financing, 2005

Source: Burrill and Company (2004)

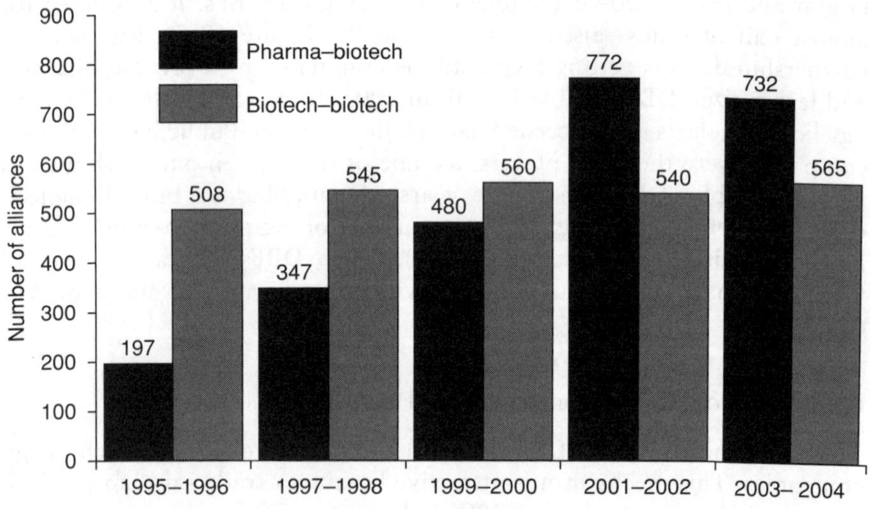

Figure 10.8 Biotechnology therapeutic alliances

Source: Thiel (2005)

Table 10.10 Top ten research partnership and licensing deals of 2005

Client/licensee	*Deal type*	*Deal value ($ million)*	*Description*
Pfizer/Incyte	Collaboration, development, license, research	803	Portfolio of CCR2 antagonists
Biogen Idec/Protein Design Labs	Codevelopment, copromotion	800	Zenapax (daclizumab) in multiple sclerosis, M200 (volociximab) and HuZAF (fontolizumab) in all indications
Novartis/Alnylam Pharmaceuticals	Codevelopment, collaboration, equity, license	700	RNAi therapeutics collaboration
Novartis/Anadys Pharmaceuticals	Codevelopment, copromotion, license, option	570	ANA975 and TLR7 oral prodrugs for hepatitis C virus and hepatitis B virus
Roche/BioCryst	Copromotion, development, license	560	BCX-4208 (PNP inhibitor) for transplantation
Emory University/Gilead, Royalty Pharma	Asset purchase	525	Royalty interest in Emtriva
Pfizer/Coley Pharmaceutical	Collaboration, development, license	515	Promume for cancers
Cephalon/Alkermes	Codevelopment, license, supply	490	Vivitrex for alcohol dependence in US
Wyeth/Progenics	Codevelopment, copromotion, license, option	417	Methylnaltrexone (MNTX) for opiold-induced bowel disorders
GlaxoSmithKline/Vertex	Development, license	405	VX-409 for pain

Source: Lähteenmäki and Lawrence (2006)

pharmas have acquired single product and strategic technology from biotechnology firms. This phenomenon is continuing to increase, including the Novartis–Chiron deal with $5.1 billion in 2005 and the recent $1.1 billion deal between CAT and AstraZeneca in 2006. Larger DBFs are becoming major M&A counterparts also; for example, Amgen acquired human therapeutic antibody company, Abgenix, with a $2.2 billion deal (Table 10.11).

Table 10.11 Largest mergers and acquisitions of 2005–2006

Acquirer/target	*Deal value ($ millions)*
Novartis/Chiron	5,100
Amgen/Abgenix	2,200
Pfizer/Vicuron	1,900
Shire Pharmaceuticals/Transkaryotic Therapies	1,600
GlaxoSmithKline/ID Biomedical	1,400
OSI Pharmaceuticals/Eyetech Pharmaceuticals	935
Genzyme/Bone Care	600
Invitrogen/Dynal	390
GlaxoSmithKline/Corixa	300
Roche/GlycArt Biotechnology	181

Source: Lähteenmäki and Lawrence (2006)

NB: To these should be added: Merck, Sharp & Bohme/Serono, $13.3 billion; Pfizer/Vicuron, $1.9 billion; Novartis/NeuTec Pharma, $1.3 billion; AstraZeneca/CAT, $1.1 billion; Merck (US)/Sirna $1.1 billion; AstraZeneca/AtheroGenics $1.0 billion; Meda/Viatris $927 million; Pfizer/Incyte, $803 million; Biogen Idec/Protein Design Labs, $800 million; Novartis/Alnylam, $700 million; Glaxo/Domantis $500 million; AstraZeneca/Kudos, $230 million; Sosei/Arakis, $200 million

Source: Urquhart and Orr (2006); Jack and Urquhart (2006)

Small start-up companies should be driving consolidation considering the benefits. But the implementation of consolidation can be very challenging. The assets might cause the company to lose a clear focus. Overall the benefits of M&A can be lost in implementation issues at certain levels.

Reverse mergers and decapitation

Now that both valuation and market performance have plummeted, many biotechnology companies have little chance to make an IPO. Some therefore turn to the concept of a reverse merger as an attractive financing option for private companies to access capital (Lawrence, 2006b). In the first four months of 2006, at least seven biotechnology reverse mergers were announced. For example, Infinity Pharmaceuticals, a private DBF based in Cambridge, Massachusetts announced a reverse merger into a public company, Discovery Partners International at San Diego. A similar deal had been made across the Atlantic. Micromet located in Munich decided to take this step to merge with California-based CancerVax. UK Dundee-based Cyclacel made the reverse-merger deal with Seattle-based Xcyte Therapies. Both of these European companies have moved their headquarters to the US and now have public listings on the US market to access more favourable US capital. This 'decapitation' issue has caused official concerns in Europe about the control of domestic companies from the US, yet the UK biotechnology industry, widely seen as the strongest in Europe, is evolving a 'conveyor belt' or 'decapitation' model of offshore innovation in the US. Thus *decapitation* means retaining *examination* knowledge capabilities in the UK while accessing *exploitation* capabilities in the US (Ward, 2005). The UK's ambitions regarding domestic commercialisation of high-grade fundamental research have been thwarted by *capability* deficiencies in commercialisation. Table 10.12 lists some key 2005 UK decapitations.

The overwhelming reasons for separation of R&D from business capabilities cited by these firms' representatives were:

- seeking US chief executives to run the business
- proximity to an experienced pool of management talent
- weak state of IPO markets for biotechnology firms in the UK
- more sophisticated US shareholder base
- better access to higher capability ('savvy') biotechnology fund managers
- proximity to the deal-making hubs of big pharma.

Thus, as may easily be seen, companies with valuable technologies that managements fail to exploit in a timely manner are likely to find themselves being acquired if for no other reason than that they will have lagging stock market valuations that will make them easy picking.

Follow-on funds, PIPEs and convertible debt

The last potential financial gap for growing biotechnology firms occurs at post-IPO stage. European financial regulation has tended to protect existing shareholders from possible dilution. Thus it makes *private investment in public equity* (PIPE) a less convenient financing vehicle for European companies than US ones. In 2005, only $306 million follow-on funding was

Table 10.12 R&D decapitations of UK biotechnology businesses, 2005

Company	*US HQ*	*Details*
BioVex	Cambridge, MA	Retains R&D in Oxford, UK
Cyclacel	Short Hills, NJ	Retains R&D in Dundee, UK Reverse merger into Xcyte Therapeutics, Seattle, WA
Domantis	Waltham, MA	Retains R&D in Cambridge, UK
Lorantis	Bloomsbury, NJ	Retains R&D in Cambridge, UK Acquired by Celldex Therapeutics
Microscience	Gaithersburg, MD	Retains R&D in Wokingham, UK Acquired by Emergent BioSolutions
Solexa	Hayward, CA	Retains R&D in Cambridge, UK Merger with Lynx Therapeutics
Zeneus	Frazer, PA	Retains R&D in Oxford, UK Acquired by Cephalon

Source: BioCentury (Ward, 2005)

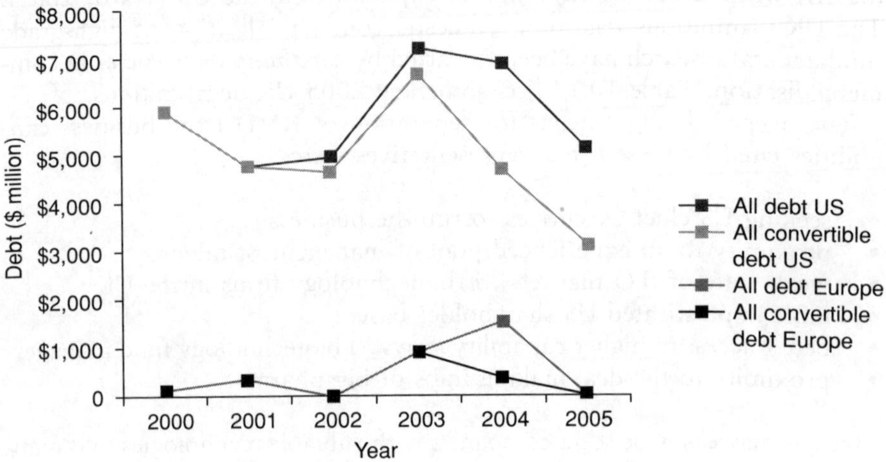

Figure 10.9 Comparision of debt in Europe and US, 2000–2005

Source: BioCentury; Mitchell (2005)

raised in Europe, less than 10 per cent of the amount raised in the US with $4 billion follow-on funding (Ward, 2006b). Although US biotechnology raised less in IPOs than Europe in 2005, the IPO does not represent the end of the financing trail in the US, at least. There mature firms require more late-stage or follow-on funding. Thus US companies continue to raise substantial amounts of money in PIPEs and debt finance.

Convertible debt is a regularly significant source of financial support for US companies, but has yet to be in demand in Europe owing to the relative immaturity of the biotechnology sector (Figure 10.9). The use of debt involves fixed payments; failure to make these payments leads to default and may potentially result in bankruptcy. Banks do not share the firm's upside potential and are therefore particularly concerned about downside risk. Banks ration credits to firms with high default and bankruptcy risk, thereby limiting firms' capacity to get additional debt (Carpenter and Petersen, 2002). Equity – especially from specialised VC investors – plays a crucial role in the growth of young, high-potential firms. Bank debt is even more important for these growth-oriented ventures (Baeyens and Manigart, 2006). Debt finance was described as the 'secret weapon' of biotechnology development in the US. Despite a decline of more than one-third, from $8.5 billion in 2004 to $5.5 billion in 2005, debt continued to be one of the top funding mechanisms. This finance vehicle was dominated by the largest biotech firms with enormous funding amounts (Table 10.13) (Lähteenmäki and Lawrence, 2006). The most notable debt offering of 2005 was a $2 billion note from Genentech in July. In February of 2006,

Table 10.13 Top ten debt financing of 2005

Company name	*Date completed*	*Amount raised ($ millions)*
Genentech	7/13	2,000
Cephalon	6/2	800
Invitrogen	6/14	325
Nektar Therapeutics	9/22	315
PDL Biopharma	2/9	250
Human Genome Sciences	8/4	230
AtheroGenics	1/7	175
CV Therapeutics	6/29	130
Alexion Pharmaceuticals	1/19	125
Nabi Biopharmaceuticals	4/14	100

Source: BioCentury; Lähteenmäki and Lawrence (2006)

Amgen announced a large $5 billion convertible debt deal, becoming the largest biotechnology debt offering in history so far.

Conclusions

This chapter has ranged widely on an important part of the reason why biotechnology firms exist, and in particular how it is that the gap between performance of US and European DBFs is to an extent conditioned by their distinctive financial environments or 'varieties of capitalism' (Hall and Soskice, 2001). The detail in the chapter is relatively up-to-date and captures relatively recent shifts since the end of the 1990s technology boom with the subsequent stock market downturn. In some respects, Europe may have become a more attractive place to get early-stage venture funding than it was or than is currently the case in the US. This is because Europe is perceived as having more varied seed capital opportunities. Venture capitalists were never very interested in this kind of funding even though they began by investing such funds into the early high-technology firms starting up in California in the 1970s. Nowadays many of them have become in effect merchant or investment banks interested mainly in safer investments than biotechnology firms. Into that space we saw big pharma becoming more of a supplicant to DBFs for innovative knowledge in the face of their drying pipelines. But we also saw how alliances often lead to acquisitions where VC has so diluted the original shares of the DBF that a trade-sale is the only way of realising some return on initial investments. Business models have shifted from services to products and back again. Outsourcing of development work to third parties that later sell the successfully trialled product on to another buyer is also a model being successfully practised. Finally, debt-financing is in its infancy for European biotechnology firms, but may have been an unexpectedly important part of the continuing success of leading US biotechnology firms.

11 Conclusions

Biotechnology's proximities, pipelines and platforms

Introduction

This book has investigated the biotechnology world from many and diverse angles, setting it within a knowledge economy context, evolving a new knowledge-focused theoretical framework against which to test empirical characteristics of biotechnology, and exploring this from scientific, industrial, financial and policy viewpoints in developed and developing countries with respect to agro-food, environmental, energy and, most of all, healthcare biotechnology. What has been conclusively shown undermines a certain bias in the perceptions of many commentators and policy makers as well as government ministers that biotechnology is a rather esoteric scientific and commercial activity that may not be enjoyed everywhere. This rests on a narrow, vertical view of it as a very specific and circumscribed, advanced technology. But it can easily be shown that while it is not as pervasive in everyday life as, for example, information and communication technology (ICT), it is rapidly taking on the appearance and influence of a general-purpose technology, widely applicable outside its own small originating genetic engineering field. For aspects of biotechnology that were not explored in the security, policing and even ICT itself are merely some of the few that are now involved in DNA fingerprinting, biometrics and bio-chips as replacements for the ubiquitous silicon chips.

In summarising the book's main findings the focus is upon three cross-cutting features. The first of these is *proximity*. The book was stated from early on to be a study in the combined and related intellectual fields of economic sociology and evolutionary economic geography, unifying once more the social and spatial dimensions of economic life. Hence it was going to be disposed to be interested in concepts central to both these overlapping fields. The notion of embeddedness in both a relational and more structural perspective, after Granovetter (1992), was utilised forensically, as were the related concepts of networks and nodes encompassing communities of practice. Conceptions of the importance of learning, knowledge, variety, notably related variety, search and selection with knowledge spillovers being facilitated by geographical, but also other kinds of proximity, entered the scene frequently. These are more commonly utilised, along with concepts like path dependence, in economic geography. Time and time again biotechnology was found to be, although a youthful endeavour, remarkably path-dependent upon established physical locations close to high-quality, usually public, bioscientific research institutes and university centres of excellence. 'Clustering', as this is called in more mainstream business school treatments such as that of Porter (1998), is the term normally used to describe this. But two observations arise. First, 'cluster' is a modern synonym for what Alfred Marshall (1918) called *industrial districts*, a term still widely used in evolutionary economic literature, particularly in Italy, where they have been and continue to be studied as living phenomena to this day. In other words, in the language of a biotechnology entrepreneur interviewed for part of the research reported on here, clusters are a retro-model that business typically does not identify very strongly with, although many policy makers and ministers are virtually inseparable from the concept and make great claims for how many their region or country possesses. Second, because of this 'clustering' in biotechnology, where it may be something of a pioneer in the sense of its path dependence on proximity to scientific research facilities, it is easy to assume that all high-tech is like that.

Other high-tech industry, notably ICT, also clusters, but in a comparative study of the two fields Cooke *et al.* (2007) found that ICT clusters closer to its customers and suppliers than to its university science laboratories. Thus it can be deduced that, in this, proximity to research is imperative and biotechnology is, if not unique, then relatively singular. However much modern economic policy places burdens on universities to be regional economy locomotives, in truth it may only be the few with world-class research that can really have a chance of being so. The book showed this to be the case in biopharmaceuticals, particularly for important university towns and some quite large but not always major or world cities. But it also showed, especially in Chapters 8 and 9, that there are platforms of other industrial and organisational activities that relate to the narrowly defined biopharmaceutical activities and many of these 'megacentres' interacted with a diverse, sometimes cross-over platform of other technologies, some of which are reprised below. Moreover, many of the other biotechnologies are found occasionally in proximity to healthcare biotechnology activity, but often are not. Agro-food biotechnology was found, not surprisingly, sometimes located in agricultural contexts, and environmental in older industrial areas in need of cleaning up. So the broad biotechnology platform consists of networks and nodes that span broad technological application platforms. Ministers need to understand this before denying regions or, in the developing world, countries the right to access the knowledge inputs and the fruits of biotechnological knowledge. Anyway, as we have seen, they cannot keep biotechnology as an exclusive club because the mechanisms by which big pharma control access, IPR and pricing have been shown to be vulnerable precisely and mainly due to the proliferation of biotechnological research and diseases that threaten suspension of international IPR conventions in developing countries.

This links to the subject of the second unifying thread in the book, which is the notion of *pipelines*. This term is utilised distinctively by biotechnologists themselves and academic observers investigating biotechnology from an economic sociology perspective. In the former case it means the status of a drug in a process of blind clinical testing, which goes from preclinical to mammalian, to small human samples and finally to large patient populations in structured samples involving placebos and the candidate drug being tested for efficacy or negative side-effects. Many of the tables in Chapter 10, it will be recalled, used this meaning of pipeline in comparing the progress and abundance of pipeline products in different countries' pipeline portfolios. But the possibly more intellectually intriguing usage of the term 'pipeline' is that introduced in a sociological study of biotechnologists in Greater Boston (Owen-Smith and Powell, 2004). They noticed that many of the firms whose owners they interviewed were legally contracted to conduct out-sourced R&D or production for 'big pharma', precisely as this book has emphasised to be the case in many chapters. Whether such principals and agents were geographically proximate or not, they were termed pipeline relationships, meaning that they were intended to be sealed or confidential agreements. Yet scientists and entrepreneur scientists, in particular in this context, still adhere to some of the Mertonian norms of open science, discussed in Chapter 1 under the acronym CUDOS. They have a public duty to 'speak truth to power' in the Jeffersonian sense, thus they talk informally with colleagues or neighbours for purposes of informal scrutiny or just plain garrulousness.

In this manner open science conventions give rise to a related concept of 'open innovation' where large firms actively seek out knowledgeable scientists or entrepreneurs to conduct their innovation for them (Chesbrough, 2003). Firms understand the value of such knowledge spillovers, weigh up the costs and benefits accordingly, and continue to increase doing it for the return benefits they get from knowledge spillovers from others. In sociological terms, knowledge spillovers are presented as coming from the channels or conduits by means of which social interaction in clusters occurs. Owen-Smith and Powell (2004) refer to leakage from these channels as normal discourse in open scientific settings. So here we gain an excellent understanding of the means by which a related variety of industry in a platform of moderately complementary activities enables high lateral absorptive capacity to thrive, as our theoretical framework and discussion of knowledge asymmetries and capabilities 'bundles' proposed, as a growth culture in its own right in the context of the knowledge economy. As was made clear in Chapter 2 and subsequently, where related variety exists, economic growth occurs with greater facility.

Finally, the book spoke about *platforms*. Biotechnologists speak in everyday language about platform technologies that enable drug discovery to be accelerated. A test for diagnosing the likelihood of a particular disease by sensing certain kinds of leakage into the bloodstream of diseased cells may be an inordinately important step in identifying the precise genetic mutation that eventually leads to discovery of an inhibitor molecule that corrects the problem. In some circles, it is argued that diagnostics of that kind are virtually part of the therapeutic treatment itself as our innovation biography of Herceptin in section three of this chapter demonstrates (Slamon *et al.*, 2006). So the briefly traditional frontier between biotechnological platform technologies and drug discovery as a kind of class barrier or gold standard privileging the latter may swiftly be dissolving. Then there is the notion, a little closer to our preferred definition, of industry platform which around biotechnology would include all and more of the nodes in the biotechnology value chain, also discussed below in this chapter, but it would bring in other sectors closer to bulk chemicals, on the one hand, and on the other, bioimaging atom colliders or synchrotrons, such as that newly built near the Oxford biotechnology cluster whose sun-like brightness enables molecular imaging to be conducted on cellular material. In other words a platform involves many related industries and is the source of related variety growth cultures not least because of the knowledge spillovers among cognate scientists, technologists and entrepreneurs.

Proximity

One of the main achievements of this book has been the consistency with which it proved possible to demonstrate how biotechnology practically always agglomerates, whether it is in developing or developed countries, in healthcare biotechnology or in the other varieties looked at such as agro-food biotechnology and so on. This is a vindication of the theoretical approach taken, which has been to have knowledge as the centrepiece of the explanatory framework of growth cultures in the knowledge economy. Of course as suggested above, biotechnology may be different from other knowledge-based industries in the precision with which its location can be predicted close to research institutes. It is known, at least for Austria and the UK, for example, that ICT firms also agglomerate, but close to customers and suppliers and often not so close to university research (Cooke *et al.*, 2007). Naturally there are exceptions, but also mis-perceptions. For example, many writers make the false connection regarding the origins of Silicon Valley and the explanatory presence of Stanford University. But the story of Silicon Valley really only dates from Shockley's arrival from Bell Labs in New Jersey accompanied by his PhD students, most of whom established semiconductor or venture capital firms in what later came to be called, journalistically, Silicon Valley. While it is true that their first employer, Fairchild Camera and Instrument, located Fairchild Semiconductor in Stanford's Industrial Park, that was a shortlived affair as the 'traitorous eight' soon left to create the likes of Intel and American Micro Devices

(Saxenian, 1994). But the clear exception that perhaps does prove the rule is the telematics cluster at Aalborg University, Denmark where some sixty telecom start-up firms can still be found in the University Technology Park and nearby, even though most have long been acquired by the likes of Texas Instruments, Motorola and Siemens. This is partly because there is high social capital in the group and a highly active cluster association, NordCom, that continues the long-term embedding process. It might be argued that Finland displays comparable characteristics for smaller ICT firms in university science parks, but they are mainly there to act as local suppliers to R&D and other plants of Nokia (Stoerring and Dalum, 2007).

Without labouring the point, proximity is important for highly knowledge-based businesses, including more services-oriented ones such as financial services, and creative industries such as new media (Lievrouw and Livingstone, 2006). This comfortably dispenses with the over-eager dismissal of the importance of face-to-face and tactile business, academic and policy contact in the Internet Age (Cairncross, 1997). But being wrong can still assist social scientific explanation since it forces thinking as to why such a prognosis is wrong about the perishing of proximity in economic life. The answer is to be found in the rather careless conflation of information with knowledge. It will be remembered how the 1990s were, again journalistically, characterised as the Information Society under the misperception that information rather than knowledge really drives progress. In the 2000s there is much more labelling of the condition in which we exist as a Knowledge Economy. This is because social science, including economics, has gained a much better grip on the generative growth powers of knowledge than it had before, as we saw in Chapter 2.

Proximity is, as we have seen, intimately tied up with two key and socially powerful developmental mechanisms: innovation and knowledge spillovers. Tackling the latter first, research by Van Ark *et al.* (2002) shows that one of the three key drivers in explaining the US lead over the EU in industrial productivity is superior knowledge spillovers in the US. This analysis assigns most of the explanation of total factor productivity in the US to aspects of ICT. Two drivers are early access to the newest technologies and high investment in such innovations, especially in fields like retail. For example, Wal-Mart, whose founder Sam Walton took radio frequency identification (RFID) classes at IBM, has what is widely considered to be one of the best retail ICT-based logistics systems in the world running its supply chains. Meanwhile, many services businesses and organisations in the EU are paper-based. The point about knowledge spillovers is that with multiple new users of advanced technologies within and beyond the walls of its purchasers, absorptive capacity rises exponentially, as there is usually someone available for informal operational advice.

This fact assisted in the solution of what was known as 'the Solow paradox', when the economist and Nobel laureate of that name noted that US ICT expenditure was rising but productivity was not. The reason was the lag between purchase and efficient and effective utilisation, which could have been shortened had systematic training in ICT use been more widespread, but it actually improved more as a consequence of the activation by workers of their social capital.

To complete this summary of the importance of proximity to modern economic and social life, one only has to think of the abiding and continuing power of *cities* to house and feed often vast populations. Periodically, the literature speaks of city crises and population shifts from urban to rural locations. But in reality these are either outliers from the statistical bell-curve or they are age-related cycles connected to family-formation or retirement. Out-migrants are normally replaced by younger in-migrants, with a significant proportion today often migrating from abroad. It is clearly because this occurs as more and more knowledge economy activity evolves, involving far more transactional relations in a more entrepreneurial, principal-agent style of service industry interaction. Thus despite their negative externalities like congestion and pollution, cities are increasingly fundamental to economies. More so than in the Industrial Age, when much industry was found outside cities, near sources of raw materials. It is the powerful element of dynamic knowledge-based capabilities that drive the current revitalisation of city life. Information by contrast is frequently merely passively available in digitised forms that can be meaningless until cognitive skills expressing knowledge application are deployed in relation to it.

In previous settings this has been illustrated in terms of the railway train timetable. It is meaningless except to the passenger who seeks his or her correct departure and destination arrival time (Cooke and Piccaluga, 2006). Moreover, with scientific or creative knowledge of the kind discussed in Chapter 2 in so much greater demand and, for example, new forms of public and private management which privilege process over substance (e.g. issuing tenders, managing consultants and contracts), there has been an enormous rise in the number of meetings, workshops, seminars and conferences at which attendees seek knowledge of consequence to their business. An indication of this fact is in the following quotation regarding the US alone:

> Many reviews and surveys reveal that meetings dominate workers' and managers' time and yet are considered to be costly, unproductive and dissatisfying. Studies show that meetings are essential and that the number of meetings and their duration has been steadily increasing. Studies of managers and knowledge workers reveal that they spend between 25%–80% of their time in meetings, suggesting that meetings are an important part of one's working life. Estimates of meeting expenses range from costs of $30 million to over $100 million per year to losses between $54 million and $3.7 billion annually! Self estimates of meeting productivity by managers in many different functional areas range from 33%–47%.
>
> (Romano and Nunamaker, 2001)

This is despite evidence that there is also a growth in the number of Internet and video conferences, but clearly although often dissatisfying, knowledge workers find meetings and conferences in geographical, albeit temporary, proximity increasingly addictive. Where cities score, therefore, is in their capability to concentrate knowledge spillovers and facilitate innovation. Even smaller knowledge cities like university towns may have this capability: consider the following two illustrations of what could be called the special category of *retro-innovation*. Despite its apparently oxymoronic connotations, this is a seriously innovative socio-cultural and economic phenomenon with major implications for certain master institutions like governments and global corporations. Being referred to here is the organic food movement. Alice Waters, once a Berkeley hippie, spent a year in France during her studies and realised that most American food was sub-standard and possibly poisonous from the growth factor, anti-insect and weedkiller chemicals she knew were liberally sprayed on the vegetables and fruits that grew in the nearby San Joaquin Valley. Ironically, on this, the most fertile farmland in the US, Silicon Valley was then rising, also contributing not a few hazardous wastes, including solvents used by the electronics industry. She evolved a philosophy to source local, fresh, seasonal, preferably organic ingredients for the food she served in Chez Panisse, the restaurant she opened at 1517 Shattuck Avenue near the Berkeley campus in 1971. She was said to lie in bed at night worrying about what to feed her customers, knowing she could make them happy. She also believed that if she could find the perfect dish to feed each person she could change the world. In a real and meaningful sense, she has begun achieving that vision, since modern cuisine such as that practised by the Slow Food movement, of which she is an International Governor, or recommended by contemporary celebrity chefs, is precisely in line with Alice's philosophy. It is widely understood in food circles that Waters initiated a food revolution, to the extent that even supermarkets have had to respond to the demand for organic food or, as it is referred to in the US, natural food, and sometimes even a little derogatorily in continental Europe as eco-food, the hippy connotation hanging onto the term.

Initially Waters simply wanted to feed her friends, not start a food revolution. Having been to France, she had seen the way a good bistro could become the heart of a neighbourhood, so she envisioned a small café, which in the early days faced financial ruin, so Chez Panisse was forced to impose financial discipline. The vision was adjusted to making the restaurant, as it was now conceived, the best one she could possibly manage, even if that meant rethinking the whole concept of what a restaurant might be. The key lay in the ingredients. Disgusted with the fish that was sold in stores, she bought a truck and sent someone down to the port to find fishermen as they docked their boats. When she could not find the baby lettuces she had loved in France, she dug up her backyard and grew her own. She found foragers to hunt for mushrooms. She persuaded farmers to let their lambs run wild through the hills. She demanded better bread. Before long, she had developed an entire network of people producing food just for her. The founding chef was Jeremiah Tower, now a leading US Celebrity Chef and owner of Stars, one of San Francisco's leading California regional cuisine restaurants. Chez Panisse served only one meal a day, but customers would habitually reserve months ahead, something no longer practised, a month now being the maximum booking time. Some with long-term reservations, from the east coast or abroad, might express disappointment at being served 'only chicken', but the first taste of the chicken would produce awe as they ate the best chicken they had ever tasted, discovering 'they never knew that food could taste so good', as Alice put it in the 'American Masters' television programme about her. Proximity thus infuses the Waters philosophy, whether by reaction against the local agro-chemicals or the quest for her own local food suppliers network or the fact that until recently Chez Panisse was her only food outlet. She has subsequently opened Fanny's Café in New York but devotes much energy to her campaign to establish the Edible Schoolyard that provides urban public school students with a one-acre organic garden and a kitchen classroom.

On its website is a quotation from Paul Cézanne to the effect that '*The day is coming when a single carrot, freshly observed, will set off a revolution*'. Using food systems as a unifying concept, pupils learn how to grow, harvest, and prepare nutritious seasonal produce. It relates also to a growing movement to transform school lunch into a vibrant expression of education for sustainability. The Centre for Ecoliteracy launched the 'Rethinking School Lunch' initiative as part of the national effort to restore the connection of farms to communities, meals to culture, and health to environment. Of course, in the UK academics like Kevin Morgan, my colleague and past co-author, and Celebrity Chef Jamie Oliver have been engaged, not unproblematically, with the same vision (Morgan and Murdoch, 2001). The key point here is that after 1945 agro-chemicals became the hegemonic agricultural technology, driven by scale, innovativeness, and technology and driving traditional methods of agricultural production to the wall. The retro-innovation of organic, natural or eco-farming demands re-scaling, associative support, common, alternative standard-setting and skills training to be undertaken by committed retro-innovators to escape, largely at their own expense, this 'common-sense' but potentially poisonous hegemony.

A different but related part of this retro-innovation concerns the rise of the organic supermarket, box-based delivery service, farms and farm shops such as the prize-winning one at Daylesford in Gloucestershire. But here, interest lies in the innovators of this complementary movement to that inspired by Alice Waters. Whole Foods Market was founded in another university town, Austin, Texas, when a few local businessmen decided that the natural foods industry was ready for a supermarket format. Its founders were John Mackey and Renee Lawson Hardy, owners of Safer Way Natural

Foods, and Craig Weller and Mark Skiles, owners of Clarksville Natural Grocery. The original Whole Foods Market opened successfully in 1980 with a staff of 19 people. At the time, there were fewer than half a dozen natural food supermarkets in the United States. Products were sourced locally as far as possible, but also globally, often from small, uniquely dedicated food artisans. The key vision is to offer the highest-quality, tastiest, least-processed, naturally preserved foods. This is based on the philosophy that pure food, unadulterated by artificial additives, sweeteners, colourings and preservatives, is the best tasting and most nutritious food available. In the UK, the equivalent, Fresh & Wild was founded in 1998 by Hass Hassan, who had worked in yet another university town, Boulder, Colorado. He acquired the long-established City Health Store in London and opened the first new Fresh & Wild in Camden Town in February 1999. Another existing health store in fashionable Notting Hill joined the emerging group in April 1999. Rapid growth and new store openings followed at Clapham Junction in October 2000, Soho in December 2000 and Stoke Newington in June 2001. Customers arrived in large numbers to experience Fresh & Wild's style of natural food retailing and the stores rapidly gained recognition, also winning awards like the Soil Association prize for best organic retailer. The first store outside of London opened in the Clifton area of Bristol in December 2003. The seven stores, together with central kitchen and warehouse, were acquired by Whole Foods Market in January 2004, showing that even organics are not immune to ownership concentration impulses. Proximity and knowledge spillovers in and around local and, as necessary, global artisanal sourcing of organic food produce epitomise this alternative, innovative food approach. It is beginning to make inroads into the domination by agri-business of advanced economy food culture, with its dependence upon agro-chemicals, food miles and large-scale production of generic, processed, flavour-reduced, generic produce.

Pipelines

As noted in the introduction to this chapter, pipeline interactions are of central importance to biotechnology. We noted two meanings, the first of which is common in the biopharmaceuticals and clinical trials fields, referring to the progress of a drug candidate through clinical trials prior to hoped-for approval by the regulatory authorities. The meaning we are more concerned with in this book is that deployed by Owen-Smith and Powell (2004) to explain constraints on knowledge diffusion by contractual connections between principal and agent firms that do not quite work as intended because of continued adhesion by scientists to Mertonian norms relating to public duties to share knowledge. Of course scientists understand contractual rules and laws, but it seems that, as Cohen (2005) discovered regarding patents, they frequently ignore such constraints in pursuing their research. Moreover, principals also benefit from such localised knowledge spillovers from embedded researchers working in neighbouring fields.

It is important also to say something about the relative importance of knowledge outsourcing above and beyond the data supplied in Chapter 10, important and useful as that is. Data from the US National Science Foundation reports changes in the size of firms conducting industrial R&D over time. Since 1981 the trend is consistent with those emphasised in this book, namely a sharp decline of in-house R&D by large corporations and a sharp rise in industrial R&D conducted by SMEs. This is presumably at least in part contracted, rather as suggested in Chapter 6, since the amounts of R&D expenditure are too large to be explained only by internal DBF R&D. It is also noteworthy that R&D outsourcing by large enterprises in general goes far more to SMEs and medium-sized firms than to universities. This is not necessarily because it is development work not interesting for universities, since Table 11.1 showed that many SMEs and mid-size firms conduct basic research. It is possibly a rising figure from a relatively small base, but the longitudinal data are not available in these two data sets to test such a hypothesis. However, in relation to the earlier conclusion about the rise of transactional work from principals to agents, these data are further general support for the thesis that there is a significant amount of contractual work being conducted by US SMEs in general, and biotechnology firms in particular, that hitherto would have been conducted in-house by big pharma.

The final pipelines question that cannot be answered with complete confidence yet is where are these SMEs? It is highly likely that most biotechnology firms in Table 11.1 are located in the Massachusetts and California bioregions centred on Cambridge and Boston in the former with San Francisco and San Diego in the latter. We can see from Table 11.4 below that 387 of the core biotechnology firms in the US were located in

Table 11.1 High-tech US industrial R&D by size of enterprise, 2001

Company size ($ million)	*Total*	*Biotech*	*Materials*	*Nanotech*
5–24	4,828	na	na	na
25–49	3,750	118	na	na
50–99	8,202	398	na	89
100–249	12,916	869	77	481
250–499	8,702	533	41	245
500–999	10,564	1,300	51	1,042
1,000–4,999	26,748	2,155	1,504	3,004
5,000–9,999	17,487	na	na	2,538
10,000–24,999	27,065	149	na	1,616
25,000+	78,244	na	593	5,592
TOTAL	198,555	7,350	5,153	14,624

Source: NSF (2005); Research and Development in Industry (2001)

Table 11.2 Outsourced US company (non-federal) industrial R&D by size of enterprise, 2001

Company size ($ million)	*Total US RDO*	*To firms*	*To universities*	*To elsewhere*
5–24	311	na	na	311
25–49	265	na	na	248
50–99	489	43	7	438
100–249	1,080	184	25	863
250–499	260	na	3	205
500–999	248	60	19	168
1,000–4,999	1,104	253	na	839
5,000–9,999	1,021	406	na	326
10,000–24,999	1,702	79	na	1,621
25,000+	2,242	1,293	68	818
TOTAL	8,723	2,378	401	5,837

Source: NSF (2005); Research and Development in Industry (2001)

those three city-regions in 2000. So hypothetically those, along with the 185 from other leading centres like Washington, Raleigh-Durham and Seattle 572 (Table 4.1, Chapter 4), mean not far short of half the US total of core healthcare biotechnology businesses are found in these six locations. A few more exist in New York and Philadelphia and that more or less summarises the geographical destination for most of the pharmaceutical R&D outsourcing to US biotechnology knowledge domains of various kinds. Given that the first three are the locations of the longest-established and highest-quality DBFs, it is likely that they receive a substantial share of the total outsourced pharmaceuticals R&D.

Two brief illustrations of pipeline relations that fostered successful innovation across proximate and non-proximate knowledge domains or industry–academe pipeline interfaces concern the proprietary products Pill Cam, which challenges the predominance of endoscopy, and Herceptin, the breast cancer drug from Roche via its subsidiary, Genentech. Although this book tends to be resolute in not privileging 'chance' or luck in innovation theorisation, unavoidable is an element of 'fortune favouring the prepared mind' – a direct quotation from Louis Pasteur, whom we last visited in Chapter 1, precisely defining Pasteur's Quadrant of 'use-inspired fundamental research' of the kind to be described. Pill Cam is most associated with Haifa, Israel-based Gadi Iddan, who had been a missile scientist at the Israeli Arms Development Authority. Afterwards he sought potential civilian applications of the video micro-camera devices utilised in rocket guidance. Visiting Boston, an acquaintance living nearby was an endoscopist who told him about the discomfort and even internal injury the mini-camera on a tube inserted into the gastro-intestinal area often induced in patients. Iddan returned to Haifa, established the Given Imaging start-up in Yoqneam Industrial Park, south of Haifa, which houses many advanced medical technology start-ups spinning out of Technion, the Israel Institute of Technology's leading incubator, and built a team that invented a gastro-intestinal video-imaging system inside a pill-sized container. It has been utilised successfully by 400,000 patients worldwide (Iddan *et al.*, 2000).

Herceptin is one of the new generation of biotechnology drugs that targets specific genetic mutations by designing antigens (Schreiber *et al.*, 1988) to inhibit the functioning of mutant genes. This antibody methodology was made possible by the Milstein-Köhler monoclonal antibodies (Mabs) discovery in Cambridge, UK in 1975. For 12 years to 1987, Dennis Slamon and his colleagues of the Division of Haematology and Oncology at the David Geffen Medical Centre, University of California, Los Angeles, conducted clinical and laboratory research that led to the development of Herceptin. The drug targets a specific genetic alteration found in about 30 per cent of breast cancer patients. In 1998, the US Food and Drug Administration approved the use of the Herceptin Mab for treatment of advanced breast cancer. Slamon's research proved a relationship between the gene HER-2/neu, which encodes a tyrosine kinase, a signalling mechanism, which if it mutates, causes overexpression of cell growth and a particularly aggressive form of breast cancer. Beginning with an $8,000 small grant from UCLA's Jonsson Comprehensive Cancer Centre, Slamon was subsequently funded by a large endowment from Revlon and funding from the Entertainment Industry Foundation of Hollywood. Fortune favoured the prepared mind when Slamon met Axel Ullrich of Genentech at Denver Airport. Genentech, in their pre-Roche days, were heavily and ambitiously involved in antibody-based cancer research. Slamon lobbied tenaciously for Genentech to fund the translational research, which they eventually did. Ullrich at Genentech and Dennis Slamon's team subsequently created Herceptin, but not until after years of delay caused by the re-shuffles and strategy redirections occasioned by Roche. Through the aforementioned innovative scientific logic involving design of a molecule aimed at the gene responsible for excessive cell growth, Ullrich, now at the Max Planck Institute for Biochemistry in Munich, accidentally cloned HER2 in 1985 (Ullrich, 2002). The intervening years saw lengthy clinical trial and much lobbying by patient interest groups to gain access to the drug (see Herceptin Timeline, Table 11.3). In November 2006, the US FDA approved Herceptin as part of a treatment for early-stage breast cancer. Eight of the antibodies on the market by 2005 were approved for the treatment of cancer. Among these were three of the six antibodies with more than $1 billion in individual sales in 2005. Genentech developed and currently markets all three of these blockbuster drugs in the US, while Roche holds marketing rights in the rest of the world. These three antibodies, Rituxan (marketed as MabThera outside the US) for treatment of non-Hodgkin's lymphoma, Herceptin for treatment of breast cancer, and Avastin for treatment of metastatic colorectal cancer, represent three of Roche's five top selling

Table 11.3 Herceptin development timeline

Date	*Event*
1975	Köhler and Milstein discover Mabs, 1985, Cambridge (UK)
1985	Axel Ullrich and Genentech team clone HER2 gene, Genentech, San Francisco
1987	Dennis Slamon and UCLA team link HER2 overexpression with metastatic breast cancer
1991	Genentech files new drug application with FDA. Phase 1 clinical trials
1993	Phase 2 clinical trials
1997	Phase 3 drug combination trials
1998	FDA gives 'Fast Track' status for metastatic breast cancer treatment. Genentech contracts DAKO for diagnostic kit. Roche–Genentech US only licensing agreement, Roche rights outside US
2000	Genentech issues warning on pulmonary side effects of treatment
2002	FDA approval for usage in adjuvant (early stage) metastatic breast cancer treatment
2005	Further side effects note from Genentech
2006	FDA approves combined therapeutic treatment regimen for Herceptin

Source: Genentech

pharmaceutical products and had combined sales greater than $6 billion worldwide in 2005.

Three more things may usefully be recalled from the parts of this book that deal with pipeline issues. First, notice that much US R&D outsourcing is probably not to extremely distant locations; relatively few go abroad although European and US pharmas have development and production facilities overseas. There has been an inclination in the economic geography literature to view pipelines as distant networking destinations, but in biotechnology clusters we also find R&D institutes like Novartis, Amgen, Abbott, Pfizer and Wyeth, to name but a few that have located sometimes major R&D headquarters, as in the case of Novartis, in platform locations like Cambridge, Massachusetts. San Francisco and San Diego do not have many pharmas in their state, let alone regions, and from the cluster diagrams in Chapter 7 it is clear that some coast-to-coast R&D outsourcing occurs. Most US and foreign pharmas are headquartered in New Jersey, once known as 'The Medicine Chest of the World', so it is logical that if the best contractor is on the east coast it is likely to be favoured. This shows up in the extremely high levels of pharma investment in the Greater Boston cluster compared to the Californian ones in Table 4.1, Chapter 4 and Table 11.4 in this chapter. In the late 1990s and early 2000s it was running at double the Californian pharma investment. So we may conclude that Cambridge, Massachusetts and Boston – where Harvard Medical School is located across the Charles River – is the epicentre of global bioscientific academic journal publishing (Figure 5.4, Chapter 5), the biggest recipient of NIH public research funding (Table 3.3, Chapter 3) and as we have seen, the destination of the largest share of big pharma investment, which is principally R&D outsourcing through milestone payments to and licensing deals with DBFs. Thus Greater Boston is the capital city region in world healthcare biotechnology.

A second noteworthy point is that, as Coenen (2006) shows, there has been a significant increase between 1990–1996 and 1997–2003 in the amount of international collaboration in that part of the innovation process dealing with intellectual property rights. Coenen looks at the co-patenting patterns from an agro-food biotechnology cluster, Saskatoon, and the mixed agro-food and healthcare biotechnology cluster in Skåne, Sweden and sees an increase in magnitude of international collaborations: from Saskatoon, mainly to the US, from Skåne to the US and Europe, although the overall ratios of domestic to overseas collaboration remained similar in this expanded total. We noted in Chapter 5 how healthcare biotechnology co-patenting was very strong among Cambridge, Boston, San Francisco and San Diego, with Cambridge and Oxford UK being junior partners occasionally, with other countries involved only rarely. Given Cambridge–Boston's dominance of healthcare biotechnology in the three crucial dimensions discussed above, it is not surprising to see a strong performance also in a fourth crucial area, patenting. IPR, of course, effectively begins the biotechnology innovation process of commercial exploitation of exploration knowledge gained in the biomedical research laboratories, and this bioregion has critical mass, particularly at Harvard, MIT and, in specific expertise areas, Boston and Tufts Universities.

Finally, we noted some interaction between developed country bio-pharmaceutical expertise and that in developing countries. Probably relatively and absolutely, Cuba was the most impressive performer in that respect, especially when it is considered that there has been a trade embargo to the logical partners in neighbouring US bioregions for between forty and fifty years. Even so, regulations that allowed trading in knowledge for healthcare purposes, which there were signs had not been policed very vigorously until the onset of the Bush administration from 2000 onwards, meant that regular interaction with the likes of established San Francisco biotechnology vaccines company Chiron and others in less rigidly politicised trade régimes like Australia and Canada, had flourished and continued to thrive. This was based on the excellence of Cuban bioscience and medical research, notably in fields such as meningitis and hepatitis. Otherwise, the beginnings of R&D outsourcing (mainly the latter) are today visible to China and India, where in the former case a rapid growth culture has prevailed since 1992 and in the latter historic excellence in medical research was now beginning to reap international returns, signifying a shifting landscape of biotechnological interactions.

Platforms

We come finally to another key element of the biotechnology landscape which has a double meaning. For bioscientists and biotechnologists a platform is a technological support system for drug discovery, including diagnostic kits and associated bioequipment. As noted earlier, there are signs that this distinction is breaking down, but it still resonates in the business field as some firms specialise more in platforms than drug discovery. But from the newest thinking in evolutionary economic geography comes a more macro-idea of a biotechnology platform, an attempt to capture key aspects of which is displayed in Figure 11.1. This draws attention to such platform industries as nanotechnology, ICT and, closer to biotechnology, bioreactors and biologics. What are these and how do they combine to present a bioregional platform?

Let us consider nanobiotechnology as a drug delivery system first. In February 2005 the first nanoparticulate drug delivery product, Abraxane for the treatment of breast cancer, was launched by Abraxis Oncology, a division of American Pharmaceutical Partners, Inc. This required the US FDA to create a new class of therapeutic products. Nanotechnologies operate at subatomic levels where nanoparticulate drug delivery systems may bind a therapeutic compound like an inhibitor molecule to a nanoparticle or wrap it in a nanoshell. Common materials utilised include gold, platinum or silicon nanoparticles. The Abraxane system uses a nanoparticulate shell constructed from albumin which can be targeted to specific cells through combination with a specific monoclonal antibody. When the nanoshell reaches the target site, it is penetrated by laser, releasing the drug. This provides high accuracy, enabling precise targeting of designer molecules which might otherwise have side effects in the rest of the body. In this new field, the market leader in 'nanoemulsions', Elan Pharmaceuticals of Ireland, has developed a nano-crystallisation system for milling drug compounds to nanometre scale particles to improve biological uptake into patients. Substantial investment in nanotechnology R&D is occurring. In 2004 from both public and private sectors was an estimated $US8.4 billion. An estimated 15 per cent of this was directed at nanobiotechnology applications. With the nanobiotechnology sector making up only around 8 per cent of the nanotechnology industry worldwide in 2004 (Figure 11.2), funding in nanobiotechnology remains plentiful.

Drug delivery is an important challenge for pharma since most drugs are delivered generically to the whole body, often producing side effects, as with Herceptin. This challenge is of utmost importance in oncology, where treatment involves the selective destruction of tumour cells. This often has non-target effects on surrounding body tissue, resulting in the side effects most commonly observed with chemotherapeutics such as alopecia and nausea.

What about platform supports like bioreactors and biologics? Bioreactors can be as small as a Petri dish with agar gel or as large as a biosynthesis

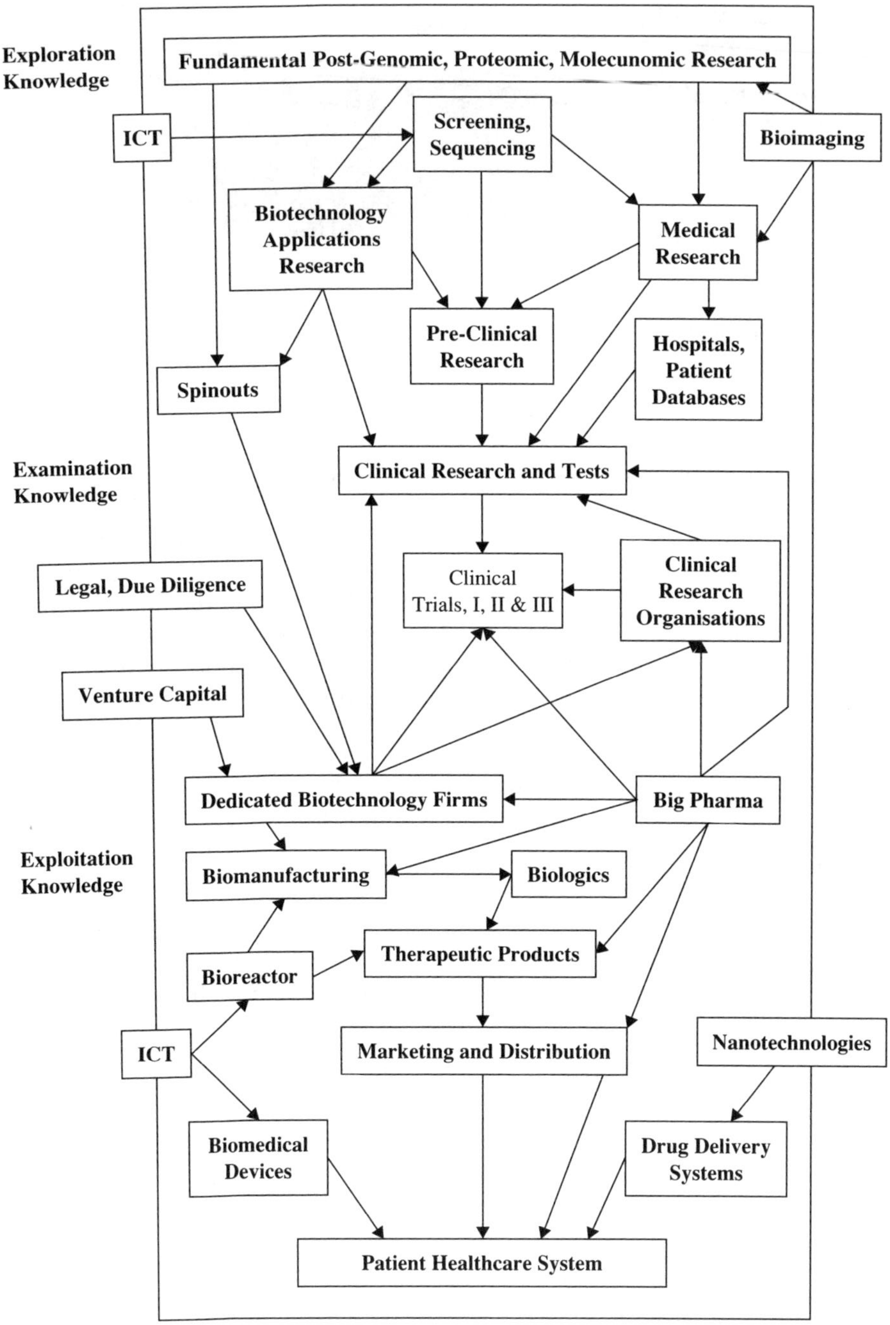

Figure 11.1 Conceptual model of the bioscientific and biotechnological value chain

Source: Cooke (2005a)

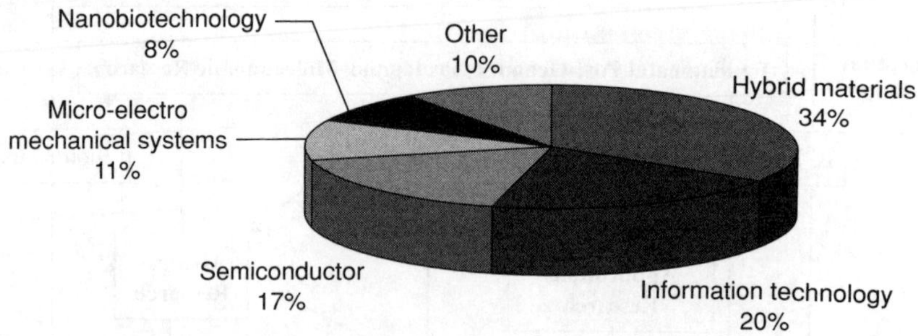

Figure 11.2 Global nanotechnology shares by application, 2004

Source: Pharmalicensing.com

plant consisting of complex programmable logic control software systems, various biosynthesising vessels and stainless steel components. Agar gel is produced from seaweed and acts as the growth culture platform in most laboratory experimentation on bacteria, cells and the like. As a platform element it has moved into the agro-food world, following the interventions of Yale literary critic Harold McGee (1984) and Oxford physicist Nicholas Kurti (1997), inventor of the culinary concept of 'molecular gastronomy' mutating into what might be termed the 'agro-pharma-biotech' platform. Agar gel's humble role in the rise of molecular gastronomy celebrity chefs like Ferran Adrià of El Bulli restaurant in Gerona, Catalonia, Heston Blumenthal of the Fat Duck, Bray, near London, and Thomas Keller of the French Laundry, Yountville, Napa, California is that it solidifies when warm, allowing such culinary innovations as hot langoustine jelly to be produced, as at Le Champignon Sauvage restaurant at Cheltenham, Gloucestershire. Nanotechnology has thus by now invaded the culinary industry, to the dismay of many for whom the fact that nanoingredients are both unregulated and can cross membranes between, for example, blood and brain or lungs, with unknown consequences, is a matter of scientific but not yet corporate concern. Biologics, by contrast, are what certain kinds of bioreactor produce through fermentation or biosynthesis of, for example, the bulk contents of biotechnological drugs that have been refined over and over again from fermented or biosynthesised cellular material. Clearly, these are extremely diverse but related varieties of bioplatform content that sustain the translation of academic research into commercialised products and services.

Ultimately, locales that have the appropriate combinations of knowledge domains ranging from biopharmaceuticals expertise of the kind evolved in San Francisco as represented by Genentech, and in Los Angeles as supported by the pipeline relations to UCLA oncologists supported by Hollywood and Revlon endowments, to innovative financial packages such as those more publicly promoted in Scotland's embryonic biotechnology platform, reach positions in global bioregional lists like Table 11.4, or on global agro-food platforms described in Chapter 9. The table contains locations not normally thought of as globally significant concentrations of bundles of knowledge domain combinations, nor do those in many developing countries. But the economic sociology and evolutionary economic geography of biotechnology is ultimately more knowledge-driven than finance-driven, so the usual 'global city' rankings get severely disturbed by their juxtaposition with global bioregions and their 'growth cultures', which are simultaneously innovative, problematic and revolutionary.

Table 11.4 Core biotechnology firms and key clusters, 2000: comparative US and European performance indicators

Location	*DBFs*	*Life scientists*	*VC*	*Big pharma funding*
Boston	141	4,980	$601.5 m.	$800 m./annum 96-01
San Francisco	152	3,090	$1,063.5 m.	$400 m./annum 96-01
San Diego	94	1,430	$432.8 m.	$320 m./annum 96-01
Toronto	73	1,149	$120.0 m.	$89 million (2002)
Montreal	72	822	$60.0 m.	$120 million (2002)
Lund-Medicon	104	5,950	$80.0 m.	$300 million (2002)
Stockholm-Upp.	87	2,998	$90.0 m.	$250 million (2002)
Cambridge	84*	2,650	$250.0 m.	$105 million (2000)
Washington DC	83	6,670	$50.0 m.	$360 million (2000)
Ral-Durham	72	910	$192.0 m.	$190 million (2000)
Zurich	70	1,236	$57.0 m.	$85 million (2002)
Jerusalem	60*	1,015	$300.0 m.	$54 million (2002)
Munich	59**	5,500	$266.0 m.	$54 million (2001)
Oxford	59*	3,250	$100.0 m.	$90 million (2002)
Paris (Evry)	58	1,800	$60.0 m.	$40 million (2002)
Berlin	55**	3,700	$122.0 m.	$30 million (2001)
Rhineland	54**	1,250	$30.0 m.	$40 million (2000)
Singapore	38	1,063	$200.0 m.	$88 million (2001)
Scotland	38*	3,600	$35.0 m.	$125 million (2002)
Rhein-Neckar	37**	3,200	$40.0 m.	$20 million (2000)
Seattle	30	1,810	$49.5 m.	$91 million (2000)

Source: Cortright and Mayer (2002); NIH; NRC; BioGenTech, Cologne; BioM, Munich; BTH Heidelberg; BioTop, Berlin; VINNOVA, Sweden; Dorey (2003); ERBI, UK; Scottish Enterprise (2003); Kaufmann *et al.* (2003); *Innovating Regions in Europe* update for Israel: http://www.innovating-regions.org/network/whoswho/regions_search.cfm?region_id=292; Oxford Bioscience Network (2003)

NB: Consistent with Chapter 7, where research for this book facilitated updated firm number statistics, these are here included. * 2004; ** 2005

Notes

Chapter 2 The knowledge economy and growth cultures

1 Karl Marx (1973) *Grundrisse*, Harmondsworth, Penguin.
2 These are concentrations of science, technology, talent, innovation and much public as well as private investment. They are thus more complex than purely market-driven than 'clusters'.
3 A further theoretical analysis of the relations between innovation capability and varieties of proximity is presented in Boschma (2005).
4 A seminal US article is Glaeser *et al.* (1992); see also for the US, Florida (2002). For Canada, see Polèse (2002). For UK, see Begg (2002). For Europe see Cheshire and Magrini (2002). For Israel, see Cooke and Schwartz (2003).
5 Despite the important insight regarding the practical and conceptual content between tacit (implicit) and codified (explicit) knowledge, these authors badly mix up 'information' and 'knowledge' as if they were identical, a problem by no means confined to them. In this chapter we will generally refer to 'knowledge' in relation to 'information' according to the following: *information* is a passive resource given *meaning* by the application of *knowledge*, which facilitates *action*. Thus if the actor has *knowledge* of her destination, *meaning* is given to the *information* in the train timetable, enabling appropriate travel *action* to be taken.
6 See also Scharmer (2001). This can take many forms, some no doubt deeply embedded and culturally as well as psychologically shaped. But an instance, discussed by Metcalfe (2005), might be the sense of unease, dissatisfaction or disagreement with the *status quo* that must be the first step in the knowledge creation process that may result in a commercial innovation.
7 Marshall (1918), p. 603. Notice how 'dissatisfaction' with the *status quo* is the stimulant of restless innovative intent by both user and producer in this telling instance.
8 The difference between disruptive and radical innovation is as follows. Following Christensen (1997) and others, disruptive innovation is revolutionary regarding technological profile, e.g. candle loses out to light bulb. Yet context, e.g. room, house, remain largely unaltered and its built environment can persist for hundreds of years. However, in the innovation systems tradition radical innovation changes both technological profile and context or environment, e.g. air travel versus all other historic forms of travel. That is, airports, air traffic control systems, skills and so on are radical, literally 'root and branch', innovations in historic terms. Equating the two, as tends to happen in some literature is less helpful in conceptual terms than distinguishing them analytically.
9 Because of current and anticipated high oil prices, hybrid fuel and fuel cells are topical innovations that have only recently, in the case of the Prius, been marketed seriously in competition with standard models. Prius hybridisation is of petrol (or

diesel) and electrical power. An innovation biography of fuel cells, if they become a feasible alternative to either conventional internal combustion or current hybrid engines, would rest upon Grove's early discoveries, and subsequently aerospace technology, especially rocketry, since that is where hydrogen fuel cells remain technologically supreme as they have since the days of Wernher Von Braun.

Chapter 4 The microbiology revolution and the crisis in pharmaceuticals

1 Parexel is the CRO responsible for administering in March 2006 the drugs test on humans whereby six men suffered multiple organ failure after taking a drug designed to treat rheumatoid arthritis, leukaemia and multiple sclerosis. The trial was being run on behalf of TeGenero, a now bankrupt German DBF start-up, on its anti-inflammatory drug TGN1412. Subsequent evaluation suggested that the drug's 'tail' could have triggered the release of a flood of inflammatory molecules called a 'cytokine storm', causing organs to shut down.

2 The 'innovation biography' of this successful biotechnologically derived drug is given in Chapter 11.

Chapter 5 Academic growth cultures

1 The 'exploration' and 'exploitation' distinction was, of course, first made in organisation science by March (1991). It is necessary today to introduce the intermediary 'examination' form of knowledge to capture the major stages of the move from research to commercialisation in biotechnology because of the long and intensive trialling process in both agro-food and pharmaceutical biotechnology. However, a moment's reflection about other sectors suggests that this kind of knowledge and its organisational processes has been rather overlooked as it applies outside biotechnology (see Cooke, 2005a, 2005b).

2 San Francisco's megacentre capability in ICT gives its biotechnology competitive advantage in bioinformatics, screening, gene sequencing and genetic engineering software and technologies.

3 As shown below (section eight) a strong public role in cluster building is also evident in Finegold *et al.*'s (2004) analysis of biopharmaceuticals developments in Singapore.

4 But disappointment has grown since the downturn at the only modest growth and drying pipeline dynamics associated with Germany's policy-induced start-up biotechnology businesses. Moreover German high-tech entrepreneurship in general suffered a heavy blow with the demise of the Neuer Markt, its now defunct take on NASDAQ.

5 Unusually, the role of 'big pharma' is rather under-emphasised in this analysis of its relation to health biotechnology. This is not because large pharmaceuticals firms are unimportant in this context, for they clearly are. However, for exploration and even, increasingly, examination and some exploitation knowledge production they practise 'open innovation' as Chesbrough (2003) demonstrates for the case of Millennium Pharmaceuticals, a leading bioinformatics supplier that re-designed itself as a biotechnological drugs manufacturer through investment of 'open innovation' contract earnings from the likes of Monsanto and Eli Lilly. These practices are now emulated by specialist suppliers in industries like ICT, automotives and household care according to the same author. This chimes with a more general hypothesis we can call 'Globalisation 2' in which in a 'knowledge economy' the drivers of globalisation become 'knowledgeable clusters' of various kinds. These exert irresistible attractions for large corporates who become 'knowledge supplicants' as their in-house R&D becomes ineffective and inefficient. They pay for, but no longer generate, leading-edge analytical knowledge for innovation.

6 In Chesbrough's (2003) book titled *Open Innovation*, these outsourcing of R&D characteristics are shown to have become common in industries outside biotechnology, notably ICT and homecare products. Accordingly, in-house R&D in the largest US firms is shown to have declined from 71 per cent in 1981 to 41 per cent in 1999. Meanwhile that conducted in small firms rose from 4 per cent to 23 per cent at the same time.
7 It is arguable smaller ones like Seattle should not stay in Table 5, but they can in any case be compared by those who are interested. The justification for including these bioregions is argued in Cooke (2004a).
8 NIH funding for medical and bioscientific research in Cambridge-Boston was in excess of $1.1 billion by 2000, $1.5 billion by 2002 and $2.1 billion in 2003. Cooke (2004b) shows it exceeded all of California by 2002 and by 2003 the gap widened to $476 billion ($2,021 compared to $1,545 billion). Interestingly, this is a recent turnaround since the 1999 total of $770 million was marginally less than the amount of National Institutes of Health funding passing through the Northern California cluster in 1999, a statistic that only increased to $893 million in 2000. Thus Greater Boston's supremacy is recent but definitive. San Diego's NIH income includes that earned by Science Applications International Corporation. This firm is based in San Diego but performs most of its NIH research outside its home base as a research agent for US-wide clients. Thus it warrants mention but is excluded from totals calculated by this author. This is not done in the Milken International report 'America's Biotech and Life Science Cluster', June 2004, which ranks San Diego as the top US cluster. This oversight seriously weakens its claims for San Diego's top US cluster position. Further reasons for rejecting the Milken Institute's ranking of San Diego first as well as inclusion of questionable research funds are that the Institute deploys a spurious methodology based on research dollars per metropolitan inhabitant to promote San Diego's ranking. Finally, the research was commissioned by local San Diego interests (Deloitte's San Diego) and excludes 'big pharma' funding, on which San Diego performs less than half as well as Boston (Table 5.1).

Chapter 6 The shifting landscape of bioscience policy

1 For the UK, results from the DTI study of clusters reveal extraordinarily high location quotients for biotechnology businesses in Cambridge and Oxford, also the main university and other public R&D locations (DTI, 2001). In the US, the Brookings Institute reports (e.g. Cortright and Mayer, 2002) consistently show there to be some nine main life sciences clusters in the US. In Canada, the work of Niosi and Bas (2001) shows high concentrations and intra-cluster innovative interaction in Montreal and Toronto. Comparable results exist for Germany (Kettler and Casper, 2000), France (Lemarié *et al.*, 2001), Sweden (VINNOVA, 2001) and Israel (Kaufmann *et al.*, 2003). The sample of one regional biotechnology 'cluster failure' written about by Orsenigo (2001) suggests Italy's *exceptionalism* regarding advanced technology activity among G7 countries, something emphasised by Tavoletti (2004) who shows the highest unemployment for all skill/qualification categories (including workers having received no formal education at any level) in all Italian regions to be that of post-doctoral students.
2 To magnify the 'multi-scalar' dimension here would require a further study. An anonymous referee suggests that Bunnell and Coe (2001) and Mackinnon *et al.* (2002) 'need to be acknowledged and cited' and this is duly done. They say it is wrong to emphasise the regional level and wrong to overlook regional specificity. This is an improvement upon positions such as that of Bathelt (2003) in the same journal that only nations have specificity and that they are also *closed* systems. The geographical scale debate seems in danger of generating more heat

than light. At a general level Cooke *et al.* (2000) showed how multi-level governance of innovation in Europe has resulted in *innovation* measures often having been taken regionally by stimulus from European Commission policies while member states devoted more attention and many more resources to science policies. The regions in question are supralocal, sub-national or meso-level entities with policy legitimacy in the relevant field. In the EU, for example, many, but not all, happen to be EU NUTS 2 regions – a statistical artifice in some cases, an isomorph of a legitimate meso-level entity in others. In yet others, NUTS 2 areas aggregate to a legitimate meso-level policy entity. Here, regions are meso-level policy entities with legitimacy for implementing innovation measures emanating supranationally, or from state-level (e.g. in England) or from within the region (e.g. N. Ireland's *Think, Create, Innovate* strategy). Economists who specialise in science, innovation and technology studies have little or no understanding of the presence of regions and their institutions as, albeit often weak, actors in innovation and, as this chapter shows, tentatively in science policy too – hence the need to reiterate it. Geographers, of course, are more fortunate.

3 As a consequence of the disaster that resulted in deformed births after mothers had taken Thalidomide to control morning sickness, trialling of candidate treatments goes through three lengthy phases. The first is usually mammalian, the second with small groups of patients with the drug's target disease, and the third a large sample of patients with controls that yield statistically valid results.

4 Science Applications International Corporation is based in San Diego but it performs most of this research outside its home base as a research agent for US-wide clients. Thus it warrants mention but is excluded from these rankings. This is not done in the Milken International report 'America's Biotech and Life Science Cluster', June 2004, thus seriously weakening its claims for San Diego's top US cluster position.

5 Definitionally, 'bioinformatics' involves capturing molecular or genetic information on databases that identify chemical compounds that may inhibit particular disease-causing molecular or genetic combinations. 'Biosciences' are those that involve study of medicine, biology, biochemistry, pharmaceutical and other treatments, medical devices and instruments, research and testing relevant to intervention with and improvement of human, animal and plant life. 'Big pharma' is a term used by academe, government and industry as an abbreviation for large pharmaceuticals companies. 'Biotechnology' involves tools like genetic engineering to create treatments that counter diseases (e.g. the example of Glivec in Table 6.1). 'Orphan drugs' are used in diseases or circumstances which occur so infrequently that there is no reasonable expectation that the cost of developing and making available a drug for such disease or condition will be recovered from sales for such drugs. Firms gain exclusivity and fee waivers for producing such drugs. 'High throughput screening' occurs when a compound interacts with a target in a productive way so that the compound then passes the first milestone on the way to becoming a drug. Compounds that fail this initial screen go back into the bioinformatics library from whence they came, perhaps to be screened later against other targets.

6 Centres of Excellence are generally considered to be more exacting, higher-quality designations than Centres of Expertise. Though numerous governments, like that of Finland, began with the former, regional politics caused the latter to replace it so that more regions could qualify. This action is criticised further on in this chapter.

Chapter 7 The cluster model in biotechnology

1 This is a widely accepted norm in most locations testified to in research by Zook (2002) and Powell *et al.* (2002) among many others. It is because of the venture

capitalist's need for a 'hands-on' relationship with his or her investment, possibly 'at the drop of a hat'. The greater the distance away from the investment the greater the uncertainty about management control. As a case in point, Kleiner Perkins Caufield, Byers, the leading US venture capitalist, has 80 per cent of its so-called 'keiretsu' investments in biotechnology and ICT within an hour's drive of its Sand Hills Road headquarters in Palo Alto.

2 Lasker Prize winner and founder of molecular studies of gene regulation Dr Mark Ptashne, is currently Ludwig Chair of Molecular Biology at the Sloan-Kettering Institute in New York. He was involved in the creation of the biotechnology industry, including as a co-founder with Tom Maniatis of the Genetics Institute, one of the first biotechnology companies. The Genetics Institute was acquired for $3 billion by American Home Products (AHP) in 1996. AHP has since changed its name to Wyeth and the Genetics Institute no longer operates as a separate business unit although Ptashne remained with Wyeth Research. In February 2004 it was announced that Ptashne and Maniatis had formed a new company, Acceleron Pharma of Cambridge, Massachusetts. The newly formed company raised $25 million in venture capital to open its initial lab in December 2003. Founding investor Polaris of Waltham, MA invested $250,000 in initial seed funding. Advanced Technology Ventures of Palo Alto, California and Waltham, MA; Flagship Ventures of Cambridge and Alameda, California; and Venrock Associates of Menlo Park, California, New York and Cambridge co-led the round. Sutter Hills Ventures in Palo Alto also contributed financing. Acceleron plans to use the money to develop protein-based treatments for a range of diseases from osteoporosis to obesity to muscle wasting-related illnesses such as Lou Gehrig's disease, to be ready for clinical trials within three years. After he left the Genetics Institute, Harvard professor Maniatis went on to found ProScript in Cambridge, which developed the drug Velcade to treat multiple myeloma, a rare form of leukaemia. That company eventually became part of Millennium Pharmaceuticals Inc. in Cambridge after a series of acquisitions, and Velcade is now Millennium's signature drug. Acceleron co-founder Wylie Vale himself launched Neurocrine Biosciences Inc. in San Diego, a $2 billion company where he remains a board member. Millennium bought LeukoSite from Cambridge, MA and COR Therapeutics from San Francisco with $76 million earned by selling its DNA sequencing technology to Eli Lilly and Monsanto. This enabled the company to fulfil Millennium's founders' vision of moving from platform technologies into therapeutic treatments. Millennium also merged with Cambridge Discovery Chemistry in 2000, giving it a UK presence and adding 100 chemists to a firm that was founded only in 1993 and which by 2000 was valued at $11 billion and employing 1,530 people. These are highly illustrative details of the strong network linkages among firms in global bioregions.

3 As discussed in Chapter 2, a distinction is being made here between 'exploration' and 'exploitation' knowledge. Clearly the first refers to basic research and the second to commercialisation. This may involve commercialisation of technology as well as therapeutics. The distinction was made by March (1991). It has recently been augmented with a third concept, 'examination' knowledge to cover the important expertise involved in trialling, testing and prototyping (Cooke, 2005b).

4 As discussed in previous chapters, this economic sociology concept, originating in Polanyi (1944), is central to the theory of clustering. It refers to the ties between firms which may be weak or strong, but proximity in clusters offers firms both kinds. However, Novartis is rather unusual in establishing very strong ties with San Diego institutions and firms. This is true also for San Francisco and, as we shall, see Boston.

5 This practice was emulated by and from other pharma companies. Currently Aventis is implementing a comparable strategy in Toronto. Meanwhile between 1999 and 2003, Pfizer, Wyeth (acquiring Genetics Institute), Amgen (acquiring Immunex), Aventis (Ariad), Abbott (BASF) and AstraZeneca were all represented in Cambridge, Massachusetts, many through acquisition. However, the embedding strategy of Novartis is both distinctive and, in relation to basic research, deeper.
6 In 2000 Syrxx was founded, as were Kalypsys and Phenomix in 2001.
7 Novartis also has a bioincubator at Zug in Switzerland.
8 This information is sourced from the *Novartis* website and the firm's strategic management study reported in Zeller (2004).
9 This was argued forcefully to be the new core competence of big pharma by Nightingale (2000).
10 This account is given in Landers (2004).
11 Thus PowderJect Pharmaceuticals was owned by Chiron, San Francisco (now owned fully by Novartis), part of British Biotech was bought by OSI Pharmaceuticals, Long Island, NY, Oxford Molecular is owned by Accelrys, San Diego, Oxford Asymmetry was bought by Evotec OAI from Hamburg, and Oxford Glycosciences is owned by Union Belgique Chemie.
12 Swedish DBFs are noted suppliers and R&D partners to leading foreign pharmaceuticals firms (e.g. Carlsson Research and Merck; BioVitrum and Amgen; Kario Bio and Wyeth). However, the official www.sweden.se website notes that: 'Both Astra-Zeneca and Pharmacia collaborate with numerous biotech companies, some of them Swedish. . . . The pharmaceutical companies Astra (now AstraZeneca) and Pharmacia (today part of Pfizer) have stimulated the growth of the Swedish biotech industry, not only in the pharmaceutical and medical subsectors but also, for example, in biotech tools and supplies. Examples are Prevas Bioinformatics contract solutions for AstraZeneca, Biovitrum and GE Healthcare while blockbuster drugs like *Losec* (AstraZeneca) and *Celebrex* (Pharmacia) were clinically trialled by Swedish clinical research organisations (CROs) as were many others originating abroad (e.g. Lipitor, Norvasc and Zoloft for Pfizer; Lipovas for Merck; and Paxil for Glaxo SmithKline).'
13 SLU is the Swedish University of Agricultural Sciences.

Chapter 9 Environmental, energy and agro-food bioregions

1 The source for these data is Omland and Ernst (2004).
2 Valentin and Lund-Jensen (2003) show how large food companies are leaders in R&D, mostly conducted in-house, in agro-food exploration, examination and exploitation. Thus Raisio and Danone have been leaders in lactobacter nutraceuticals like Benecol and Actimel, while global corporates like Unilever and Nestlé conducted biological research in-house from their earliest days. Thus this industry does not display as many DBFs as biopharmaceuticals. Sweden's leader ProViva was discovered in academia (Lund University) and developed by local firm Probi.
3 The role of synchrotrons in the future of bioregions has been noted. In the UK, government invested $500 million with partners in a replacement for its nuclear age synchrotron near Manchester. The new Diamond Synchrotron is being constructed near Oxford's biosciences cluster. The old one is being adapted for bioimaging uses. In March 2004 the UK's Birmingham University announced its strategy of purchasing an upgraded synchrotron.
4 I am grateful to my colleague, Kevin Morgan, an expert on regional food service innovation, for this insight (see Morgan *et al.*, 2006).
5 This is similar to the strategy of the city of Copenhagen in neighbouring Denmark, whose Dogme Organic Project is based on a set of criteria that designate a

number of Danish municipalities to be 'locomotives' in developing environmentally friendly practices and protecting the environment. One of the rules in the project is to assure that at least 75 per cent of the foodstuff used in the municipal kitchens and canteens is organic before 2009. In order to achieve this goal the city of Copenhagen has launched the Dogme Organic Project.

6 Greenomics (a Plant Research International affiliate) conducted in 2000 the shotgun cloning and high-quality sequencing of *Lactobacillus plantarum*. A first automated annotation of the genome was provided using bioinformatics tools and Greenomics provided Wageningen Centre for Food Sciences (WCFS) with a primary functional classification of all genes. This was the largest project WCFS had outsourced since its foundation in 1997 in the 'Wageningen Food Valley'. WCFS, a research alliance of government, food industry and research organisations, performs long-term strategic and fundamental research in food sciences. The partners include DSM, the Netherlands Dairy Industry Association, Unilever NV, Avebe BA, Cosun, and the Cebeco group while the research organisations include Wageningen University and Research Centre, TNO Nutrition and Food Research Institute, and NIZO food research. See http://www.plant.wageningen-ur.nl/news/2000-12_en.htm.

7 Table 9.2, excluding the two large research firms, gives 15 as the mean firm size in the Wageningen–Nijmegen bioregion.

References

Academy of Finland (2002) *Biotechnology in Finland*, Helsinki, Academy of Finland

Adamec, V., Cassell, B., Smith, E. and Pearson, R. (2006) Effects of inbreeding in the dam on dystocia and stillbirths in US Holsteins, *Journal of Dairy Science*, 89, 307–314

Adams, M. (2004) Functional foods gain in popularity, but health claims on most brand-name groceries mislead consumers, *Natural Health Newsletter*, 19 July

Akerlof, G. (1970) The market for 'lemons': quality uncertainty and the market mechanism, *Quarterly Journal of Economics*, 84, 488–500

American Association for the Advancement of Science (1999) *US R&D Budget and Policy Programme*, Washington, AAAS

Andermann, A. (1996) Physicians, fads and pharma: a history of Aspirin, www.mjm.mcgill.ca/issues/v02n02/aspirin

Ashcroft, B., Dunlop, S. and Love, J. (1995) UK innovation policy: a critique, *Regional Studies*, 29, 307–311

Audretsch, D. and Feldman, M. (1996) Knowledge spillovers and the geography of innovation and production, *American Economic Review*, 86, 630–640

Baeyens, K. and Manigart, S. (2006) Follow-on financing of venture capital backed companies: the choice between debt, equity, existing and new investors, *Vlerick Leuven Gent Working Paper Series 2006/05*, University of Gent and KU Leuven

Bagchi-Sen, S. and Scully, J. (2004) The Canadian environment for innovation and business development in the biotechnology industry; a firm-level analysis, *European Planning Studies*, 12, 961–984

Barnett, A. (2003) Revealed: how drug firms hoodwink medical journals, *Observer*, 7 December

Bathelt, H. (2003) Growth regimes in spatial perspective 1: innovation, institutions and social systems, *Progress in Human Geography*, 27, 789–804

Bazdarich, M. and Hurd, J. (2003) *Anderson Forecast: Inland Empire and Bay Area*, Los Angeles, Anderson Business School

Becattini, G. (2001) *The Caterpillar and the Butterfly*, Florence, Felice de Monnier

Becattini, G. and Dei Ottati, G. (2006) The performance of Italian industrial districts and large enterprise areas in the 1990s, *European Planning Studies*, 14, 1139–1162

Begg, I. (ed.) (2002) *Urban Competitiveness*, Bristol, Policy Press

Benner, M. and Sandström, U. (2000) Internationalising the triple helix: research funding and norms in the academic system, *Research Policy*, 29, 291–301

Best, M. (2000) Silicon Valley and the resurgence of Route 128, in J. Dunning (ed.) *Globalization, Regions and The Knowledge Economy*, Oxford, Oxford University Press

Best, M. (2001) *The New Competitive Advantage*, Oxford, Oxford University Press

Bioability (2005) European biotechnology venture capital database, Atlanta, Thomson Bioworld

BioGenTec (1998) *BioGenTec Atlas*, Cologne, BioGenTec

Bioscience Innovation and Growth Team (2004) *Bioscience 2015: Improving National Health, Increasing National Wealth*, London, BIGT

Black, B. and Gilson, R. (1998) Venture capital and the structure of capital markets: banks versus stock market, *Journal of Financial Economics*, 47, 243–277

Boschma, R. (2005) Proximity and innovation: a critical assessment, *Regional Studies*, 39, 61–74

Boschma, R. and Frenken, K. (2003) Evolutionary economics and industry location, *Review of Regional Research*, 23, 183–200

Bottazzi, G., Dosi, G. and Fagiolo, G. (2002) On the ubiquitous nature of agglomeration economies and their diverse determinants: some notes, in A. Curzio and M. Fortis (eds) *Complexity and Industrial Clusters*, Heidelberg, Physica-Verlag

Breschi, S. and Lissoni, F. (2001) Localised knowledge spillovers versus innovative milieux: knowledge 'tacitness' reconsidered, *Papers in Regional Science*, 80, 255–273

Brown, J. and Duguid, P. (2001) *The Social Life of Information*, Boston, Harvard Business School Books

Brumfiel, G. (2002) Bell Labs inquiry spreads to superconductors, *Nature*, 401, 5

Bunnell, T. and Coe, N. (2001) Spaces and scales of innovation, *Progress in Human Geography*, 25, 569–589

Burnell, S. (2002) Biomed researchers applaud NIH budget, *UPI Science News*, 31 January, 12

Burrill and Company (2004) Strategic partnering, *Quarterly Newsletter*, 2, 2

Burrill and Company (2006) Life Science: a changing prescription, *Bio2006 Conference*, Chicago, 10 April

Burton-Jones, A. (1999) *Knowledge Capitalism*, Oxford, Oxford University Press

Cairncross, F. (1997) *The Death of Distance: How the Communications Revolution Will Change Our Lives*, Boston, Harvard Business School Press

Caniëls, M. and Romijn, H. (2005) Localised knowledge spillovers: the key to innovativeness in industrial clusters? *Cambridge Journal of Economics*, 29, 497–515

Cantwell, J. and Iammarino, S. (2003) *Multinational Corporations and European Regional Systems of Innovation*, London, Routledge

Carlsson, B. (ed.) (2001) *New Technological Systems in the BioIndustries: An International Study*, London, Kluwer

Carpenter, R. and Petersen, B. (2002) Capital market imperfections, high-tech investment, and new equity financing, *Economic Journal*, 112, F54–F72

Casper, S. and Karamanos, A. (2003) Commercialising science in Europe: the Cambridge biotechnology cluster, *European Planning Studies*, 11, 805–822

Cassidy, J. (2002) *dot.con*, London, Allen Lane

Chandler, A. (1988) *Scale and Scope*, New York, Free Press

Chaya, D. (2005) Survival strategies for start-ups, *Journal of Commercial Biotechnology*, 11, 130–133

Chesbrough, H. (2003) *Open Innovation*, Boston, Harvard Business School Press
Cheshire, P. and Magrini, S. (2002) The distinctive determinants of European urban growth: does one size fit all? *Research Papers in Spatial and Environmental Analysis*, 73, Dept. of Geography, London School of Economics
CHI/PWC (California Healthcare Institute and PriceWaterhouseCoopers) (2002) Biomedicine: the new pillar in Northern California's economy, La Jolla and San José, CHI/PWC
Christensen, C. (1997) *The Innovator's Dilemma*, Boston, Harvard Business School Press
Clarke, L. (1998) Setting the pace, *Pharmaceutical Forum*, 46, 2–6
Coase, R. (1937) The nature of the firm, *Economica*, 4, 386–405
Coe, J. (2002) Networked pharma: innovative strategies to overcome margin deterioration, www.contactpharma.com/June021.htm
Coenen, L. (2006) *Faraway, So Close! The Changing Geographies of Regional Innovation*, Lund, Lund University Press
Coenen, L., Moodysson, J. and Asheim, B. (2004) Nodes, networks and proximities: on the knowledge dynamics of the Medicon Valley biotech cluster, *European Planning Studies*, 12, 1003–1018
Cohen, W. (2005) Patents and appropriation: concerns and evidence, *Journal of Technology Transfer*, 30, 57–71
Cohen, W. and Levinthal, D. (1989) Innovation and learning: the two faces of R&D, *Economic Journal*, 99, 569–596
Cooke, P. (1988) Flexible integration, scope economies, and strategic alliances, *Environment & Planning D: Society & Space*, 6, 233–256
Cooke, P. (2001a) Biotechnology clusters in the UK, *Small Business Economics*, 17, 43–59
Cooke, P. (2001b) New economy innovation systems: biotechnology in Europe and the USA, *Industry & Innovation*, 8, 267–289
Cooke, P. (2001c) Regional innovation systems, clusters and the knowledge economy, *Industrial & Corporate Change*, 10, 945–974
Cooke, P. (2002) *Knowledge Economies*, London, Routledge
Cooke, P. (2003a) Towards regional science policy? The rationale from biosciences, in D. Wolfe (ed.) *Clusters Old and New*, Montreal, McGill Queens University Press
Cooke, P. (2003b) Economic globalisation and its future challenges for regional development, *International Journal of Technology Management*, 26, 401–420
Cooke, P. (2003c) The evolution of biotechnology in three continents: Schumpeterian or Penrosian? *European Planning Studies*, 11, 757–764
Cooke, P. (2004a) Globalisation of bioregions: the rise of knowledge capability, receptivity and diversity, *Regional Industrial Research Report 44*, Cardiff, Centre for Advanced Studies
Cooke, P. (2004b) The accelerating evolution of biotechnology clusters, *European Planning Studies*, 12, 915–920
Cooke, P. (2005a) Rational drug design, the knowledge value chain and bioscience megacentres, *Cambridge Journal of Economics*, 29, 325–342
Cooke, P. (2005b) Regionally asymmetric knowledge capabilities and open innovation: exploring 'Globalisation 2' – a new model of industry organisation, *Research Policy*, 34, 1128–1149
Cooke, P. (2006) Global bioregions: knowledge domains, capabilities and innovation system networks, *Industry & Innovation*, 13, 437–458

Cooke, P. and Davies, S. (1993) *An Economic Profile of North Rhine-Westphalia*, Report to the Welsh Development Agency, Cardiff, Centre for Advanced Studies

Cooke, P. and Piccaluga, A. (eds) (2006) *Regional Development in the Knowledge Economy*, London, Routledge

Cooke, P. and Schwartz, D. (2003) Regional knowledge economy variations: an EU–Israel comparison, *Regional Industrial Research Report 45*, Cardiff, Centre for Advanced Studies

Cooke, P. and Wills, D. (1999) Small firms, social capital and the enhancement of business performance through innovation programmes, *Small Business Economics*, 13, 219–234

Cooke, P., Boekholt, P. and Tödtling, F. (2000) *The Governance of Innovation in Europe*, London, Pinter

Cooke, P., Etxebarria, G. and Uranga, M. (1998) Regional systems of innovation: an evolutionary perspective, *Environment and Planning A*, 30, 1563–1584

Cooke, P., Heidenreich, M. and Braczyk, H. (eds) (2004) *Regional Innovation Systems*, 2nd edition, London, Routledge

Cooke, P., Manning, C. and Huggins, R. (2000) Industrial liaison and academic entrepreneurship in Wales, *Zeitschrift für Wirtschaftsgeografie*, 44, 246–260

Cooke, P., De Laurentis, C., Tödtling, F. and Trippl, M. (2007) *Regional Knowledge Economies*, Cheltenham, Edward Elgar

Cooke, P., Kaufmann, D., Levin, C. and Wilson, R. (2006) The biosciences knowledge value chain and comparative incubation models, *Journal of Technology Transfer*, 31, 115–129

Cortright, J. and Mayer, H. (2002) *Signs of Life: The Growth of Biotechnology Centres in the US*, Washington DC, The Brookings Institute

Curzio, A. and Fortis, M. (eds) (2002) *Complexity and Industrial Clusters*, Heidelberg, Springer

De la Mothe, J. and Niosi, J. (eds) (2000) *The Economics and Spatial Dynamics of Biotechnology*, London, Kluwer

Dicken, P. (2001) Globalising processes and local economies, paper to Institute for Regional Innovation (Invent) conference 'Global-local: Regional Economic Development in an era of Globalisation and Localisation', Vienna, December

Dimov, D. and Murray, G. (2006) An examination of the determinants of the incidence and scale of seed capital investments by venture capital firms, 1962–2002, http://www.exeter.ac.uk/sobe, forthcoming in *Small Business Economics*

Dohse, D. (2001) Technology policy and the regions: the case of the BioRegio contest, *Research Policy*, 29, 1111–1133

Dorey, E. (2003) Emerging market Medicon Valley: a hotspot for biotech affairs, *BioResource*, March, www.investintech.com

Dosi, G. (1988) Sources, procedures and microeconomic effects of innovation, *Journal of Economic Literature*, 26, 1120–1171

DTI (1999) *Biotechnology Clusters*, London, Department of Trade and Industry

DTI (2001) *Business Clusters in the UK: A First Assessment*, London, Department of Trade and Industry

DTI (2005) *Comparative Statistics for the UK, European and US Biotechnology Sectors*, London, Department of Trade and Industry

DTI/FCO (1998) *Biotechnology in Germany: Report of an ITS Expert Mission*, London, Department of Trade and Industry and Foreign and Commonwealth Office

Dunning, J. (ed.) (2000) *Globalization, Regions and the Knowledge Economy*, Oxford, Oxford University Press

Dyer, G. (2002) Testing times for the pharmaceuticals industry, *Financial Times*, 15 February, 17

Dyer, G. and Griffith, V. (2001) Growing together, *Financial Times*, 19 December, 20

Edling, C. and Stern, C. (2003) Scandinavian rational choice sociology, *Acta Sociologica*, 46, 5–16

EMEA (2002) *Medicinal products with a Community marketing authorisation, the European Agency for the Evaluation of Medicinal Products*, London, EMEA

Ernst & Young (2006) *Beyond Borders*, http://www.ey.com

Etzkowitz, H. and Leydesdorff, L. (eds) (1997) *Universities and the Global Knowledge Economy*, London, Pinter

European Commission (2003) *Third European Report on Science and Technology Indicators*, Brussels, European Commission

European Investment Fund (2004) Technology Transfer Accelerator (TTA) Final Report, Luxemburg, EIF/EIB

Farrar, S. (2002) Sinn Fein backer pays for centre, *The Times Higher Education Supplement*, 8 February, 3

Fazeli, S. (2005) The European biotech sector: could it achieve more? *Journal of Commercial Biotechnology*, 12 (1), 10–19

Feldman, M. and Audretsch, D. (1999) Innovation in cities: science-based diversity, specialisation and localised competition, *European Economic Review*, 43, 409–429

Feldman, M. and Francis, J. (2003) Fortune favours the prepared region: the case of entrepreneurship and the Capitol Region biotechnology cluster, *European Planning Studies*, 11, 757–764

Ferrer, M., Thorsteinsdóttir, H., Quach, U., Singer, P. and Daar, A. (2004) The scientific muscle of Brazil's health biotechnology, *Nature Biotechnology*, (Supplement), 8–12

Finegold, D., Wong, P. and Cheah, T. (2004) Adapting a foreign direct investment strategy to the knowledge economy: the case of Singapore's emerging biotechnology cluster, *European Planning Studies*, 12, 921–942

Florida, R. (2002) *The Rise of the Creative Class*, New York, Basic Books

Freeman, C. (1987) *Technology Policy and Economic Performance: Lessons from Japan*, London, Pinter

Fuchs, G. (ed.) (2003) *Biotechnology in Comparative Perspective*, London, Routledge

Galison, P. (1997) *Image and Logic: A Material Culture of Microphysics*, Chicago University Press

Gans, J. and Stern, S. (2003) The product market and the market for 'ideas': commercialisation strategies for technology entrepreneurs, *Research Policy*, 32, 333–350

Gereffi, G. (1999) International trade and industrial upgrading in the apparel commodity chain, *Journal of International Economics*, 48, 37–70

Giaccaria, P. (2006) *Territorial Competitiveness in a Systemic Perspective. Evidence from Turin's Territorial Productive Systems*, unpublished PhD thesis, London School of Economics

Gibbon, P. (2006) *Decoding Organic Standard-Setting and Regulation in Europe (1991–2005)*, Vienna, UNIDO

Gibbons, J. (2000) The role of Stanford University, in C. Moon Lee, W. Miller, M. Hancock and H. Rowen (eds) *The Silicon Valley Edge: A Habitat for Innovation and Entrepreneurship*, Stanford, Stanford University Press

Gibbons, M., Limoges, H., Nowotny, S., Schwartzman, S., Scott, P. and Trow, M. (1994) *The New Production of Knowledge*, London, Sage

Giesecke, S. (2000) The contrasting roles of government in the development of biotechnology industry in the US and Germany, *Research Policy*, 29, 205–223

Glaeser, E., Kallall, H., Scheinkman, J. and Shleifer, A. (1992) Growth in cities, *Journal of Political Economy*, 100, 1126–1152

Granovetter, M. (1992) Problems of explanation in economic sociology, in N. Nohria and R. Eccles (eds) *Networks and Organisations: Structure, Form and Action*, Boston, Harvard Business School Press

Gregersen, B. (1992) The public sector as a pacer in National Systems of Innovation, in B. Lundvall (ed.) *National Systems of Innovation*, London, Pinter

Griffith, V. (2002) Biotech groups stuck on a rollercoaster ride, *Financial Times*, 17 February, 29

Griffith, V. (2003) Idec Pharmaceuticals and Biogen agree $6.8 bn merger, *Financial Times*, 24 June, 25 and 32

Griliches, Z. (1992) The search for R&D spillovers, *Scandinavian Journal of Economics*, 94, 29–47

Haas, P. (1992) Introduction: epistemic communities and international policy coordination, *International Organisation*, 46, 1–37

Hall, B. (2005) The financing of innovation, University of California at Berkeley, Department of Economics

Hall, P. and Soskice, D. (2001) *Varieties of Capitalism: The Historical Foundations of Comparative Advantage*, Oxford, Oxford University Press

Hall, P. and Markusen, A. (eds) (1985) *Silicon Landscapes*, London, Allen & Unwin

Harmaakorpi, V. (2006) Regional Development Platform Method (RDPM) as a tool for regional innovation policy, *European Planning Studies*, 14, 1085–1114

Harrison, B. (1994) *Lean and Mean: The Changing Face of Corporate Power in the Age of Flexibility*, New York, Basic Books

Hassink, R. and Lagendijk, A. (2001) The dilemmas of interregional institutional learning, *Environment & Planning C: Government & Policy*, 19, 65–84

Hegel, G. (1830) *Encyclopaedia of the Physical Sciences*, Frankfurt, Suhrkamp Verlag

Helpman, E. (ed.) (1998) *General Purpose Technologies and Economic Growth*, Cambridge, MIT Press

Henderson, R., Orsenigo, L. and Pisano, G. (1999) The pharmaceutical industry and the revolution in molecular biology: interactions among scientific, institutional and organisational change, in D. Mowery and R. Nelson (eds) *Sources of Industrial Leadership*, Cambridge, Cambridge University Press

Henderson, V. (2003) Marshall's scale economies, *Journal of Urban Economics*, 53, 1–28

Hodgson, G. (1993) *Economics and Evolution; Bringing Life Back Into Economics*, Cambridge, Polity

Hodgson, J. (2006) Private biotech 2004 – the numbers, *Nature Biotechnology*, 24, 635–641

Howard, K. (2003) Biotechs sue Columbia over fourth Axel patent, *Nature Biotechnology*, 21, 955–956

Howell, M., Trull, M. and Dibner, M. (2003) The rise of European venture capital for biotechnology, *Nature Biotechnology*, 21, 1287–1291

Humphrey, J. (2006) *Global Value Chains in the Agrifood Sector*, Vienna, UNIDO

Humphrey, J. and Schmitz, H. (2000) Governance and Upgrading: Linking Industrial Cluster and Global Value Chain Research, *Working Paper 120*, Sussex University, Institute of Development Studies

Huston, L. and Sakkab, N. (2006) Connect and develop: inside Procter & Gamble's new model of innovation, *Harvard Business Review*, March, 58–66

Iddan, G., Meron, G., Glukhovsky, A. and Swain, P. (2000) Wireless capsule endoscopy, *Nature*, 405, 417, 25 May

Invest Skåne (2004) *Food Skåne*, Malmö, Skåne Inward Investment

Irvine, J. and Martin, B. (1984) *Foresight in Science: Picking the Winners*, London, Pinter

Jack, A. (2005) Focus turns to question of drugs advertising, *Financial Times*, 14 January, 3

Jack, A. (2006) The drug companies: a new mood of co-operation, *Financial Times*, Special Report on Business and AIDS, 4 December, 3

Jack, A. (2007) Sanofi chairman hits out at generic drug manufacturers, *Financial Times*, 2 January, 15

Jack, A. and Urquhart, L. (2006) Change of culture: how big pharma is picking the best of biotech as a sector starts to mature, *Financial Times*, 12 January, 17

Jacobs, A. and Wager, E. (2005) European Medical Writers Association guidelines on the role of medical writers in developing peer-reviewed publications, *Current Medical Research and Opinions*, 21, 317–321

Jacobs, J. (1969) *The Economy of Cities*, New York, Random House

Jaffe, A., Trajtenberg, M. and Henderson, R. (1993) Geographic localization of knowledge spillovers as evidenced by patent citations, *Quarterly Journal of Economics*, 108, 577–598

Johnson, J. (2002) 'Valley' in the Alps, *Financial Times*, 27 February, 10

Kaiser, R. (2003) Multi-level science policy and regional innovation: the case of the Munich cluster for pharmaceutical biotechnology, *European Planning Studies*, 11, 841–858

Kaiser, R. and Liecke, M. (2006a) Recent developments in the four leading German biotechnology clusters, Munich, Ludwig Maximilians University (mimeo)

Kaiser, R. and Liecke, M. (2006b) Sector review of the European biotechnology industry: Germany, Switzerland and France (EURODITE Project Report D3b), Brussels, DG Research

Kaufmann, D., Schwartz, D., Frenkel, A. and Shefer, D. (2003) The role of location and regional networks for biotechnology firms in Israel, *European Planning Studies*, 11, 823–840

Kazmin, A. and Jack, A. (2006) Merck to offer Aids drug price cut to Thais, *Financial Times*, 4 December, 10

Kellogg, D. (2006) Toward a post-academic science policy: scientific communication and the collapse of the Mertonian norms, *International Journal of Communications Law & Policy*, autumn, available at SSRN: http://ssrn.com/abstract=900042

Ketels, C. (2006) Knowledge clusters, Presentation to Nordic Council of Ministers Conference Investing in Research and Innovation, Copenhagen, 16–18 October

Kettler, H. and Casper, S. (2000) *The Road to Sustainability in the UK and German Biotechnology Industries*, London, Office of Health Economics

Klepper, S. (2002) Capabilities of new firms and the evolution of the US automobile industry, *Industrial and Corporate Change*, 11, 645–666

Kosonen, K. (2006) Linking less-favoured Finnish regions to the knowledge economy through university filial centres, in P. Cooke and A. Piccaluga (eds) *Regional Development in the Knowledge Economy*, London, Routledge

Krugman, P. (1995) *Development, Geography and Economic Theory*, Cambridge, MIT Press

Kumar, N., Quach, U., Thorsteinsdóttir, H., Somsekhar, H., Daar, A. and Singer, P. (2004) Indian biotechnology – rapidly evolving and industry led, *Nature Biotechnology* (Supplement), 31–36

Kurti, N. (1997) *But the Crackling is Superb: An Anthology of Food and Drink by Fellows and Foreign Members of the Royal Society*, London, Royal Society

Lagendijk, A. (2001) Scaling knowledge production: how significant is the region? In M. Fischer and J. Fröhlich (eds) *Knowledge, Complexity and Innovation Systems*, Berlin, Springer-Verlag

Lagnevik, M., Sjöholm, I., Lareke, A. and Östberg, J. (2003) *The Dynamics of Innovation Clusters: A Study of the Food Industry*, Cheltenham, Edward Elgar

Lähteenmäki, R. and Lawrence, S. (2006) Public biotechnology 2005 – the number, *Nature Biotechnology*, 24, 625–634

Landers, P. (2004) Testing machines were built to streamline research – but may be stifling it, *Wall Street Journal*, 24 February, wsj.com

Latour, B. (1998) From the world of science to the world of research? *Science*, 280, 208–209

Law, J. (2006) *Big Pharma: How the World's Biggest Drug Companies Control Illness*, London, Constable

Lawrence, S. (2001) *Health Funding Update*, New York, The Foundation Centre

Lawrence, S. (2006a) State of biotech sector – 2005, *Nature Biotechnology*, 24 (6), 603

Lawrence, S. (2006b) Reverse mergers attract top-tier biotechs in sluggish IPO market, *Nature Biotechnology*, 24, 598–599

Lawton Smith, H. (2004) The biotechnology industry in Oxfordshire: enterprise and innovation, *European Planning Studies*, 12, 985–1002

Lawton Smith, H. and Bagchi-Sen, S. (2004) Guest Editorial: Innovation geographies; international perspectives on research, product development, and commercialisation of biotechnologies, *Environment & Planning C: Government & Policy*, 22, 159–160

Le Fanu, J. (1999) *The Rise and Fall of Modern Medicine*, London, Abacus

Lee, C. and Walshok, M. (2001) *Making Connections*, Report to University of California, Office of the President

Lee, D. and Dibner, M. (2005) The rise of venture capital and biotechnology in the US and Europe, *Nature Biotechnology*, 23, 672–676

Lemarié, S., Mangematin, V. and Torre, A. (2001) Is the creation and development of biotech SMEs localised? Conclusions from the French case, *Small Business Economics*, 17, 61–76

Lerner, J. (1995) Venture capitalists and the oversight of privately-held firms, *Journal of Finance*, 50, 301–318

Lerner, J. (1999) The government as venture capitalist: the long-run effects of the SBIR program, *Journal of Business*, 72, 285–318
Lievrouw, L. and Livingstone, S. (eds) (2006) *The New Media Handbook*, 2nd Edition, London, Sage
Lumme, A., Kauranen, I., Autio, E. and Kaila, M. (1993) New technology-based companies in Cambridge in an international perspective, *Working Paper 35*, Cambridge University, Small Business Research Centre
Lundvall, B. (1988) Innovation as an interactive process, in G. Dosi, C. Freeman, R. Nelson, G. Silverberg and L. Soete (eds) *Technical Change and Economic Theory*, London, Pinter
Lundvall, B. (ed.) (1992) *National Systems of Innovation*, London, Pinter
Lundvall, B. and Borrás, S. (1997) *The Globalising Learning Economy*, Luxembourg, European Commission
McGee, H. (1984) *On Food and Cooking: The Science and Lore of the Kitchen*, New York, Scribner
McKelvey, M. (1996) *Evolutionary Innovation: The Business of Biotechnology*, Oxford, Oxford University Press
McKelvey, M., Alm, H. and Riccaboni, M. (2003) Does co-location matter for formal knowledge collaboration in the Swedish biotechnology-pharmaceuticals sector? *Research Policy*, 32, 485–501
Mackinnon, D., Cumbers, A. and Chapman, K. (2002) Learning, innovation and regional development: a critical appraisal of recent debates, *Progress in Human Geography*, 26, 293–311
Maillat, D. (1998) Interactions between urban systems and localised productive systems: an approach to endogenous regional development in terms of innovative milieu, *European Planning Studies*, 6, 117–130
Malerba, F. and Orsenigo, L. (2002) Innovation and market structure in the dynamics of the pharmaceutical industry and biotechnology: towards a history-friendly model, *Industrial & Corporate Change*, 11, 667–703
Malmberg, A. and Maskell, P. (2002) The elusive concept of localization economies: towards a knowledge-based theory of spatial clustering, *Environment and Planning A*, 34, 429–449
March, J. (1991) Exploration and exploitation in organisational learning, *Organization Science*, 2, 71–87
March, J., Sproull, L. and Tamuz, M. (1991) Learning from samples of one or fewer, *Organisation Science*, 2, 1–13
Marshall, A. (1918) *Industry and Trade*, London, Macmillan
Marx, K. (1973) *Grundrisse*, Harmondsworth, Penguin
Maskell, P., Eskelinen, H., Hannibalsson, I., Malmberg, A. and Vatne, E. (1998) *Competitiveness, Localised Learning and Regional Development: Specialisation and Prosperity in Small, Open Economies*, London, Routledge
Mayer, C., Schoors, K. and Yafeh, Y. (2005) Sources of funds and investment activities of venture capital: evidence from Germany, Israel, Japan and the United Kingdom, *Journal of Corporate Finance*, 11, 586–608
Merton, R. (1973) *The Sociology of Science: Theoretical and Empirical Investigations*, Chicago, University of Chicago Press
Metcalfe, S. (2005) Knowledge and Enterprise: What Makes Capitalism Work? Plenary presentation to 5th Triple Helix Conference 'The Capitalisation of Knowledge', Turin, 18–21 May

Milbank Memorial Fund (2000) *State Healthcare Expenditures*, New York, Milbank Memorial Fund

Mihell, D., Kingham, D. and Stott, M. (1997) *The Development of the Biotechnology Sector in Oxfordshire: Implications for Public Policy*, Oxford, Oxford Innovation

Mitchell, P. (2005) Could bank loans solve Europe's biotech financing slump? *Nature Biotechnology*, 23, 1459–1460

MJNER (French Ministry for Youth, Education and Research) (2003) *The R&D Intensity of French Bioregions*, Paris, MJNER

Morgan, K. (2004) The exaggerated death of geography, *Journal of Economic Geography*, 4, 3–21

Morgan, K. and Murdoch, J. (2001) Organic vs. conventional agriculture: knowledge, power and innovation in the food chain, *Geoforum*, 31, 159–173

Morgan, K., Marsden, T. and Murdoch, J. (2006) *Worlds of Food: Place, Power and Provenance in the Food Chain*, Oxford, Oxford University Press

Mowery, D. and Ziedonis, A. (2000) Numbers, quality and entry: how has the Bayh-Dole Act affected US university patenting and licensing? In A. Jaffe, J. Lerner and S. Stern (eds) *Innovation Policy and the Economy*, Cambridge, MIT Press

Murray, F. and Stern, S. (2005) Do Formal Intellectual Property Rights Hinder the Free Flow of Scientific Knowledge? An Empirical Test of the Anti-Commons Hypothesis, *NBER Working Paper 11465*

Murray, F. and Stern, S. (2006) When ideas are not free: the impact of patents on scientific research, Presented at NBER 'Innovation Policy and the Economy Conference', Washington, May

Mytelka, L. (2006) Pathways and policies to biopharmaceutical innovation systems in developing countries, *Industry & Innovation*, 13, 415–436

Nelson, R. (ed.) (1993) *National Innovation Systems*, Oxford, Oxford University Press

Nightingale, P. (2000) Economies of scale in experimentation: knowledge and technology in pharmaceutical R&D, *Industrial & Corporate Change*, 9, 315–359

Niosi, J. and Bas, T. (2001) The competencies of regions – Canada's clusters in biotechnology, *Small Business Economics*, 17, 31–42

Niosi, J. and Bas, T. (2003) Biotechnology megacentres: Montreal and Toronto regional systems of innovation, *European Planning Studies*, 11, 789–804

Nonaka, I. and Reinmöller, P. (1998) The legacy of learning: toward endogenous knowledge creation for Asian economic development, *Wissenschaftszentrum Berlin Jahrbuch*, Berlin, Wissenschaftszentrum

Nonaka, I. and Takeuchi, H. (1995) *The Knowledge-Creating Company*, Oxford, Oxford University Press

Nothlings, U. (2005) Processed meats and pancreatic cancer, Cancer Research Centre, University of Hawaii, Honolulu

Nowotny, H., Scott, P. and Gibbons, M. (2001) *Re-Thinking Science*, Cambridge, Polity

NSF (2005) *Research and Development in Industry, 2001*, Washington, National Science Foundation

OECD (1999) *S&T Indicators: Benchmarking the Knowledge-Based Economy*, Paris, Organisation for Economic Cooperation and Development

OECD (2000) *Health Data*, Paris, Organisation for Economic Co-operation and Development

Omland, N. and Ernst, H. (2004) Vitalisation of industry through the regional promotion of knowledge-intensive new firms – the case of German biotechnology, Paper to Japanese Institute for Labour Policy and Training conference on 'Globalisation and Revitalisation of Regional Employment: Comparison of Germany and Japan', 26 March, Tokyo

Orlikowski, W. (2002) Knowing in practice: enacting a collective capability in distributed organising, *Organisation Sciences*, 13, 249–273

Orsenigo, L. (1989) *The Emergence of Biotechnology*, New York, St Martin's Press

Orsenigo, L. (2001) The (failed) development of a biotechnology cluster: the case of Lombardy, *Small Business Economics*, 17, 77–92

Orsenigo, L., Pammolli, F. and Riccaboni, M. (2001) Technological change and network dynamics: lessons from the pharmaceutical industry, *Research Policy*, 30, 485–508

Owen-Smith, J. and Powell, W. (2004) Knowledge networks as channels and conduits: the effects spillovers in the Boston biotechnology community, *Organization Science*, 15, 5–21

Owen-Smith, J., Riccaboni, M., Pammolli, F. and Powell, W. (2002) A comparison of U.S. and university-industry relations in the life sciences, *Management Science*, 48, 24–43

Pammolli, F., Riccaboni, M. and Orsenigo, L. (2000) Variety and irreversibility in scientific and technological systems, in A. Nicita and U. Pagano (eds) *The Evolution of Economic Diversity*, London, Routledge

Parkes, C. (2004) Job losses in Silicon Valley worse than first feared, *Financial Times*, 25 March, 9

Penrose, E. (1959) *The Theory of the Growth of the Firm*, Oxford, Oxford University Press

Piore, M. and Sabel, C. (1984) *The Second Industrial Divide*, New York, Basic Books

Polanyi, K. (1944) *The Great Transformation*, New York, Rinehart

Polanyi, M. (1966) *The Tacit Dimension*, London, Routledge

Polèse, M. (2002) *The Periphery in the Knowledge Economy*, http://www.inrs-ucs.uquebec.ca

Porter, M. (1990) *The Competitive Advantage of Nations*, New York, Free Press

Porter, M. (1998) *On Competition*, Boston, Harvard Business School Press

Porter, M. (2002) *Clusters of Innovation: Regional Foundations of US Competitiveness*, Washington, Council on Competitiveness

Porter, M., Sachs, J. and Warner, J. (2000) *The Global Competitiveness Index*, Boston, Harvard Business School Press

Powell, W., Koput, K. and Smith-Doerr, L. (1996) Interorganizational collaboration and the locus of innovation: networks of learning in biotechnology, *Administrative Sciences Quarterly*, 41, 116–145

Powell, W., Koput, K., Bowie, J. and Smith-Doerr, L. (2002) The spatial clustering of science and capital: accounting for biotech firm-venture capital relationships, *Regional Studies*, 36, 291–305

Raagmaa, G. and Tamm, P. (2004) An emerging biomedical business in a low capitalised country, *European Planning Studies*, 12, 943–960

Rallet, A. and Torre, A. (1998) On geography and technology: proximity relations on localised innovation networks, in M. Steiner (ed.) *Clusters and Regional Specialisation*, London, Pion

Rehfeld, D. (1995) Disintegration and reintegration of production clusters in the Ruhr Area, in P. Cooke (ed.) *The Rise of the Rustbelt*, London, UCL Press

Richardson, G. (1972) The organisation of industry, *Economic Journal*, 82, 883–896

Rochepeau, A. (2004) *The Pharmaceutical Biotechnology Industry in France*, Paris, Ministry of Research

Romano, N. and Nunamaker, J. (2001) Meeting analysis: findings from research and practice, *Proceedings of the 34th Annual Hawaii International Conference on System Sciences*, 1–13

Rorty, R. (1980) *Philosophy and the Mirror of Nature*, Oxford, Basil Blackwell

Ryan, C. and Phillips, P. (2004) Knowledge management in advanced technology industries: an examination of international agricultural biotechnology clusters, *Environment & Planning C: Government & Policy*, 22, 217–232

Sabel, C. (2002) Diversity, not specialisation: the ties that bind the (new) industrial districts, in A. Curzio and M. Fortis (eds) *Complexity and Industrial Clusters*, Heidelberg, Springer

Sapsed, J., Gann, D., Marshall, N. and Salter, A. (2005) From here to eternity? The practice of knowledge transfer in dispersed and co-located project organisations, *European Planning Studies*, 13, 831–852

Sassen, S. (1994) *Cities in a World Economy*, Thousand Oaks, Pine Forge Press

Sassen, S. (forthcoming) Electronic networks, power, and democracy, in C. Avgerou, R. Mansell, D. Quah and R. Silverstone (eds) *The Oxford Handbook of ICTs*, Oxford, Oxford University Press

Saxenian, A. (1994) *Regional Advantage*, Cambridge, Harvard University Press

Scharmer, C. (2001) Self-transcending knowledge: organising around emerging realities, in I. Nonaka and D. Teece (eds) *Managing Industrial Knowledge: Creation, Transfer and Utilisation*, London, Sage

Schlosser, E. (2001) *Fast Food Nation*, London, Penguin

Schreiber, H., Ward, P., Rowley, A. and Strauss, H. (1988) Unique tumour-specific antigens, *Annual Review of Immunology*, 6, 465–483

Schumpeter, J. (1975) *Capitalism, Socialism and Democracy*, New York, Harper & Row

Scottish Executive (2001) *A Science Strategy for Scotland*, Edinburgh, Scottish Executive

Senker, J. and Van Zwanenberg, P. (2001) *European Biotechnology Innovation Systems*, Final Report to EU-TSER Programme, University of Sussex, Science Policy Research Unit

Shane, S. (2004) *Academic Entrepreneurship*, Cheltenham, Edward Elgar

Shapiro, S. (2005) Agency theory, *Annual Review of Sociology*, 31, 263–284

Shorett, P., Rabinow, P. and Billings, P. (2003) The changing norms of the life sciences, *Nature Biotechnology*, 21, 123–125

Slamon, D., Romond, E., Perez, E. and CME Consultants (2006) Advances in adjuvant therapy for breast cancer, *Clinical Advances in Haematology & Oncology*, 4, (Supplement), 4–9

Small Business Economics, Special Issue (2001) editor G. Fuchs, *Small Business Economics*, 17, 1–153

Stankiewicz, R. (2002) The evolving design space of technology and the public R&D system, Paper to Conference on 'Rethinking Science Policy: Analytical Frameworks for Evidence-Based Policy', Science Policy Research Unit, University of Sussex, 21–23 March

Sternberg, R. and Litzenberger, T. (2004) Regional clusters in Germany – their geography and relevance for entrepreneurial activities, *European Planning Studies*, 12, 767–791

Stewart, T. (2001) *The Wealth of Knowledge*, London, Nicholas Brealey

Stoerring, D. and Dalum, B. (2007) Cluster emergence: a comparative study of two cases in North Jutland, Denmark, in P. Cooke and D. Schwartz (eds) *Creative Regions: Technology, Culture and Knowledge Entrepreneurship*, London, Routledge

Stokes, D. (1997) *Pasteur's Quadrant: Basic Science and Technological Innovation*, Washington, Brookings Institution

Suchman, M. (2000) Dealmakers and counsellors: law firms as intermediaries in the development of Silicon Valley, in M. Kenney (ed.) *Understanding Silicon Valley*, Stanford University Press

Svensson-Henning, M. (2003) *The Scanian Food Cluster and Innovation System: An Introduction to the Case Studies on Food-Oriented Biotechnology Firms*, Lund, University of Lund

Swann, P., Prevezer, M. and Stout, D. (eds) (1998) *The Dynamics of Industrial Clustering*, Oxford, Oxford University Press

Tavoletti, E. (2004) Higher education and high unemployment: does education matter? An interpretation and some critical perspectives, in P. Cooke and A. Piccaluga (eds) *Regional Economies as Knowledge Laboratories*, Cheltenham, Edward Elgar

Teece, D. and Pisano, G. (1998) The dynamic capabilities of firms: an introduction, in G. Dosi, D. Teece and J. Chytry (eds) *Technology, Organisation and Competitiveness*, Oxford, Oxford University Press

Teich, A. (1999) The science of health and the health of science, *Information Impacts*, 12 November

Thiel, K. (2005) A very firm handshake: biotech's growing negotiating power, *Nature Biotechnology*, 23, 1221–1226

Thorsteinsdóttir, H., Sáenz, H., Quach, U., Daar, A. and Singer, P. (2004) Cuba – innovation through synergy, *Nature Biotechnology*, (Supplement), 19–24

Törnquist, G. (2002) *Science at the Cutting Edge: The Future of the Øresund Region*, Copenhagen, Copenhagen Business School Press

Tulkki, P., Järvenisvu, A. and Lyytinen, A. (2001) *The Emergence of Finnish Life Sciences Industries*, Helsinki, Sitra

Uglow, J. (2003) *The Lunar Men: The Friends Who Made the Future*, London, Faber & Faber

Ullrich, A. (2002) Molecular targets in cancer therapy and their impact on cancer management, *Oncology*, 63 (Supplement 1), 1–5

Urquhart, L. and Orr, R. (2006) Valuing pharmas is not an exact science, *Financial Times*, 8 June, 22

US Department of Agriculture (1999) *Food Consumption Survey*, Washington, USDA

Valentin, F. and Lund-Jensen, R. (2003) Science-driven discontinuities and the organisation of distributed innovations: the case of biotechnology in food

processing technologies, Paper to DRUID Summer Conference 'Creating, Sharing and Transferring Knowledge', Copenhagen, 12–14 June
Van Ark, B., Melka, J., Mulder, N., Timmer, M. and Ypma, G. (2002) *ICT Investment and Growth Accounts for the European Union, 1980–2000*: Final Report on 'ICT and Growth Accounting' for the DG Economics and Finance, Brussels, European Commission
Van Reenen, J. (2002) Economic issues for the UK biotechnology sector, *New Genetics & Society*, 21, 109–130
VINNOVA (2001) *The Swedish Biotechnology Innovation System*, Stockholm, VINNOVA
VINNOVA (2003) *Swedish Biotechnology-Scientific Publications, Patenting and Industrial Development*, Stockholm, VINNOVA
Ward, M. (2005) Strategy: decapitation, *BioCentury*, 13, A1–A4
Ward, M. (2006a) Europe's iceberg, *BioCentury*, 14, A1–A4
Ward, M. (2006b) CAT nipped, *BioCentury*, 14, A1–A4
Warde, A. and Martens, L. (1998) Eating out and the commercialisation of mental life, *Journal of Food Studies*, 100, 147–153
Webster, A. and Etzkowitz, H. (1991) *Academic-Industry Relations: A Second Academic Revolution?* London, Science Policy Support Group
Wellcome Trust (2002) Clinical research facilities: an introduction, *Wellcome News*, 12 February
Wickelgren, I. (2002) *The Gene Masters*, New York, Times Books
Williamson, O. (1985) *The Economic Institutions of Capitalism*, New York, Free Press
Wolter, K. (2002) Can the US experience be repeated? The evolution of biotechnology in three European regions (mimeo), Duisburg University, Germany (published in P. Cooke and A. Piccaluga [eds] 2004, *Regional Economies as Knowledge Laboratories*, Cheltenham, Edward Elgar)
Woolcock, M. (1998) Social capital and economic development: towards a theoretical synthesis and policy framework, *Theory & Society*, 27, 151–208
Zeller, C. (2002) Regional and North Atlantic knowledge production in the pharmaceutical industry, in V. Lo and E. Schamp (eds) *Knowledge – the Spatial Dimension*, Münster, Lit-Verlag
Zeller, C. (2004) North Atlantic innovative relations of Swiss pharmaceuticals and the proximities with regional biotech areas, *Economic Geography*, 80, 83–111
Zhenzhen, L., Jiuchun, Z., Ke, W., Thorsteinsdóttir, H., Quach, U., Singer, P. and Daar, A. (2004) Health biotechnology in China – reawakening of a giant, *Nature Biotechnology* (Supplement), 13–18
Zitt, M., Ramanana-Mahary, S., Bassecoulard, E. and Laville, F. (2003) Potential science-technology spillovers in regions: an insight on geographical co-location, *Scientometrics*, 56, 259–282
Zook, M. (2002) Grounded capital: venture financing and the geography of the Internet industry, 1994–2000, *Journal of Economic Geography*, 2, 151–177
Zucker, L., Darby, M. and Armstrong, J. (1998) Geographically localised knowledge: spillovers or markets, *Economic Inquiry*, 36, 65–86
Zucker, L., Darby, M. and Armstrong, J. (1999) Intellectual Capital and the Firm: The Technology of Geographically Localized Knowledge Spillovers, *National Bureau of Economic Research Reprint 2217*, Cambridge, Massachusetts, NBER

Index